Coral Reefs of Australia

Perspectives from Beyond the Water's Edge

Editors: Sarah M. Hamylton, Pat Hutchings and Ove Hoegh-Guldberg

CSIRO

PUBLISHING

This book is dedicated to Australia's reefs and the people who live with them.

A catalogue record for this book is available from the National Library of Australia.

ISBN: 9781486315482 (pbk)
ISBN: 9781486315499 (epdf)
ISBN: 9781486315505 (epub)

How to cite:
Hamylton SM, Hutchings P, Hoegh-Guldberg O (Eds) (2022) *Coral Reefs of Australia: Perspectives from Beyond the Water's Edge*. CSIRO Publishing, Melbourne.

Published by:

CSIRO Publishing
Private Bag 10
Clayton South VIC 3169
Australia

Telephone: +61 3 9545 8400
Email: publishing.sales@csiro.au
Website: www.publish.csiro.au
Sign up to our email alerts: publish.csiro.au/earlyalert

Front cover: The reef at Lord Howe Island (image credit: Matthew Curnock)
Back cover: Watson Island, Great Barrier Reef (image credit: Sarah Hamylton)

Edited by Joy Window (Living Language)
Cover design by Cath Pirret
Typeset by Envisage Information Technology
Index by Bruce Gillespie
Printed in China by 1010 Printing International Ltd

CSIRO Publishing publishes and distributes scientific, technical and health science books, magazines and journals from Australia to a worldwide audience and conducts these activities autonomously from the research activities of the Commonwealth Scientific and Industrial Research Organisation (CSIRO). The views expressed in this publication are those of the author(s) and do not necessarily represent those of, and should not be attributed to, the publisher or CSIRO. The copyright owner shall not be liable for technical or other errors or omissions contained herein. The reader/user accepts all risks and responsibility for losses, damages, costs and other consequences resulting directly or indirectly from using this information.

CSIRO acknowledges the Traditional Owners of the lands that we live and work on across Australia and pays its respect to Elders past and present. CSIRO recognises that Aboriginal and Torres Strait Islander peoples have made and will continue to make extraordinary contributions to all aspects of Australian life including culture, economy, and science. CSIRO is committed to reconciliation and demonstrating respect for Indigenous knowledge and science. The use of Western science in this publication should not be interpreted as diminishing the knowledge of plants, animals and environment from Indigenous ecological knowledge systems.

The views expressed in this publication are those of the independent author(s) and do not necessarily represent those of, and should not be attributed to, the Australian Coral Reef Society or the lead editors.

This publication has been supported by the Australian Coral Reef Society and the Australian Academy of Sciences.

Australian
Coral Reef Society

The paper this book is printed on is in accordance with the standards of the Forest Stewardship Council® and other controlled material. The FSC® promotes environmentally responsible, socially beneficial and economically viable management of the world's forests.

MIX
Paper from responsible sources
FSC® C016973

Jul22_01

FOREWORD

Sylvia A. Earle

For thousands of years, the warm waters along Australia's coasts have gleamed with the rainbow hues of healthy coral reef systems, home to hundreds of kinds of corals, anemones, jellyfish, echinoderms, algae and seabirds, thousands of species of sponges and molluscs, two dozen kinds of reptiles, more than a hundred kinds of sharks, at least 1500 bony fish species, 30 kinds of mammals and members of more than 30 other of the major categories of life, including myriad microbes. In the 21st century, one kind of mammal – *Homo sapiens* – looms large as both an agent of unprecedented destruction and as the best hope for an enduring future for these ancient metropolises of life.

As a young scientist in the 1950s, I was led to believe that the ocean was a realm apart from life on the land, a mysterious space so vast that nothing humans do could alter its nature and that no one could really predict, let alone change, the weather, the climate, the chemistry – or the composition of life on Earth. But around the world, rapid loss of terrestrial wildlife and wild places had inspired nations to protect special places as parks and reserves. By the 1970s, concerns were growing about signs of ocean decline, and in 1975 Australia responded by establishing the Great Barrier Reef Marine Park Authority in an effort to safeguard the health of that treasured region.

Throughout decades of exploration on, around and under the ocean I have witnessed the most profound time of discovery -- and the greatest era of loss – in all of human history. The coral reefs I observed along Australia's eastern coast and beyond in the Coral Sea in the 1970s and 1980s have been largely transformed into ghost reefs, where far fewer sharks, grouper, snappers, lobsters and even parrotfish swim among hauntingly pale branches and mounds of coral.

Armed with evidence gathered by thousands of scientists recording data from spacecraft and satellites high in the sky, submersibles and instruments deployed deep within the sea, and measurements and observations in and across the globe, it is clear that humans are, indeed, changing the nature of nature. Their behaviours are warming the Earth, diminishing the abundance and diversity of life and changing planetary chemistry. In a few decades, human actions are unravelling the basic living systems that have shaped Earth's rocks and water into an oasis of life, a mostly blue miracle in an otherwise inhospitable universe. Our life support systems are in trouble, and therefore, so are we.

The good news is that it is not too late to reverse much of the harm and turn from decline to recovery. Nature-based solutions that protect and restore wildlife and wild places are achieving results, with more than 70 nations pledging to protect at least 30 per cent of the land and sea by 2030. In this thoughtful, authoritative, magnificently illustrated volume, the authors share their insights not only about the wondrous nature of Australia's legendary coral reefs, but also about actions that inspire hope for a long and prosperous future for them and for all of life on Earth, humans very much included.

CONTENTS

ACKNOWLEDGEMENTS

This book has been curated to mark the Hundred Year Anniversary of the Australian Coral Reef Society and the council and members are thanked for their support and contributions.

The text has benefited from the insights of several expert chapter reviewers, including Professor Maria Byrne, Dr Michelle Dyer (GBRMPA), Dr Zena Dinesen, Catheline Froehlich, Dr Saskia Jurriaans (AIMS), Professor Richard Kenchington, Brett Lewis, Professor David Suggett, Dr Gergely Torda, and Professor Colin Woodroffe.

We would like to thank the Australian Academy of Sciences for providing financial support through their Regional Collaboration Programme towards book research and publishing costs. The Regional Collaborations Programme is supported by the Australian Government under the National Innovation and Science Agenda. We are also grateful for the ongoing support of the Australian Research Council for coral reef science through their grant programs and large initiatives such as the Centre for Excellence in Coral Reef Studies, which has been critically important to Australian coral reef science. Similarly, many organisations like the Australian Institute of Marine Science, the Great Barrier Reef Marine Park Authority and the Great Barrier Reef Foundation have provided key input and financial support to broaden the research, education and communication necessary to build solutions and a better future for Australia's coral reefs in a stable, zero carbon world. The long list of individuals and organisations who have supported science and management projects over the past century of coral reef science cannot be comprehensively described.

The efforts of several people in searching out archival information and images are much appreciated: Professor Tom Spencer and Professor Barbara Brown at the Natural History Museum, Royal Society and Royal Geographical Society in London, Joanna Ruxton at the Great Barrier Reef Marine Park Authority, Trisha Fielding at the James Cook University Library, Jack Ennis Butler at the National Library of Australia, Patricia Egan at the Australian Museum, Joy Wheeler at the Royal Geographical Society and Dr Anne Hoggett at Lizard Island Research Station.

Members of the Australian Coral Reef Society are thanked for supplying images throughout the book. Matthew Curnock and Martin Colognoli deserve a special mention for generously providing their spectacular photographs. Brett Lewis's excellent graphic design suggestions were invaluable.

Thank you to the following people for providing the photograph at the start of each chapter: Matthew Curnock (Chapters 1 and 8), Paul Jones (Chapter 2), Prithvi Bhattacharya (Chapters 3 and 5), Brett Lewis (Chapter 4), Steph Gardner (Chapter 6), Julia Sumerling (Chapter 7).

Our editors at CSIRO Publishing, Mark Hamilton, Briana Melideo, Tracey Kudis and Joy Window, are thanked for their sharp eyes, useful suggestions and patient guidance in pulling the manuscript together.

PREFACE

Sarah Hamylton, Pat Hutchings and Ove Hoegh-Guldberg

Australia's tropical coastline supports mangroves, seagrass and coral reefs, runs for thousands of kilometres and takes on distinct regional characteristics as it crosses from the arid deserts of Western Australia to the Wet Tropics of Queensland. No single person has dived the entire vastness of Australia's coral reefs, many of which are seldom visited by humans. Only recently have many of these reefs become accessible to visitors, tourists, scientists and those seeking to make a living from them.

Coral reefs may be contingent on geology and biology, but, as Iain McCalman writes in his book *The Reef: A Passionate History*, they are also products of human perception that have been imagined into existence down the millennia. In Australia, it is an imaginative picture that extends back, uniquely, at least 60 000 years. Aboriginal and Torres Strait Islander Australians regard coral reefs as part of a sea country that incorporates all living people, animals, plants, and creation spirits, about which stories have been told for many generations. More recently, our perceptions of reefs have been shaped by discovery and exploration. Reefs have transformed from mortal danger to natural mystery, from bountiful expanse of untapped resources to unique ecosystems worthy of protection.

As people have become increasingly familiar with these seascapes, helped by aquaria, SCUBA technology, books and documentaries, their curiosity has deepened about how such a wealth of biodiversity thrives in remote and often low-nutrient oceans. Much of this biodiversity remains undescribed. This book offers a range of perspectives on the relationship between coral reefs and humans. It is a relationship shaped by hours spent floating above reefs, walking reef flats, dangling instruments from boats, and making observations at field sites and in laboratories. Australian scientists have gone beyond unlocking the geological and biological secrets of reefs to occupy a broader role as advocates, communicators and advisors, driven by their profound admiration for, and desire to protect reefs and their associated ecosystems.

What began for most as a passion is now deadly serious. As local and global threats have multiplied, our understanding of coral reefs in the Anthropocene is more important than ever, with humans directly changing planetary temperatures. Corals are now the 'canary in the coalmine' for the impacts of global climate change, having suffered more frequent and intense mass coral bleaching and mortality, dramatic species range shifts, ocean acidification, and more. Australia's coral reefs have sustained a global interest for centuries and the diverse perspectives in this book trace our evolving relationship with them, telling inspiring stories of scientific discovery and ocean governance that have put Australia at the forefront of environmentally responsible coral reef management.

Curating this collection has been an organic process in which we included as many voices as possible, although they inevitably encompass only a small portion of all the people who interact with Australia's reefs. Ideas and insights are presented in distinctive styles; they focus on different ways in which people interact with reefs and are sometimes contradictory. We bring together different worldviews to emphasise the differences between perspectives, from industry to Indigenous, from mesophotic to mass spawning. It is the interplay between them that offers a deeper understanding of what it means to work with and live alongside Australia's reefs.

AUTHOR AFFILIATIONS

Maria Fernanda Adame	Australian Rivers Institute, Griffith University, Brisbane, Qld
Robin Beaman	College of Science and Engineering, James Cook University, Cairns, Qld
Giglia A. Beretta	School of Life Sciences, University of Technology, Sydney, NSW
David J. Booth	School of Life Sciences, University of Technology Sydney, NSW
John Bradley	Monash Indigenous Studies Centre, Monash University, Melbourne, Vic
Barbara Brown	School of Natural & Environmental Sciences, University of Newcastle upon Tyne, UK; Environmental Research Unit, University of the Highlands and Islands, Caithness, Scotland
Nicola Browne	School of Molecular and Life Sciences, Curtin University, Bentley, WA
Maria Byrne	School of Life and Environmental Sciences, University of Sydney, Sydney, NSW
Emma Camp	Climate Change Cluster, University of Technology Sydney, Sydney, NSW
Mel Cowlishaw	Great Barrier Reef Marine Park Authority, Brisbane, Qld
Graeme Cumming	ARC Centre of Excellence for Coral Reef Studies, James Cook University, Townsville, Qld; Australian Coral Reef Society Council member
Matthew I. Curnock	CSIRO Land and Water, Townsville Qld
Jon C. Day	ARC Centre of Excellence for Coral Reef Studies, James Cook University, Townsville, Qld; formerly Great Barrier Reef Marine Park Authority, Townsville, Qld
Christopher Doropoulos	CSIRO Oceans & Atmosphere, St Lucia, Qld
Sophie Dove	ARC Centre of Excellence for Coral Reef Studies and School of Biological Sciences at the University of Queensland, St Lucia, Qld
Michelle Dyer	Great Barrier Reef Marine Park Authority, Townsville, Qld; Australian Coral Reef Society Council member
Michael J. Emslie	Australian Institute of Marine Science, Townsville, Qld
Gal Eyal	ARC Centre of Excellence for Coral Reef Studies, School of Biological Sciences, University of Queensland, St Lucia, Qld; Mina and Everard Goodman Faculty of Life Sciences, Bar-Ilan University, Israel
Ann Elias	Department of Art History, University of Sydney, Sydney, NSW
Leanne Fernandes	Senior Research Fellow, School of Marine Biology, College of Science and Engineering, James Cook University, Townsville, Qld
Maoz Fine	The Alexander Silberman Institute of Life Science, The Hebrew University of Jerusalem, Israel
Catheline Froehlich	School of Earth, Atmospheric and Life Sciences, University of Wollongong, Wollongong, NSW; Australian Coral Reef Society Council member
James Gilmour	Australian Institute of Marine Science, Indian Ocean Marine Science Centre, University of Western Australia, Perth, WA; Oceans Institute, University of Western Australia, Crawley, Western Australia, Australia
Howard Gray	Maritime historian, Batavia Coast, WA
Jack Greenwood	Department of Zoology, The University of Queensland (1961–2002)
Valerie Hagger	School of Biological Science, University of Queensland, St Lucia, Qld
Sarah M. Hamylton	School of Earth, Atmospheric and Life Sciences, University of Wollongong, Wollongong, NSW; President, Australian Coral Reef Society
Melanie Hava	Melanie Hava is a Mamu Aboriginal artist. Her mother is from the Dugul-barra and Warii-barra family groups of the Johnstone River catchment

Harold Heatwole	Department of Zoology, University of New England, Armidale, NSW
Andrew Heyward	Australian Institute of Marine Science, Perth, WA; Oceans Institute, University of Western Australia, Crawley, Western Australia, Australia
Ove Hoegh-Guldberg	ARC Centre for Excellence for Reef Studies, Great Barrier Reef Foundation (Chief Scientist) and School of Biological Sciences at the University of Queensland, St Lucia, Qld
Andrew S. Hoey	ARC Centre of Excellence for Coral Reef Studies, James Cook University, Townsville, Qld
Thomas H. Holmes	Marine Science Program, Department of Biodiversity, Conservation and Attractions, Kensington, Perth; Oceans Institute, University of Western Australia, Crawley, WA
David Hopley	Emeritus Professor, College of Science and Engineering, James Cook University, Townsville, Qld
Terry Hughes	ARC Centre of Excellence for Coral Reef Studies, James Cook University, Townsville, Qld
Pat Hutchings	Senior Fellow, Australian Museum Research Institute, Australian Museum, Sydney, NSW
Alan Kendrick	Marine Science Program, Department of Biodiversity, Conservation and Attractions, Kensington, Perth, WA
Emma Kennedy	Australian Institute of Marine Science, Townsville, Qld
Michael J. Kingsford	Marine Biology and Aquaculture and Centre of Excellence for Coral Reef Studies, College of Science & Engineering, James Cook University, Townsville, Qld
Richard Leck	Head of Oceans, World Wide Fund for Nature, Brisbane, Qld
Brett Lewis	Earth and Atmospheric Science, Queensland University of Technology, Brisbane, Qld; Australian Coral Reef Society Council member
Michelle Linklater	Coastal and Marine Unit, NSW Department of Planning and Environment, Sydney, NSW
Rohan Lloyd	College of Arts, Society & Education, James Cook University, Townsville, Qld
Jennifer Loder	Director of Community Partnerships, Great Barrier Reef Foundation, Brisbane, Qld
Catherine E. Lovelock	School of Biological Sciences, University of Queensland, St Lucia, Qld
Leah Lui-Chivizhe	School of Communication and Centre for the Advancement of Indigenous Knowledges (CAIK), University of Technology, Sydney, NSW
Damien Maher	School of Environment, Science and Engineering, Southern Cross University, Lismore, NSW
Jennie Mallela	The Research School of Biology, The Australian National University, Canberra, ACT
Helene Marsh	College of Science & Engineering, James Cook University, Townsville, Qld
Iain McCalman	Institute of Humanities and Social Sciences, Australian Catholic University, Sydney, NSW
Eva McClure	ARC Centre of Excellence for Coral Reef Studies, James Cook University, Townsville, Qld
Anthony McKnight	Awabakal, Gumaroi, Yuin man and School of Education, University of Wollongong, Wollongong, NSW
Roger McLean	School of Science, University of New South Wales Canberra, Canberra, ACT
Efrat Meroz-Fine	Faculty of Law, the Hebrew University of Jerusalem, Israel
Tiffany H. Morrison	ARC Centre of Excellence for Coral Reef Studies, James Cook University, Townsville, Qld
Nurjannah Nurdin	Research and Development Center for Maritime, Coast and Small Islands, Hasanuddin University, Makassar, Indonesia
John M. Pandolfi	ARC Centre of Excellence for Coral Reef Studies, School of Biological Sciences, University of Queensland, St Lucia, Qld
Marji Puotinen	Australian Institute of Marine Science, Perth, WA
John Quiggin	School of Economics, University of Queensland, St Lucia, Qld
Kate Quigley	Australian Institute of Marine Science, Townsville, Qld; Minderoo Foundation
Russell Reichelt	Formerly Australian Institute of Marine Science and Great Barrier Reef Marine Park Authority, Townsville, Qld

Zoe Richards	Coral Conservation and Research Group, School of Molecular and Life Sciences, Trace and Environmental DNA Laboratory, Curtin University; Collections and Research, Western Australian Museum, Perth, WA
David Ritter	Greenpeace Australia Pacific, Sydney, NSW
Chris Roelfsema	School of Earth and Environmental Sciences, University of Queensland, St Lucia, Qld; Australian Coral Reef Society Council member
Barbara Robson	Australian Institute of Marine Science, Townsville, Qld
Claire L. Ross	Department of Biodiversity, Conservation and Attractions, Kensington, Perth, WA
Peter Sale	Emeritus Professor, University of Windsor, Canada; formerly University of Sydney, Sydney, NSW
Verena Schoepf	Institute for Biodiversity and Ecosystem Dynamics, University of Amsterdam, Amsterdam, The Netherlands
Colin Simpfendorfer	College of Science and Engineering, James Cook University, Townsville, Qld
Gavin Singleton	Yirrganydji Djabugay Traditional Owner from the Cairns to Port Douglas region in North Queensland, Australia; Project Manager, Dawul Wuru Aboriginal Corporation/Coordinator, Yirrganydji Land and Sea Ranger Program, Qld
Carrie Sims	Australian Institute of Marine Science, Townsville, Qld; Australian Coral Reef Society Council member
Scott Smithers	College of Science and Engineering, James Cook University, Townsville, Qld
Brigitte Sommer	School of Life Sciences, University of Technology Sydney, NSW; School of Life and Environmental Sciences, University of Sydney, NSW
Tom Spencer	Cambridge Coastal Research Unit, Department of Geography, University of Cambridge, UK
David Suggett	Climate Change Cluster, University of Technology, Sydney, NSW; Australian Coral Reef Society Council member
Frank Talbot	Director of the Australian Museum (1966–75), Sydney, NSW
Gergely Torda	Centre of Excellence for Coral Reef Studies, James Cook University, Townsville, Qld
Tali Treibitz	Hatter Department of Marine Technologies, Charney School of Marine Sciences, University of Haifa, Haifa, Israel
Vinay Udyawer	Australian Institute of Marine Science, Darwin, NT
Madeleine Van Oppen	Australian Institute of Marine Science, Townsville, Qld; School of BioSciences, University of Melbourne, Melbourne, Vic
Charlie Veron	Coral Reef Research, former Chief Scientist, AIMS, Townsville, Qld
Nathan Waltham	Marine Data Technology Hub, College of Science and Engineering, James Cook University, Townsville, Qld
Selina Ward	School of Biological Sciences, University of Queensland, St Lucia, Qld; Australian Coral Reef Society Council member
Gregory Webb	School of Earth and Environmental Sciences, University of Queensland, St Lucia, Qld
Jody Webster	Geocoastal Research Group, School of Geosciences, University of Sydney, Sydney, NSW
Shaun Wilson	Marine Science Program, Department of Biodiversity, Conservation and Attractions, Kensington, Perth & Oceans Institute, University of Western Australia, Crawley, WA
Tim Winton	Novelist and patron of the Australian Marine Society
Kennedy Wolfe	Marine Spatial Ecology Laboratory, University of Queensland, St Lucia, Qld; Australian Coral Reef Society Council member
Colin Woodroffe	School of Earth, Atmospheric and Life Sciences, University of Wollongong, Wollongong, NSW
Matan Yuval	Hatter Department of Marine Technologies & the Department of Marine Biology, Charney School of Marine Sciences, University of Haifa, Haifa, Israel; the Inter-University Institute of Marine Sciences, Eilat, Israel
Imogen Zethoven	Blue Ocean Consulting; formerly Australian Marine Conservation Society

ABBREVIATIONS

A$ Australian dollars
ACRS Australian Coral Reef Society
AE Assisted evolution
AIMS Australian Institute of Marine Science
AMSA Australian Maritime Safety Authority
AO Order of Australia
ARC Australian Research Council
AUV Autonomous underwater vehicle
BP Before Present
CITES Convention on Trade in Endangered Species
COP Conference of the Parties
COREMAP Coral Reef Management and Rehabilitation Program
COTS Crown-of-thorns starfish
CSMP Coral Sea Marine Park
CTI Coral Triangle Initiative
DNA Deoxyribonucleic acid
ENCORE Elevated Nutrients on Coral Reef Experiment
ENSO El Niño Southern Oscillation
EAC East Australian Current
EEZ Exclusive Economic Zone
EPA Environment Protection Agency
EPBC Environment Protection and Biodiversity Conservation
GBR Great Barrier Reef
GBRC Great Barrier Reef Committee
GBRF Great Barrier Reef Foundation
GBRMP Great Barrier Reef Marine Park
GBRMPA Great Barrier Reef Marine Park Authority
HIRS Heron Island Research Station
ICRS International Coral Reef Society
IUCN International Union for Conservation of Nature
IPCC Intergovernmental Panel on Climate Change
ITF Indonesian Throughflow
LED Light emitting diode

LiDAR Light Detection and Ranging
LNG Liquefied natural gas
LNP Liberal National Party
LTMP Long-Term Monitoring Program
LIRS Lizard Island Research Station
MCRMP Millennium Coral Reef Mapping Project
MODIS Moderate resolution imaging spectroradiometer
NOAA National Oceanic and Atmospheric Administration (USA)
NSW New South Wales
NT Northern Territory
PADDD Protected area downgrading, downsizing and degazettement
Qld Queensland
RNA Ribonucleic acid
ROV Remotely operated vehicle
SLR Sea-level rise
SCUBA Self-contained underwater breathing apparatus
UNEP-WCMC United Nations Environment Programme-World Conservation Monitoring Centre
UQ University of Queensland
U-Th Uranium-thorium
UNESCO United Nations Educational, Scientific and Cultural Organization
UNFCCC United National Framework Convention on Climate Change
US$ United States dollars
UV Ultraviolet radiation
Vic Victoria
WA Western Australia
WH World Heritage
WHS World Heritage site (also referred to as World Heritage Property)
WPSQ Wildlife Preservation Society of Queensland
WWF World Wildlife Fund

Cultural sensitivity warning

Readers are warned that there may be words, descriptions and terms used in this book that are culturally sensitive, and which might not normally be used in certain public or community contexts. While this information may not reflect current understanding, it is provided by the author in a historical context.

This publication may also contain quotations, terms and annotations that reflect the historical attitude of the original author or that of the period in which the item was written, and may be considered inappropriate today.

Aboriginal and Torres Strait Islander peoples are advised that this publication may contain the names and images of people who have passed away.

1
Australia's coral reefs

Australia's coral reefs fringe thousands of kilometres of tropical coastline. Here, we describe the distinctive character of Australia's reef regions, from the isolated reefs of Western Australia's shelf, including the Kimberley coastline and Ningaloo, extending offshore to the Indian Ocean territories of Cocos (Keeling) Atoll and Christmas Island and through the Timor and Arafura seas of Australia's northern coastline, to the Torres Strait, Great Barrier Reef and Coral Sea. Further offshore, the seamount reefs of Elizabeth and Middleton and Lord Howe Island are some of Australia's eastern coral outposts in the Pacific Ocean, the easternmost being Norfolk Island, some 7500 km from the westernmost reef of Cocos (Keeling) Atoll.

Australia's coral regions are subject to unique environmental conditions that have enabled corals to grow in different assemblages of species to form reef platforms of varying sizes and shapes. These reefs support a diverse community of marine life, some of which stays close to the coral reef throughout its lifetime, while others such as seabirds, whale sharks, turtles, red crabs and dugong migrate through reef environments, connecting to land and other coastal waters, including those of neighbouring countries.

Humans have lived alongside Australian reefs for many thousands of years, since well before the Sahul Shelf connected Australia to New Guinea some 20 000 years ago. In the Torres Strait, turtle and dugong are central to ceremonial life and important cultural resources to Islanders. For hundreds of years, Indonesian fishers have harvested clams, shark and bêche-de-mer, or trepang, from Australia's northern coastline. With the arrival of Europeans, marine industries grew in pearling in the Torres Strait and Coral Sea, oyster farming and lobster fisheries in the Houtman Abrolhos Islands, phosphate mining on Christmas Island and tourism on the Great Barrier Reef and Lord Howe Island. This reliance on marine resources brought humans into an intimate relationship with Australia's reefs.

Since Charles Darwin sailed to Cocos (Keeling) Atoll in 1836 and proposed his global theory of coral reef formation, our scientific understanding of Australia's coral reefs has expanded to include the broader marine life they support, while the ways humans have interacted with them have varied. Here, we provide a brief overview of Australia's reefs, their environmental characteristics and their human histories.

Coral reefs around Australia
Sarah Hamylton

Australia's continent extends from the Indian Ocean in the West to the Pacific Ocean in the East. Much of its tropical coastline is lined with some of the longest, oldest, most biodiverse and pristine coral reefs on the planet. Australia presides over around 50 000 km² of coral reefs, or 17 per cent of the world's coral reefs, inside its exclusive economic zone (EEZ) [1]. After Indonesia, Australia has the largest coral reef area of any nation. From the most extensive coral reef in the world (the Great Barrier Reef (GBR) in the north-east), to the world's longest fringing reef in the West (Ningaloo Reef), Australia's reefs formed under different conditions and are marked by unique, often very remote, environments today (Fig. 1.1).

Many individual corals combine to form limestone framework reef platforms. In turn, these platforms modify the waves reaching a shoreline and provide sediments that build beaches and islands. Environmental conditions for reef development vary substantially around Australia with regional water quality, rainfall, sea surface temperatures and oceanographic characteristics, such as tidal range and currents, all of which determine whether or not a reef will grow and thrive.

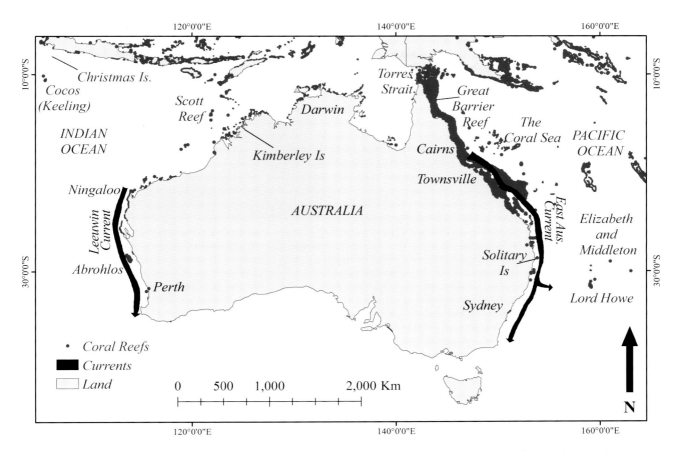

Fig. 1.1. Australia's coral reefs (red dots), with significant currents shown in black. (Image credit: Sarah Hamylton)

In relation to other regions of the world, the coral reefs of Australia have developed over a geologically stable continental shelf [2], evolving over hundreds of thousands of years in response to changing sea levels. The form of modern reefs around the coastline is inherited from older reefs that have persisted through patterns of sea-level change since the last ice age, around 11 000 years ago. From a long-term standpoint, rising sea levels from the last ice age flooded the Great Barrier Reef lagoon and created the maze of reef patches that we see today. At the same time, some of the higher, previously mountainous peaks became rocky islands (e.g. Lizard Island and Magnetic Island) that are fringed by substantial reef platforms. Fringing reefs, barrier reefs and atolls are all found in Australia, often made up of a variety of smaller submerged reef platforms that reflect

their underlying foundations. Over shorter, decadal timescales, reefs are shaped by other factors such as waves, tides, currents and rivers. Coral reefs typically grow from small patches of reef, expanding in the direction of dominant wind and currents into a continuous, often crescentic, reef. The upper shallow surface will eventually become a large platform on which sand or shingle can accumulate to form coral islands [3]. This sequence is probably responsible for the distinctive shape of many reefs in Western Australia (WA), Torres Strait and the GBR.

Over the wide and shallow continental shelf, the distribution of corals is controlled mainly by light, turbidity, and temperature (Fig. 1.2). Typical tropical reef-building corals can survive with a minimum average temperature of ~18°C in the coldest months [4]. Along the western coastline, the southward flowing Leeuwin Current brings warm tropical waters that enable reefs to grow at relatively high latitudes, such as the Houtman Abrolhos Islands at 28°S. Likewise, along the eastern coastline, the East Australian Current (EAC) carries warm waters south along the length of the GBR until it meets the Tasman Front, where it diverts offshore to support Australia's southernmost reef growth around Lord Howe Island at 31.5°S, about 780 km offshore north-east of Sydney. Although these sites mark the southernmost development of Australia's coral reefs, the presence of non-reef forming coral communities extends much further, notably to Rottnest Island near Perth on the West Coast and to the Solitary Islands on the East Coast, with some species occurring around Sydney Harbour and further south.

Currents along the shore consistently move water over reef surfaces and influence the arrival of the larvae of marine organisms that colonise reefs. Where fast-flowing currents drive high water circulation, reefs are characterised by soft corals, gorgonian sea fans and filter-feeding invertebrates such as clams. As corals are a keystone and habitat-forming species on most Australian reefs, they also support a diverse array of fish and invertebrates. In biogeographical terms, many coral species are observed in the reefs of northern Australia, particularly on reefs close to the Indo-Pacific regional hotspot of marine biodiversity known as the Coral Triangle.

Australia's coral reefs are subject to different management regimes that reflect the local pressures they experience. Recreational fishing pressures are lower for offshore reefs than fringing reefs that are more accessible to coastal towns and communities. Similarly, the nearshore East Australian reefs lie adjacent to freshwater rivers that influence regional water quality, whereas the WA reefs line arid coastlines. The seasonal patterns of cyclones also vary markedly around Australia, with more frequent and intense cyclone activity and associated rainfall close to the equator.

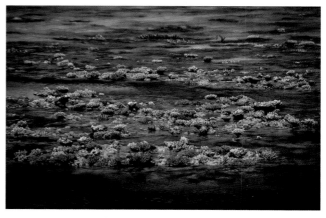

Fig. 1.2. Exposed corals at Dynamite Pass, Ribbon Reef 10, northern Great Barrier Reef. (Image credit: Matt Curnock)

Australia's reef corals and fish are on the move due to anthropogenic climate change (see 'Coral reefs on the move?' in Chapter 8). Over a hundred tropical species of fish, such as damselfish and surgeonfish, typically associated with northern coral reefs, are now also being observed as far south as Sydney Harbour, signifying a broader poleward expansion of coral reef biodiversity, most likely due to increasing sea surface temperatures in the north [5].

Whether they traverse rugged volcanic islands or secluded lagoonal coral cay settings, coral reefs impart a distinctive character to much of Australia's shoreline.

The coral reefs of Western Australia
Shaun K. Wilson, Thomas H. Holmes, Alan Kendrick and Claire L. Ross

The coral reefs of WA are morphologically diverse, with high levels of biodiversity and endemism. Carbonate reefs occur along more than 1500 km of WA coastline, incorporating oceanic atolls in the north to fringing reefs that surround the Houtman Abrolhos Islands off the midwest coast (Fig. 1.3). Across this latitudinal gradient, coral reefs are along the mainland coast and continental islands, including extensive fringing reef at Ningaloo and,

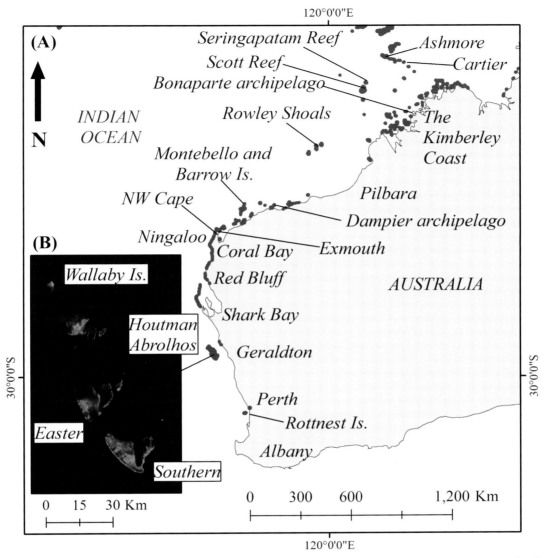

Fig. 1.3. (A) The coral reefs of Western Australia. (B) The three island groups of the Houtman Abrolhos Islands. (Image credits: Sarah Hamylton)

to a lesser extent, at Shark Bay, which are both World Heritage sites. Coastal and offshore marine reserves recognise the important role that coral reefs play in supporting biodiversity and providing ecosystem services. Reserves cover the Kimberley, Ashmore, Cartier, Rowley Shoals, waters surrounding the Montebello and Barrow islands, Ningaloo Reef and Shark Bay.

The Kimberley is the northernmost region of the WA mainland. The reefs of this region remain poorly studied due to inaccessibility and large tidal ranges of over 10 m (horizontal falls), which generate currents of up to 30 kn and highly turbid waters. As the Kimberley coastline flooded during the Holocene (i.e. approximately the last 10 000 years), a complex coastline was created with numerous islands upon which fringing carbonate reefs developed. Coral cover on inshore fringing reefs of the Kimberley is generally highest (15–25 per cent) within the shallow waters of the outer reef flat margins [6]. Tidal pools provide some relief from aerial exposure during spring low tides, often accommodating coral assemblages characterised by acroporids, while dome-shaped faviids and merulinids corals characterise other intertidal areas (Fig. 1.4A) [6].

Plateaus, terraces, and banks rise from 200–600 m along the broad northern continental shelf and slope to the west of the Kimberley coast, providing the foundations for offshore oceanic reefs such as Scott, Ashmore, Seringapatam and Rowley Shoals. On offshore reefs, corals proliferate in the clear, warm oligotrophic waters, where reef growth has kept pace with sea-level rise throughout the Holocene, creating lagoons with sandy floors and small patch reefs (Fig. 1.4B). Lagoon size, depth, density of patch reefs, and connection to the ocean differ among atolls and profoundly influence the faunal assemblages associated with each. Coral diversity is typically high, and community composition is influenced by larvae supplied by Indonesian reefs to the north and other factors. Ecological connectivity among offshore reefs is limited, though stronger connections are likely to occur within reef systems, such as the Rowley Shoals.

The continental shelf from the Kimberley to the Pilbara is wide, with fringing reefs and continental rocky islands being common along this coast, notably around the offshore Montebello and Barrow islands and the coastal Dampier Archipelago. The Pilbara coast is arid with only a few major rivers that periodically discharge into coastal waters under the influence of cyclones or high rainfall. The inshore waters of the Pilbara are turbid, partially due to tidal movements, which range from 2–6 m, decreasing in amplitude southwards. Cyclones have a profound effect on coral reefs in this region, making turbidity an important driver of coral assemblages in the region [7]. The highly turbid reefs are characterised by corals from the genus *Turbinaria*. Other taxa such as *Favites*, *Porites* and *Pavona* are more prominent in less turbid waters, while *Acropora* are typically more common in the clearer waters offshore

Fig. 1.4. Coral reefs of Western Australia. (A) Intertidal reefs of the inshore Kimberley marine reserves. (B) Patch reefs at Rowley Shoals. (C) Inshore reef in turbid waters of the Dampier Archipelago. (D) *Acropora* dominated reef flat at Ningaloo Marine Park. (E) *Turbinaria* dominated reef at Shark Bay. (F) *Acropora* dominated reef at Houtman Abrolhos Islands. (Image credits: Will Robbins (A, B), Richard Evans (C), Department of Biodiversity, Conservation and Attractions (D, E), Sahira Bell (F))

(Fig. 1.4C). The Pilbara is also a focus of major coastal and marine resource industry development, which includes port facilities, shipping channels to offshore oil and gas platforms and marine pipelines. Large-scale dredging and dumping of sediments to build and maintain shipping channels close to port areas near the Dampier Archipelago and Barrow Island have caused localised coral mortality.

The continental shelf narrows at North West Cape, south of the Pilbara Region, where a prevailing southerly wind facilitates upwelling and attracts annual migrations of whale sharks in the deep water beyond the Ningaloo Reef slope. Ningaloo Reef extends 260 km from North West Cape to Red Bluff, making it one of the longest fringing coral reefs in the world. The reef protects the coastline from oceanic swells and forms a sheltered lagoon that is up to 5 km wide at some places. Like the Pilbara, the hinterland adjacent to Ningaloo Reef is arid and terrestrial discharge is limited. Unlike further north, the tidal range is moderate (≤ 2 m) and inshore waters are relatively clear. The clearer waters of Ningaloo promote high rates of coral growth within the shallow lagoon, which is dotted with numerous patch reefs, separated by fields of canopy-forming macroalgae (seaweeds). These macroalgal fields provide habitat for juvenile fishes, including spangled emperor and cods, which are targeted by recreational fishers as adults [8]. The lagoon is bordered by a back reef and reef flat that are dominated by plating and corymbose *Acropora* (Fig. 1.4D). The reefs along much of the exposed and windward slope are characterised by encrusting, massive and corymbose growth forms that are robust to the high wave energy of this coast [9]. The proximity of this remarkable coral reef and protected lagoon to an accessible coast and the reliable seasonal presence of megafauna, like turtles, whale sharks and cetaceans, make Ningaloo Reef a very popular tourist destination with visitors staying at the small towns of Exmouth (pop. 2500) and Coral Bay (pop. 350).

Coral reefs can also be found on the lee side of Bernier, Dorre and Dirk Harthog islands, which form the western boundary of Shark Bay, and create a transitional zone between the tropical and temperate reefs. The most diverse coral communities occur around the passages and northern/southern ends of these islands, where there are clear oceanic waters flowing south. Shallow water within the bay is generally more turbid and subject to large temperature fluctuations. Coral communities here have typically low diversity and are dominated by *Turbinaria* (Fig. 1.4E). They are also surrounded by expansive seagrass meadows, a conspicuous feature of Shark Bay.

The southernmost coral reefs of WA and of the entire Indian Ocean fringe the Wallabi, Easter and Pelsaert islands, in the Houtman Abrolhos group (Fig. 1.3). These islands are ~60 km from the mainland town of Geraldton, between 28 and 29°S. As they are spread along 100 km of the continental margin, the islands are home to a diverse array of temperate and tropical marine species. Indeed, almost 200 coral species have been recorded from the islands, often occurring in proximity to temperate macroalgae, such as *Eklonia*. Branching and plating *Acropora* are among the most common corals (Fig. 1.4F), although the diversity and coverage of corals is highest on the deeper reef slopes, lagoons, and leeward sides of the islands.

Corals can also be found in the temperate waters south of Houtman Abrolhos islands, most notably around Dongara, Fisherman Island, Rottnest Island, Hall Bank, Geographe Bay and Albany, although these corals do not form carbonate reefs and the diversity of coral taxa is low. The presence of corals at such high latitudes (35°S) is partially attributable to poleward-flowing currents that transport warm water to more southern reefs of Australia. Most notable is the Leeuwin Current, which flows from North West Cape, bringing warm water to the south-west and southern coasts of Australia (Fig. 1.1). The influence of the Leeuwin Current on corals is strongest offshore, such that the Houtman Abrolhos and Rottnest islands have greater coral coverage than the adjacent mainland. Further south, the Leeuwin Current is closer to shore as it rounds Cape

Leeuwin and creates conditions favourable for stands of *Turbinaria* coral in Ngari Capes Marine Park. The Leeuwin Current is partially fed by the Holloway Current, which links to the Indonesian Throughflow (ITF) supplying tropical fish and invertebrate larvae to WA reefs.

Like many of the coral reefs around the world, WA reefs have experienced more frequent extreme warm water events since the 1980s, leading to coral bleaching and mortality. The remote nature of many WA coral reefs means that coral bleaching is less visible and seldom reported. One of the first major bleaching events was documented at Scott Reef in 1998 [10]. Major coral bleaching has subsequently been recorded at locations along the WA coastline, one of the most severe events occurring in the summer of 2010/11. La Niña conditions over the 2010/11 summer strengthened the Leeuwin Current which delivered warm water to southern reefs, and contributed to sea surface temperature anomalies of up to 5°C. During this event, coral bleaching and mortality were recorded across 1200 km of coastline, from the Montebello Islands in the north to the Houtman Abrolhos Islands and Hall Bank near Perth in the south [11]. Coral bleaching has also been recorded on inshore reefs along the Pilbara coast in 2013 and 2014 [12]. A recent assessment of the status of WA reefs found that average coral cover is currently at an all-time low at several locations where there has been long-term monitoring [13]. Undoubtedly the extent of coral cover and types of coral communities along the coast have waxed and waned over the past millennia. The onset of the Anthropocene does, however, raise questions about the resilience of WA coral reefs in the future.

The isolated reefs of Australia's north-west shelf

James Gilmour and Andrew Heyward

Three reef systems along the edge of Australia's north-west shelf emerge from oceanic waters hundreds of metres deep (Fig. 1.5). The Ashmore (Ashmore, Cartier, Hibernia), Scott (North, South, Seringapatam) and Rowley Shoals (Mermaid, Clerke, Imperieuse) reefs developed along the continental margin 5–6 million years ago [14]. Having persisted through changing sea levels and a subsiding continental margin over thousands of years, these reef systems are now hundreds of kilometres from the mainland and from each other.

Being far from the coast, the reefs are bathed in warm oceanic waters that are low in inorganic nutrients. The oceanography of these reefs is strongly influenced by the ITF, linking them over geological time to equatorial reefs at the centre of marine biodiversity in the Coral Triangle. These north-west reef systems and their sandy cays are visited seasonally by migratory species making their way along the vast WA coastline, including whales, turtles and seabirds. By comparison, human presence has been limited. Europeans have visited the reefs since the early 1800s, usually harvesting trochus (mother-of-pearl) shells and guano from the sand cays. Competition for resources led to the reefs initially being claimed by Britain, and then declared part of WA in the early 1900s. Indonesian fishers have been visiting for hundreds of years, mainly harvesting trochus, giant clam, shark fin and sea cucumber (see 'Bêche-de-mer: the cornerstone of Australian fisheries' in Chapter 2). The 1974 Australia–Indonesia Memorandum of Understanding regarding the Operations of Indonesian Traditional Fishermen in Areas of the Australian Fishing Zone and Continental Shelf allowed restricted fishing with traditional methods following the declaration of marine protected areas managed by Australian agencies.

There has been a relatively short history of scientific research on the three reef systems. Early expeditions occurred in the 1970s aboard American and Russian research vessels, but since the 1980s, most surveys of marine biodiversity have been conducted by the Western Australian Museum [e.g. 15]. Monitoring programs focusing on corals and fishes were established in the 1990s by the Australian Institute of Marine Science and WA Department of Biodiversity, Conservation and Attractions, with over two decades of regular surveying at the Scott and Rowley Shoals reefs.

Physical environment and reef habitats

Each of the Rowley, Scott and Ashmore reef systems consists of three atolls, varying in size and structure; some have fully enclosed lagoons while others are relatively open to the ocean (Figs 1.5 and 1.6A). The western flanks are exposed to the strongest waves, evident in the wider reef flat that supports a lower abundance and diversity of corals. The leeward (eastern) flank of most reefs has abundant and diverse coral communities, which extend from the reef flat down the steep slope to depths of ~30 m (Fig. 1.6B). The lagoons are mostly less than 20 m deep and have a sandy seabed, patches of staghorn corals and networks of isolated bommies with mixed communities (Fig. 1.6C). The massive, deep lagoon at South Scott Reef is open to the ocean, spanning some 300 km² and reaching depths of 30–70 m [16]. This lagoon supports extensive communities of fleshy and calcareous algae, hard corals, filter feeders and some seagrasses (Fig. 1.6D), even where light levels are < 5 per cent of those in the shallows. Experiments with some of these deep-water corals species indicate that they still rely on photosynthesis for much of their nutrition, adapting their growth form,

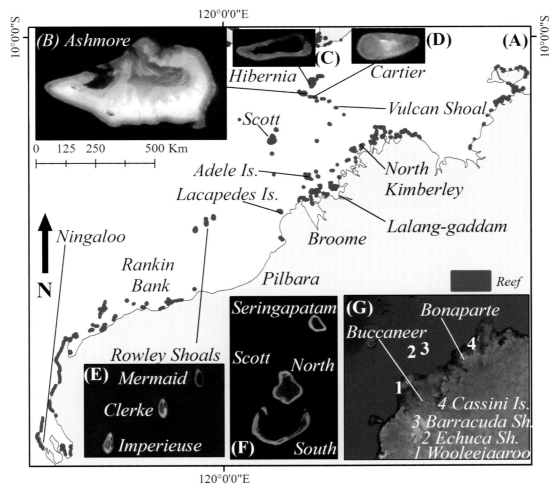

Fig. 1.5. The Ashmore, Scott and Rowley Shoals reef systems consist of three reefs (B, F, E), located near the edge of Australia's north-west shelf, hundreds of kilometres from the Kimberley coastline (G) and from each other. (Image credit: Sarah Hamylton)

Fig. 1.6.   (A) Seringapatam Reef, and the other reef atolls, rise from hundreds of metres depth, having steep outer-reef slopes and lagoons of varying sizes and shapes. (B) Coral communities on the outer eastern (leeward) slope of each reef are the most abundant and diverse, such as at Imperieuse Reef. (C) The shallow (< 20 m) lagoon at Clerke Reef has fields of staghorn corals surrounding massive bommies, often covered in corals that grow to the low tide mark. (D) The deep (50 m) lagoon at South Scott Reef has extensive communities of sponges, calcareous algae, hard and soft corals, well adapted to the low light levels. (Image credits: Nick Thake (A–C), Andrew Heyward (D))

symbionts and photosynthetic pigments to the low light [17].

Disturbances and reef resilience

Because the reef systems are isolated, they lack the chronic local pressures affecting many others around the world, such as destructive fishing, terrestrial runoff and pollution. But, as with all WA reefs, they are exposed to seasonal storms or cyclones and an increasing frequency of mass coral bleaching and mortality due to climate change [13]. The exposure of WA coral reefs to these acute disturbances varies regionally and with global weather patterns. Cyclone impacts are very rare at reefs south (> 25°S) of Ningaloo and uncommon at reefs in the far north (< 13°S), such

as the Ashmore Reef, and Cocos (Keeling) and Christmas islands. By comparison, cyclone impacts are common at the Scott and particularly Rowley Shoals reefs.

Mass coral bleaching and mortality, due to severe heat stress, have become more frequent and intense across WA, including the Ashmore and Scott reef systems. The Scott reefs bleached during all of the global coral bleaching events (1998, 2010, 2016) and at other times. During the first and third global bleaching events, coral cover decreased by 70–80 per cent across all shallow reefs (Fig. 1.7A, B). Monitoring of coral recovery following the first mass coral bleaching has increased our understanding of how isolated coral reefs respond to climate change pressures.

Fig. 1.7. (A) Severe heat stress in 1998 and 2016 caused mass bleaching and the death of most shallow water (< 20 m) corals across the Scott Reef system. (B) The heat stress in 2016 also affected the deeper (30 m) coral communities that had largely escaped bleaching in 1998, but not those in the deepest (50 m) parts of the South Reef lagoon (C). (Image credits: James Gilmour)

Such isolated reef systems do not exchange coral larvae in sufficient numbers to aid each other's recovery following disturbances [18, 19]. Recovery at the Scott reefs relied on high rates of growth and survival in the corals remaining after the mass coral bleaching, favourable habitat conditions, high water quality, and abundant fish stocks.

The isolated reef systems of Australia's northwest are increasingly affected by severe cyclones and mass coral bleaching. As they are essentially 'closed' systems, the reefs provide a grim indication of the plight of the world's reefs with ongoing climate change, especially considering many other reefs also suffer from additional pressures. The Scott reefs have been impacted by multiple bleaching events in recent years and are unlikely to recover to their previous condition if ocean temperatures increase as predicted. In contrast, the Rowley Shoals are yet to be affected by mass bleaching and provide a spectacular reminder of the natural beauty and economic value of healthy coral reefs.

Kimberley corals exposed
Zoe Richards

Australia's North West Kimberley Bioregion is one of the world's last great wilderness areas given its isolation from urban centres and agricultural influences. The Kimberley Bioregion covers ~476 000 km^2 of reefs, including over 2500 islands. This vast marine realm has been the homeland of numerous Traditional Owner groups for tens of thousands of years, and these groups have deep, ongoing spiritual connections with their sea country.

Coral reefs of the Kimberley region fall into two distinct groups: the large atolls, platform reefs, banks and shoals that occur in the offshore bioregion (including Scott Reef, Rowley Shoals and Ashmore Reef), and the fringing and submerged patch reefs that occur in the inshore bioregion. Together, the *marnany* (reefs – in the Bardi, Jawi, and Mayala languages), *waddaroo* (coral reefs – in

Dambeemangarddee), *warrurru* (reefs – in Wunambal Gaambera) and *warrirr* (reefs – in Balanggarra) form a vast network of coral reef resources that, until recently, had rarely been studied by Western scientists.

The Kimberley reefs are uniquely characterised by their biogeographical and oceanographic setting. The emergent offshore shelf atolls and reefs rise from depths of between 300 and 700 m, directly in the path of the ITF current. Between ~100–200 m water depth, many submerged banks, including Rankin Bank, Echuca Shoal, Vulcan Shoal and Barracouta Shoal, grow to within 10–30 m of the sea surface (Fig. 1.5). Above the 100 m depth contour, inshore reef communities are shaped by extremely large semi-diurnal tides (up to 11 m), which interact with the shallow, complex bathymetry and island archipelagos to produce powerful multidirectional currents and high levels of turbidity from tidal sediment re-suspension and terrigenous input during monsoonal flooding [20].

A particularly noteworthy reef in the region is *Wooleejaaroo* (Montgomery Reef), the world's largest inshore reef (total area of 400 km²). On a spring low tide, *Wooleejaaroo* sits almost 8 m above the surface of the ocean and water cascades down the sides of the reef, creating a spectacular natural phenomenon (Fig. 1.8A). For ~7000 years, the dugongs, turtles, birds, fish, molluscs and crustaceans of *Wolleejaaroo* have sustained the Yawijibaya people (Fig. 1.9).

A similar but smaller geological phenomenon can be found at *Jalan* (Tallon Island), where a coralline algae bank has coalesced to form a single 2 m high and 200 m long terrace that impounds water to feed a series of cascades over low tide (Fig. 1.7B). Turtle Reef in Talbot Bay is another unique geomorphological structure. Classified as an inter-island fringing reef, the 25 km² reef is estimated to be up to 9000 years old, rises ~4.5 m above spring low tide level, and is formed by two coalesced fringing reefs attached to the north and south Molema Islands [21]. Turtle Reef is formed by large depositions of carbonate material, and water is impounded on top of the reef by coralline algal ridges, resulting in large expanses of low tide lagoonal reef surface inhabited by corals, anemones, aggregations of tridacnid molluscs, *Rochia niloticus* (Trochus) and seagrass.

Corals are vital components of tropical ecosystems, playing an important role in carbon cycling, primary productivity and providing critical habitat for marine plants and animals. Until recently, the extent to which the Kimberley reefs were coral dominated was debated. Shallow-water marine benthic surveys were undertaken between 2009 and 2014 across the Kimberley as part of the Western Australian Museum Kimberley Woodside Collection Project to characterise the composition and structure of reefs in this little-known region for the first time. The Kimberley region was found to

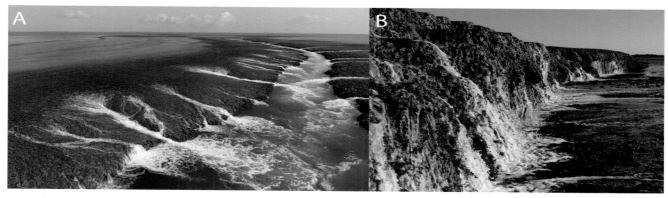

Fig. 1.8.   (A) *Wooleejaaroo* (Montgomery Reef) channel and cascades. (B) Crustose coralline algal terraces at *Jalan* (Tallon Island) impound raised lagoonal habitat that forms 2 m high cascades at low tide. (Image credits: Will Robbins (A), Zoe Richards (B))

Fig. 1.9. Yawijibayas were a self-sufficient clan who lived on the seafood resources of *Wooleejaaroo* (Montgomery Reef) for more than 7000 years. (Image credit: Rebel Films)

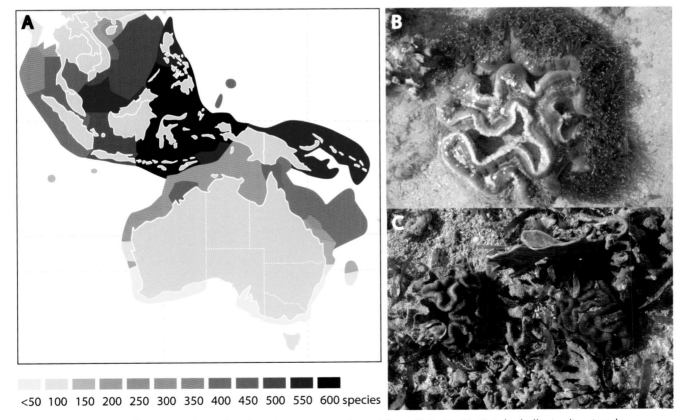

Fig. 1.10. (A) Australasian coral species diversity patterns adapted from [24]. (B, C). *Catalaphyllia jardineri* and *Trachyphyllia geoffreyi* are known only from the inshore Kimberley. (Image credits: Zoe Richards)

support a complex mosaic of highly productive and biologically diverse coral reef habitats. The average regional level of hard coral cover was 23 per cent, but this varied dramatically between stations, with 76 per cent cover recorded north-east of Cassini [22].

A synthesis of new hard coral specimens obtained during the 2009–2014 expeditions, along with historical specimens from the region, revealed that well over 400 species of scleractinian (hard) coral occur in Kimberley [23] (Fig. 1.10A). Cross-shelf gradients are apparent with many species occurring only at offshore atolls, including species not previously been recorded from Australia (e.g. *Acropora elegans* from Scott Reef, and *A. retusa* from the Rowley Shoals). However, some species such as *Catalaphyllia jardineri* and *Trachyphyllia geoffreyi* were only found at inshore locations (Fig. 1.10B, C). This biodiversity is substantially higher than the 350 species that were predicted to occur in the region (Fig. 1.10A).

The central inshore Kimberley and, more specifically, a cluster of fringing and platform reefs in the Bonaparte Archipelago support the most diverse intertidal coral communities in tropical Australia, with 225 species of scleractinian coral recorded here [25]. This level of diversity is remarkable, given that inshore reef habitats are extremely dynamic, and intertidal coral communities can be directly exposed to extreme ambient temperature

Fig. 1.11. (A) On-country meeting between *Dambeemangarddee* Traditional Owner and marine park manager, North Lalang-Garram Marine Park, Kimberley. (B) Newly discovered intertidal reefs at Cape Londonberry, North Kimberley Marine Park. (C) *Acropora aspera* exposed at low tide. (D) *Acropora hyacinthus* excreting mucous at low tide to prevent desiccation. (Image credits: Zoe Richards)

and light conditions for up to 3 hours at a time over spring tides.

A biodiversity impact study was undertaken in the Bonaparte Archipelago after the 2016 El Niño-associated thermal stress event of sustained high water temperatures during an exceptional underwater heat wave. The study found no evidence to suggest that a mass coral bleaching and mortality event occurred in the central Kimberley. This was surprising, given that bleaching events were recorded in the Western Kimberley and Scott Reef (see the previous section). The intertidal and subtidal reefs in the central Kimberley were postulated as providing a regional refuge for photo-symbiotic benthic fauna [26]. New surveys undertaken by researchers from Curtin University, the Department of Biodiversity, Conservation and Attractions and Indigenous rangers in the Lalang-gaddam, North Lalang–Garram and North Kimberley Marine Parks from 2018 to 2020 found extensive new intertidal reefs, and no evidence of recent climatic or physical disturbance (Fig. 1.11A, B).

The capacity for the coral reefs within the Kimberley region to act as refugia under future climate scenarios requires further investigation. Nevertheless, naturally extreme reef environments can provide insight into mechanisms that enable resistance to the environmental conditions that are predicted under future climate change. To ensure that the spectacular coral reefs of the Kimberley have the best chance of survival in an uncertain future, it is vital they receive protection through the creation of no-take sanctuary zones. Fortunately, such protections have been granted via a network of regional marine parks.

Traditional Owners play a key role in protecting Kimberley corals by keeping their sea country rich, alive and healthy. Enabling Traditional Owners, and especially younger generations, to access sea country for cultural expression and learning, while also supporting and enhancing the capacity of Traditional Owners and Indigenous rangers to monitor and report on the condition of sea country, will augment the conservation outcomes for the region (Fig. 1.12).

Fig. 1.12.   Emerging Balanggarra elder Marcus Maraltadj participating in a coral survey in the North Kimberley Marine Park. (Image credit: Zoe Richards)

The Cocos (Keeling) Islands

Sarah Hamylton

The Cocos (Keeling) Islands are an Australian territory in the East Indian Ocean, some 2100 km offshore from continental Australia. The islands are composed of a single horseshoe-shaped atoll and a small, isolated island, Pulau Keeling, ~25 km to the north (Fig. 1.13). At 96°E and 12°S, the nearest land masses are Christmas Island, 900 km to the north-east, and the Indonesian city of Java, 1200 km to the north. The Cocos population is concentrated in the village of Bantam on Home Island.

This is largely made up of Malaysian descendants of indentured workers, brought to the island to work on coconut plantations in the early 1800s.

Cocos (Keeling) atoll has more than 20 low-lying sandy islands around a near-continuous rim that borders a central lagoon. Several passages cut through the reef rim of the atoll, particularly in the north. Both of the Cocos (Keeling) islands rise from an ocean floor that is around 5000 m deep. The atoll has a total area of 225 km with a large, shallow central lagoon that reaches a maximum depth of 15 m. The peripheral coral reef is a gently sloping terrace outside of the atoll, and it runs to a depth of 15 m before declining steeply to the abyssal ocean floor. Inside the lagoon, distinctive blue holes and a mosaic of reticulate reefs have formed in the deeper areas, bordered by large sand sheets around the shallower intertidal flats inside the channels [27]. Cocos (Keeling) was the only coral atoll that Charles Darwin visited during the voyage on the HMS *Beagle*, and it has played a central role in his theory, and our understanding today, of reef formation.

Charles Darwin's theory of coral reef formation

While Darwin's most famous work from the *Beagle* voyage was his theory of evolution, his formal role on the voyage focused on geological matters, including coral reefs. Supported by the British Admiralty, Darwin had been tasked with seeking a better understanding of how coral reefs grow, as they were causing significant mortality from ship wrecks (see 'Encountering and charting the hazardous reefs of Australia, 1622–1864' in Chapter 2). This was a subject about which he proposed a ground-breaking theory of reef formation in a monograph titled *The Structure and Distribution of Coral Reefs*, published 7 years after he returned from the *Beagle* voyage in 1842.

The problem of coral formation concentrated specifically on the major scientific question of how different types of coral reefs, including atolls, barrier reefs and fringing reefs, formed across the world's oceans. At the time, atolls were viewed as tranquil harbours in the centre of vast open oceans where

navigators could safely stop on long sea-going voyages to take lengthy astronomical observations that were needed in order to determine their whereabouts. How could these huge and remarkable rings of coral rock composed entirely of the skeletons of tiny animals that only survive in shallow water rise from the deepest depths of the world's oceans?

It was a question that had given rise to much speculation and would go on to inspire decades of controversy. The renowned geologist Charles Lyell had already suggested that reefs grow up from a volcanic foundation. Darwin took this suggestion forward in two significant ways. First, after studying the configuration of coral reefs across the seas, Darwin proposed that many of the world's reefs have formed on top of very slowly subsiding, inactive volcanoes. Second, Darwin proposed that because coral reefs continued to grow upwards while their underlying volcanic foundations slowly subsided, a sequence existed in which fringing reefs progressively transformed to barrier reefs and then to atolls. Very gradually, fringing reefs that hug the shore would grow further away from the coastline and their lagoons would transform from paddling shallows into deeper waters. Given the passage of enough time, the volcanic landmass on which the whole thing was founded would itself sink down beneath the water surface and become an atoll.

Fringing-reefs are thus converted into barrier-reefs, and barrier-reefs, when encircling islands, are thus converted into atolls, the instant the last pinnacle of land sinks beneath the surface of the ocean. [28, p. 147]

This transition would occur so slowly as to be imperceptible to humans, but over hundreds of thousands of years would lead to the world's oceanic atolls that can be seen today.

The HMS Beagle's visit to Cocos (Keeling)

The *Beagle* glided into the channel at Cocos (Keeling) Atoll on 1 April 1836 and stayed for 12 days. The ship's surveyors took observations from all over the islands, charting water depths from the

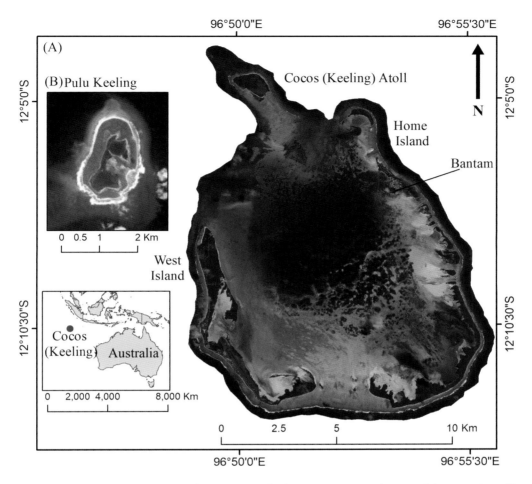

Fig. 1.13. The Cocos (Keeling) Islands, East Indian Ocean. The larger Cocos (Keeling) atoll has a series of low-lying sandy islands around the atoll rim surrounding lagoonal reticulate reefs, and the smaller Pulu Keeling island lies 25 km to the north. (Image credit: Sarah Hamylton)

internal lagoon to the external ocean with a 'lead line'. The end of the lead cone was stuffed with tallow (sticky animal fat), which would gather clues as to what was on the seafloor. If the bottom was covered in sand, this would stick to and cover the tallow. Corals, on the other hand, would leave an impression and sometimes become embedded in the tallow. Many repeat observations were taken by manoeuvring the survey boat and lowering the line into deep and progressively shallower waters, leading Darwin to conclude that Cocos (Keeling) corals do not flourish at depths greater than 20–30 fathoms (36–55 m) and rarely at depths greater than 15 fathoms (27 m).

Above the water, Darwin surveyed from the outer seaward coast of the atoll, across the islands and into the lagoon (Fig. 1.14). His diary entries reveal that he walked the island shorelines and interiors with a rock hammer and a sample bag, picking up and examining fragments of coral from the beaches and collecting specimens of flora and fauna. From the heights of the coconut trees, Darwin speculated that channels around the atoll that had previously linked the outside ocean to the internal lagoon had recently closed.

Darwin formed the opinion that the atoll had subsided by a small amount, probably owing to three earthquakes that had occurred in the

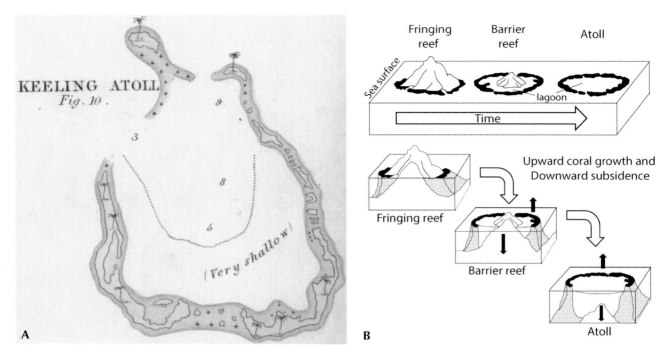

Fig. 1.14. (A) Cocos (Keeling) Atoll, mapped in 1836 by Charles Darwin for his monograph *The Structure and Distribution of Coral Reefs*. (B) The sequence of coral reef formation proposed by Charles Darwin after he visited Cocos (Keeling) atoll aboard the HMS *Beagle*. (Diagram in (B) by Sarah Hamylton)

previous 10 years. He pieced together evidence of shoreline erosion from dead coconut trees and the remains of shed foundations, which must have previously been on dry land but now languished in the shallows, shown to him by the *Beagle*'s Captain Fitzroy. His observations of atoll subsidence along with the upward growth of corals in the shallow water around the atoll both accorded with his theory, which remains the best account of atoll reef formation today.

Christmas Island
Jennie Mallela

Christmas Island is an Australian territory in the East Indian Ocean that spans an area of 135 km² with 73 km of rugged coastline (Fig. 1.15). It formed 60 million years ago when a volcanic seamount rose 5000 m from the seabed to the highest point on Christmas Island, now called Murray Hill, 361 m above sea level. This limestone and basaltic island is surrounded by rocky shores and spectacular nearshore fringing reefs that plunge steeply downwards to a depth of greater than 60 m. The reefs are bathed by the oceanic South Equatorial Current and eddies formed by the ITF and the South Java Current. At the base of the seamount, a tectonic plate shifts the island northwards by a few centimetres every year [29].

On the land, tropical rainforests, freshwater wetlands and steep sea cliffs provide a home to the islands flora and fauna. The national park covers 64 per cent of the island and expands from inland rainforests to protect coastal wetlands, and then extends 50 m offshore from the low tide mark out across the nearshore fringing reef.

Life on the reef

Christmas Island's fringing reefs are among Australia's most remote coral reef systems. They are geographically isolated, and are situated 1500 km west of the Australian mainland and 350 km south of its nearest neighbour, Indonesia.

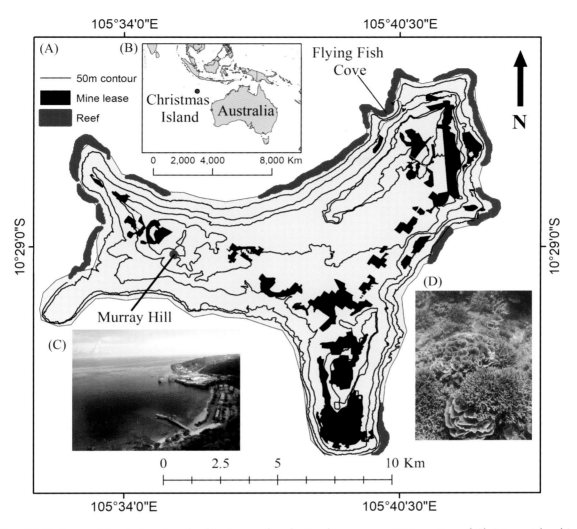

Fig. 1.15. (A) Christmas Island, showing the fringing reef and mine lease areas, (B) Location of Christmas Island in the East Indian Ocean. (C) Flying Fish Cove above the water: the phosphate loading dock. (D) Flying Fish Cove below the water: the reef. (Image credits: Sarah Hamylton and Jennie Mallela)

The narrow reef shelf surrounding the island varies in width from 20 to 100 m [30]. In September and October, oceanic upwelling brings cool, nutrient-rich water that increases planktonic productivity around the island. This attracts large megafauna such as whale sharks (*Rhincodon typus*) that migrate annually to Christmas Island to feast in the nutrient-rich waters [31] (Fig. 1.16).

The coral reefs are a high priority for biodiversity conservation as they provide some of the last safe refuges for globally threatened hard corals, including the rounded, reef-building hard coral *Acanthastrea brevis* and branching colonies of *Acropora papillare*. These species have been identified as vulnerable to extinction this century by the International Union for Conservation of Nature (IUCN) Red List [32].

Millions of endemic red crabs (*Gecarcoidea natalis*) spend most of the year in the inland tropical forests and venture to the reefs once a year to breed (Fig. 1.17). The timing of this mass march is triggered by wet season rains that typically begin to fall in November. During the march, waves of millions of red crabs scuttle over the ground

Fig. 1.16. An inquisitive whale shark, *Rhincodon typus*. (Image credit: Kelly Hoppen © Great Barrier Reef Marine Park Authority, supplied with kind permission)

turning the sandy beaches into a moving red mass.

Upon reaching the ocean, the crabs rehydrate after their long march by immersing themselves in the ocean, after which they mate. The males return to the forest and the females wait in burrows until the high tide starts to turn during the last quarter of the moon, at which point they release their eggs into the water before returning home to their inland forests [33]. The red crab larvae hatch from their eggs upon contact with the ocean and are swept onto the reef by the ebbing tide. An estimated 10 000 eggs are produced by each female. The majority of larvae are usually eaten by reef-dwelling fish and migratory animals. Any juvenile crabs that survive will emerge from the ocean ~4 weeks later and make their way to the inland forests.

The discovery of Christmas Island

Christmas Island was one of the last islands in the Indian Ocean to be discovered by European

explorers. Seafaring charts from early English and Dutch navigators record sightings of the island in the early 1600s. In 1643, Captain William Mynors, from the British East India Company, named the island after sighting it on Christmas Day. In 1886, Captain Maclear of the HMS *Flying Fish* discovered a safe anchorage at a place they named Flying Fish Cove on the north-eastern tip of Christmas Island.

Over the next couple of years, exploring parties used this bay as a base from which to collect flora and fauna samples from across the island. One of these samples, a piece of coral, changed the island's history forever. John Murray, a Scottish scientist who participated in the *Challenger* expedition (1872–76), collected coral samples from Christmas Island. In 1888, he discovered a pebble of highly sought-after pure phosphate of lime in a piece of coral that had been collected by the naval admiralty from Christmas Island [34, 35]. A concerted effort was then made by the admiralty to rapidly collect more samples, and extensive beds of highly valuable phosphate rock were soon

Fig. 1.17. The Christmas Island red crab (*Gecarcoidea natalis*) annual migratory march to the coast where they take a well-deserved dip in the ocean before mating. (Image credit: Jennie Mallela)

discovered on the island, which were mined for fertiliser.

By 1888, Christmas Island and its highly valuable phosphate reserves had been claimed by the British. The island was leased to John Murray and George Clunies-Ross, the owner of nearby Cocos (Keeling) Islands, who jointly formed the Christmas Island Phosphate Company. By the late 1890s, they had populated the island with a community of workers: 200 indentured Chinese labourers, eight European managers, five Sikh policemen and a small number of Malay workers. Phosphate mining began shortly afterwards, with the first major phosphate shipment of 36 000 tons successfully exported in 1900 [34, 35]. During World War II, the islands phosphate reserves were targeted by the Japanese, who took control of the island from 1942 until their surrender in 1945. The island was then reoccupied by the British, with the Australian and New Zealand governments buying the Christmas Island Phosphate Company in 1949. Australia gained sovereignty over Christmas Island on 1 October 1958.

In the 1980s, Indonesian asylum seekers began fleeing by boat into Australian waters, making landfall at Cocos (Keeling) atoll and Christmas Island. The government at the time, presided over by John Howard, excised these offshore islands from Australia's migration zone, thereby preventing the influx of asylum seekers arriving by boat from applying for refugee status. Temporary accommodation for asylum seekers was provided on Christmas Island at Phosphate Hill in 2001, with the later construction of a permanent immigration detention centre in the north-western section of the Island. At its peak in 2013, the facility detained 2960 asylum seekers, massively increasing the island's resident population of 1402. The centre has had an environmental impact, producing a higher volume of waste than the island is equipped to manage. It has also been marked by controversy with claims of inhumane living conditions, riots, self-harm and a court case against the Australian government by asylum seekers.

The environmental legacy of the Christmas Islands phosphate mine

Since 1895, more than 3200 ha of native tropical forest have been cleared for phosphate mining activities. This corresponds to roughly a quarter of the island area and has caused significant environmental damage both to the inland forests and to the reefs. Alongside terrestrial habitat loss, seabird nesting and chick survival have declined for species such as the endangered Abbott's booby, which only nests on Christmas Island's coastal wetlands. By the 1970s, concern over Abbott's booby triggered the formation of the first National Park declared in 1980 across the south-western portion of the island. This was expanded in 1986 and 1989 to form the Christmas Island National Park. Increased numbers of visitors to Christmas Island during colonisation introduced non-native species, including rats (*Rattus rattus*), feral cats (*Felis catus*), yellow crazy ants (*Anoplolepis gracilipes*) and wolf snakes (*Lycodon capucinus*), all of which pose a significant threat to the island's native wildlife. Additional environmental threats include pollution from land-based mining, which runs off the island onto the adjacent nearshore reefs, particularly following heavy rains [36]. Excavated soil runs off the land, clouding the coastal waters and settling on the reef, where it smothers and kills immobile reef organisms. Damage to the nearshore reefs is most obvious down-stream from the mining drying facilities (Fig. 1.18A and B).

Climate change has caused large-scale ecosystem stress in recent decades, including mass coral bleaching and mortality. In 2016, more than half of the shallow, reef-building corals turned white and bleached in response to protracted and elevated seawater temperatures. Over the next few years, many of the reefs partially recovered, with new coral colonies growing at damaged sites, and hard coral dominating much of the reef floor. In contrast, the reef sites most heavily polluted by mining runoff have been slow to recover. Phosphate mining activities are now coming to an end and efforts are underway to rehabilitate the terrestrial environment. It is hoped that the island's coral reefs and unique flora and fauna will survive long into the future.

Torres Strait
Colin Woodroffe

The Torres Strait separates the northernmost point of Australia at the tip of Cape York Peninsula and the south coast of Papua New Guinea by ~150 km. The strait is an important route for international shipping between the Pacific and Indian Oceans, having supported a lengthy history of trade between Australia and South-East Asia, by linking the Coral Sea in the east and the Arafura Sea in the west (Fig. 1.19). It contains granitic islands to the west (e.g. Badu, Prince of Wales Island) and younger volcanic islands fringed by

Fig. 1.18. (A and B) The reef site below the phosphate-drying facility showing clear evidence of sediment damage from mining runoff: fine sediment, bleaching and turf algae overgrowth. (Image credits: Jennie Mallela)

coral reefs to the east (e.g. Murray Island), as well as low-lying sand cays on several reefs in the central region.

There are several distinctive regions of coral reefs in Torres Strait [37]. The central region comprises a maze of scattered coral reef platforms that rise abruptly from shallow depths and submerged *Halimeda* banks. The eastern region comprises numerous shelf-edge and dissected reefs that are comparable to the outer ribbon reefs of the northern GBR, while many of the reefs in the western region display a distinctive east–west orientation due to the strong tidal currents they experience as a result of large volumes of water funnelling from the Coral Sea into the Gulf of Carpentaria.

European exploration of Torres Strait

In August 1770, the *Endeavour* sailed past Cape York, and Lieutenant James Cook landed on a small island, where he raised a flag in a gesture intended to take possession of the land in the name of King George III. Cook had seen no evidence of settled use of the land by the Indigenous inhabitants, despite the fact that Joseph Banks recorded observing several people on the shore, some with necklaces made from mother-of-pearl.

The *Endeavour* sailed through Endeavour Strait, the southernmost of the western entrances to Torres Strait, between Possession Island and Prince of Wales Island. We now know that the Strait had been navigated almost 164 years previously, when the Spaniard, Luis Váez de Torres, after whom it is now named, had found a route between the reefs in the *San Pedrico* in 1606.

Isthmus to islands

Neither of these explorers could have known that their transit between Australia and the large island to its north would not have been possible 10 000 years previously. Torres Strait is relatively shallow, with much of the central area no more than 25 m deep and most of the western part of

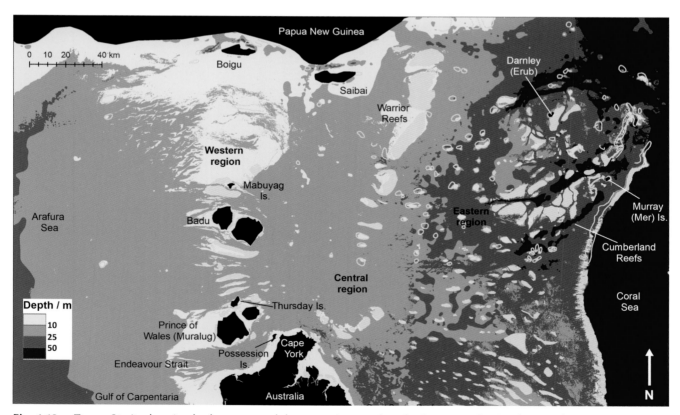

Fig. 1.19.   Torres Strait, showing bathymetry and the extensive coral reefs. (Image credit: Sarah Hamylton)

the strait less than 10 m deep (Fig. 1.20). For much of the preceding 2 million years, substantial volumes of water had been locked up in high-latitude ice sheets, meaning that the sea level was lower. During this time, Australia and New Guinea were connected across the Sahul shelf (Fig. 1.20). The area flooded as ice sheets melted, and the rising sea encroached along the eastern margin of Australia, inundating again what is now the GBR, including Torres Strait.

Torres Strait Islanders

Torres Strait covers 48 000 km², adjoining the northern GBR, but with the important difference that it is the home of Torres Strait Islanders. It seems likely that the isthmus was inhabited in the late Pleistocene when Australia and New Guinea were connected because land both to north and south was occupied; it may have been a route by which people had migrated [38]. It is unclear whether the isthmus was inhabited by Aboriginal communities or people of Melanesian descent, such as the current Torres Strait Islanders.

The first radiocarbon-dated evidence for human use of the islands comes from the Badu and Mabuyag islands and coincides with the time that these western islands became separated, with the sea level reaching close to its present level around 7000 years ago. Evidence from fossil corals on many of the reef flats indicates that the sea rose to a height that was slightly higher than present [39]. During this period, between 3000 to 6000 years ago, the islands appear to have been intermittently visited by sea-going peoples. Over the past 3000 years, several of the reef platforms have accumulated sandy reef islands, and archaeological evidence indicates the existence of a more widespread marine-focused Torres Strait cultural complex [40].

A labyrinth of reefs

Extensive reef platforms have developed across the shallow shelf of the strait over the past few millennia [39]. A near-continuous chain of outer reefs extends to the north of GBR, forming a chain of reef crests that poses a challenge for mariners. The complex of dendritic reefs, called Cumberland Reefs, exhibit the deltaic morphology that is also found in the northern GBR. Coral reefs fringe the eastern volcanic islands, reef platforms have developed over topographic remnants of previous interglacial reefs within the centre of the strait, and there are many linear reefs in the west as well as

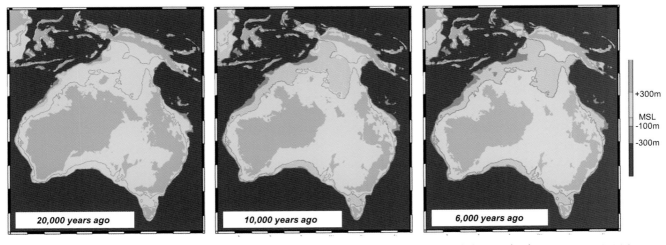

+300m

MSL
-100m

-300m

Fig. 1.20. Flooding of the Sahul shelf. Australia and New Guinea were connected during the last ice age 20 000 years ago; the sea had flooded into the Gulf of Carpentaria by 10 000 years ago as the ice melted, and Torres Strait had been breached by 6000 years ago as sea level rose to its present level. Maps prepared by the Universities of Durham and Toronto and accessed from the Permanent Service for Mean Sea Level (https://www.psmsl.org). (Image credit: Glenn Milne)

fringing reefs around the numerous granitic islands. Reefs cover more than 2400 km² of seabed that Cook referred to as a labyrinth. Isolated reefs vary from small reef patches, often with a crescentic windward (eastern) margin, to lagoonal reefs, which are surrounded by a prominent reef crest. Several of the larger reef platforms have adopted a planar form, with a reef flat that almost completely dries at low tide.

Navigating a safe passage through this maze of reefs, avoiding the shallows and shoals, and battling unpredictable weather and oceanographic conditions has remained a major challenge. Today, large vessels require pilotage through the strait. One of the earliest and most remarkable feats was by William Bligh. In 1789 he managed to steer the open launch, in which he and a handful of loyal crewmen had been set adrift following the mutiny on the *Bounty,* between the reefs. Simultaneously, in appalling conditions, he charted a new route north of Badu. Bligh made a second passage through the straits 3 years later in the *Providence,* as he carried breadfruit from Tahiti to the Caribbean. During that voyage, on 10 September 1792, well-armed Islanders approached Bligh's ship in their canoes, and a violent exchange of arrows and musket fire occurred, resulting in deaths on both sides, which gave rise to the name Warrior Reefs (Fig. 1.19) [41].

On board the *Providence* was a young Matthew Flinders who drew, perhaps under Bligh's supervision, what is considered his first chart, entitled *A Chart of the Passage between New Holland and New Guinea as seen in His Majesty's ship Providence in 1792.* Flinders said of the strait: 'Perhaps no space of 3 1/2° in length, presents more dangers than Torres Strait'. As Bligh completed his second transit through the strait, he did not realise that in 1791, the *Pandora,* which had been sent to round up some of the *Bounty* mutineers, had hit a reef near the Murray Islands and was wrecked with considerable loss of life; several survivors struggled for a week to make their way through the maze of reefs and reach Batavia (now Jakarta) in Indonesia.

In 1802, Flinders returned in the *Investigator,* following a new west–east route through the strait between the routes taken by Cook and Bligh. He discovered that Bligh's mapping had been considerably more accurate than Cook's, which suffered a discrepancy in longitude which might be explained by the poor health and hence imprecise astronomical fixes of the astronomer Charles Gréen [41]. Flinders again passed through the straits in the *Cumberland* in 1803 and wrote sailing directions for navigating from the Pacific to the Indian Ocean, which he recommended was best embarked on between April and the end of October, when favourable south-east trade winds prevailed. This represented a significant short cut compared with sailing instead around the east and north coast of New Guinea. The intricate network of reefs continued to pose hazards for shipping. In 1819, Phillip Parker King undertook detailed charting of northern GBR and Torres Strait waters, initially in the *Mermaid* and subsequently in the *Bathurst.* Nevertheless, there were numerous shipwrecks, with at least 120 recorded up until 1920.

Strong tidal currents result from contrasting tidal systems in the Coral Sea and the Gulf of Carpentaria. Many reefs are elongated, reflecting strong currents, which can reach speeds of over 2 m s^{-1} in the narrower channels. Entry through the outer reef is challenging, and a three-storey circular beacon tower, painted red and white, was constructed as a guide just to the south on Raine Island in 1844 [42].

In contrast to the crowded sailing vessels of the late 17th and early 18th centuries, the double-outrigger dugout canoes built by the Torres Strait Islanders, using trees from New Guinea, were ideal for travelling in these shallow waters.

Marine ecosystems and a sea-going culture

The rich resources of Torres Strait supported marine-focused communities on many of the islands, which were interlinked by trade and intermarriage, although also, on occasion, by inter-island conflict. The sea was, and continues to be, central to ceremonial life and contains a great diversity of marine animals, including sharks, saltwater crocodiles and seven species of turtle. Particularly noteworthy are the populations of dugong, which possibly exceeded 10 000 individuals in the 1980s. Dugong and turtle remain an important cultural resource for Torres Strait Islanders, and traditional hunting is permitted as part of a sustainable and culturally appropriate management program (Fig. 1.21). In the mid-19th century, collection of bêche-de-mer (sea cucumbers, also called trepang) commenced, followed by a prosperous pearling industry that initially used free-divers from Japan and other places in the Asia-Pacific region and subsequently followed by helmeted divers operating from the specially designed Torres Strait pearling luggers. This formed the basis for other marine-focused commercial enterprises, such as fishing and crayfish collection.

Further exploration of Torres Strait was undertaken in the mid-19th century by HMS *Rattlesnake*, captained by Owen Stanley. Thomas Huxley was aboard as assistant surgeon, although he engaged primarily in natural history, especially a study of marine invertebrates. Huxley subsequently encouraged Alfred Cort Haddon to make Torres Strait the focus of an expedition in 1898. Initially, with a biogeographical focus, Haddon turned his attention to the people, particularly in the Murray Islands. The sociological studies and ethnographic collections made by researchers on this expedition changed the course of British anthropology. The collection of ~1000 artefacts, film recordings and photographs formed the basis for the development of social anthropology and was nurtured when Haddon took up a position at the University of Cambridge.

Coincidentally, the remote Murray Islands were the catalyst for further change, when Eddie Mabo and four other Torres Strait Islanders went to the High Court of Australia in 1982; they asserted that their island (Mer) had been continuously inhabited

Fig. 1.21. *Mulungu*, a linocut by Torres Strait artist Alick Tipoti, showing two hunters returning to their island after capturing a dugong (Dhangal) and a turtle (Waru) at sea. (Image courtesy of Alick Tipoti and the Australian Art Network)

and exclusively possessed by them, and accordingly claimed land rights. In 1992 they were vindicated, and native title was recognised, finding that the Islanders had maintained traditional law. This over-turned the principle of *terra nullius*, the idea that the land belonged to no-one, which had been used as a principle in international law to justify land claims by right of occupation by the British in the absence of other peoples. In the fullness of time, the Mabo decision would prove a landmark case in the broader recognition of the rights of Aboriginal and Torres Strait Islander peoples as the Traditional Owners of their land and recognising native rights in their seas.

As is seen in most other reef areas, climate change, particularly the ongoing sea-level rise that is already occurring at much higher rates than other parts of the world, will have significant impacts on the Torres Strait. The low-lying areas of many of the islands are already experiencing coastal erosion and inundation, which is threaten-ing the communities that live on them [43].

Coral Sea
Andrew S. Hoey

The Coral Sea is situated off Australia's north-east coast bounded by Papua New Guinea to the north, the Solomon Islands, Vanuatu and New Caledonia to the east, and the Tasman Sea to the south. Australia's Coral Sea marine estate falls within the Coral Sea Marine Park (CSMP), which extends from the eastward margin of the Great Barrier Reef Marine Park (GBRMP) to the outer extent of Australia's EEZ, some 1200 km offshore (Fig. 1.22). The CSMP is among the world's largest and most isolated marine parks, encompassing an area of 989 836 km². Within the CSMP there are ~56 islets and cays as well as 20 widely separated shallow reef systems, ranging from Ashmore and Boot reefs adjacent to the Torres Strait in the north to Cato Reef in the south, and Mellish Reef (> 1000 km east of Cairns) in the far east. These shallow reef systems, including Lihou Reef as one of the world's largest atolls (~2500 km²), have a combined reef area of 15 024 km².

The reefs of the Coral Sea are largely shaped by the geomorphic, oceanographic and environmen-tal conditions of the region. Reefs within the Coral Sea are separated by oceanic waters up to 4000 m deep, rising from seamounts on four major deep-water plateaus: the Eastern Plateau in the north, the Queensland Plateau in the central region, and the Marion and Kenn plateaus in the south [44]. Potential connectivity between these isolated reefs occurs through major ocean currents, namely the west-flowing Southern Equatorial Current (Fig. 1.22). This current system strengthens during the summer months and bifurcates on the Australian continental shelf to form the south-flowing EAC and its eddies, and the Hiri Gyre in the Gulf of Papua to the north.

The Coral Sea has limited exposure to direct human pressures (e.g. fishing, runoff) relative to more accessible coastal reefs. These reefs are charac-terised by clear waters (underwater visibility > 50 m is common) and low nutrients. Coral cover on many of the reefs within the Coral Sea, especially those in the central Coral Sea, has historically been relatively low as a result of oceanic swells and frequent cyclones. These shape exposed reef faces into a rela-tively featureless habitat dominated by crustose coralline algae and few corals, while sheltered back-reef environments have higher coral cover and greater habitat complexity [44]. Moreover, reefs in the central Coral Sea have been repeatedly exposed to severe tropical cyclones [44, 45]. These distur-bances, coupled with the limited connectivity and supply of coral larvae from other sources, appear to be major determinants of coral cover on these reefs.

Coral Sea reefs support unique coral and reef fish communities that are distinct from those of the adjacent GBR and share many species with reefs in the Tasman Sea to the south (i.e. Elizabeth and Middleton reefs and Lord Howe Island), and nations to the east (New Caledonia, Vanuatu and the Solomon Islands). A striking feature of these reefs is the sheer diversity of reef fish (> 600 spe-cies) and the high abundance and biomass of sharks and other large predatory fishes, which is comparable to other isolated reef systems, such

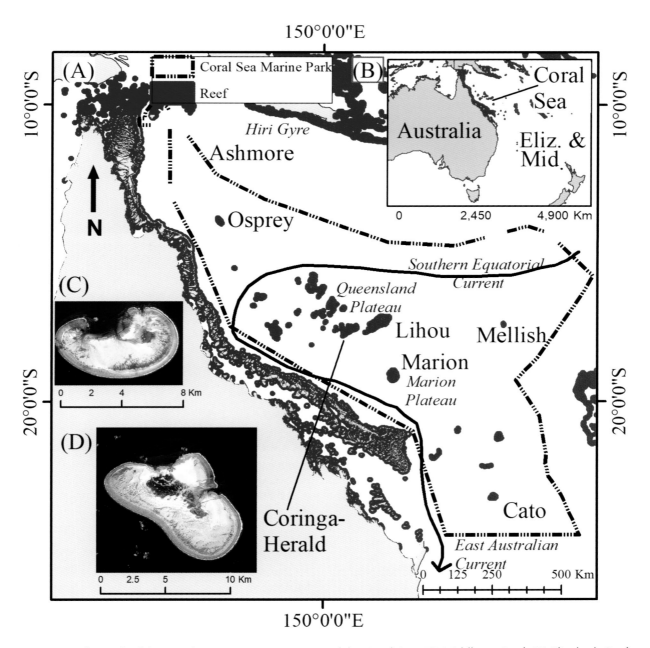

Fig. 1.22.   (A) The reefs of the Coral Sea region. (B) Location of the Coral Sea. (C) Middleton Reef. (D) Elizabeth Reef. (Image credits: Sarah Hamylton)

as the Chagos Archipelago in the central Indian Ocean [45] (Fig. 1.23). This is likely to reflect the limited fishing that occurs on these reefs.

Despite the isolated nature and hence limited direct human pressures on Coral Sea reefs, they are increasingly being exposed to the effects of climate change. Indeed, three major coral bleaching events were recorded between 2016 and 2021 (2016, 2017 and 2020) for the Coral Sea, with the 2020 event being the most severe and widespread [45]. The effects of these events, and other major disturbances, may be particularly pronounced on isolated reefs, such as those in the Coral Sea, due to the reliance on self-recruitment of coral larvae (i.e. larvae spawned from adult corals on the same reef rather than those nearby) for replenishing coral populations.

Fig. 1.23.   The Coral Sea. (A) Shallow lagoonal patch reefs with sand cay in the background. (B) Diverse coral assemblage viewed from above. (C) School of trevally (*Caranx sexfasciatus*) on the reef slope. (D) The longfin parrotfish (*Scarus longipinnis*) is one of the most common parrotfish in the Coral Sea, but is extremely rare on the adjacent Great Barrier Reef. (Image credits: Tane Sinclair-Taylor)

Elizabeth and Middleton reefs

Elizabeth and Middleton reefs are ~800 km south of Cato Reef (the southernmost reef in the Coral Sea), 170 km north of Lord Howe Island and 570 km east of Coffs Harbour on the Australian mainland coast. These reefs rise from a deep-water plateau some 2000 m deep on the Lord Howe Rise. Both reefs are of similar size (40–50 km²) with extensive spur and groove reef formations characteristic of high wave energy environments on their eastern and southern aspects. The extremely shallow rim encloses Elizabeth Reef and semi-encloses Middleton Reef to form extensive shallow lagoonal habitat interspersed with deeper blue holes.

Due to the lack of any visible landmarks and the proximity to historical shipping routes to and from

Sydney, Newcastle and Brisbane, more than 30 ships have been wrecked on Elizabeth and Middleton Reefs, several of which can be seen around the reef perimeters (Figs 1.24 and 1.25). These wrecks include the *Runic*, grounded on Middleton Reef in 1961, which at 13 500 tonnes was the world's largest refrigerated cargo liner at the time, and the Japanese tuna boat *Fuku Maru*, which came aground on Middleton Reef in 1963.

Elizabeth and Middleton reefs are high-latitude reefs and support a mix of tropical and subtropical coral, fish and algae species, including several endemic species or species that are generally rare over most of their range (Fig. 1.25) [46]. More than 100 species of scleractinian (hard) corals and > 310 species of fish have been recorded across both

Fig. 1.24. The bulk carrier 'Kyoten Maru' wrecked on Lihou Reef in the Coral Sea in 1982. (Image credit: P. Howorth © Great Barrier Reef Marine Park Authority, supplied with kind permission)

reefs, including several species that may be unique to these reefs (Fig. 1.26). For example, the double header wrasse *Coris bulbifrons*, the three-striped butterflyfish *Chaetodon tricinctus*, and McCulloch's anemonefish *Amphiprion mccullochi* are endemic to Elizabeth and Middleton reefs, Lord Howe Island and Norfolk Island. Elizabeth and Middleton reefs also support many top predators, in particular the Galápagos shark *Carcharinus galapagensis*, and the Black Cod *Epinephelus daemelii*, which is listed as vulnerable under the Commonwealth *Environment Protection and Biodiversity Conservation Act 1999* [46]. Populations of the black cod have been overfished elsewhere in the south-western Pacific, with Elizabeth and

Fig. 1.25. Elizabeth and Middleton reefs. (A) The *Environment Protection and Biodiversity Conservation Act*-listed black cod (*Epinephelus daemelii*). (B) The endemic three-striped butterflyfish (*Chaetodon tricinctus*) in shallow lagoon habitat. (C) A Galápagos shark (*Carcharhinus galapagensis*). (D) The rusting hull of the *Runic* that was wrecked on the western perimeter of Middleton Reef in 1961. (Image credits: Andrew Hoey)

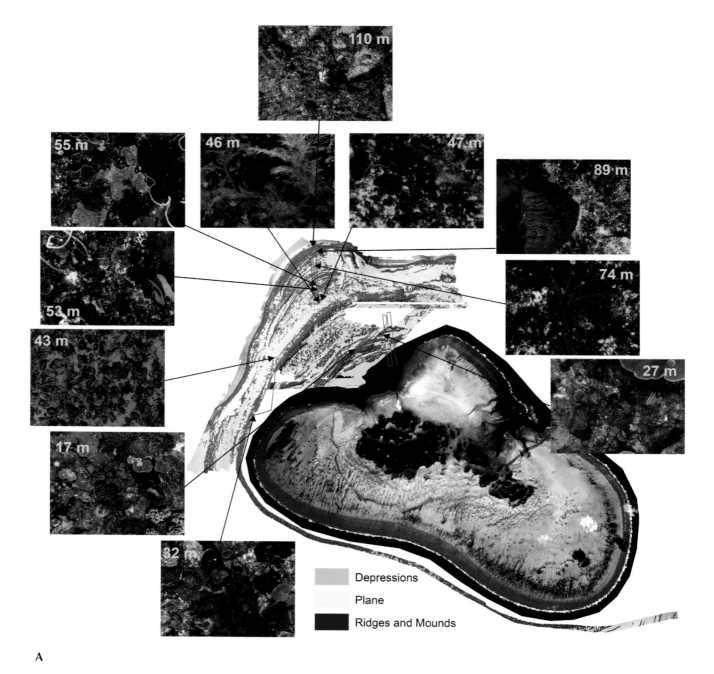

A

Fig. 1.26.   (A) Elizabeth Reef and (B) Middleton Reef mapped using multibeam sonar and autonomous underwater vehicles (AUV *Sirius* and AUV *Nimbus*) by Geoscience Australia (2020). Geomorphological features of the shelf include depressions, plane and ridges and mounds. Autonomous underwater vehicle images show the benthic habitats. Ridges and mounds are colonised by dense hard and soft corals, with a mix of barren sediments and moderate cover of turfing algae on planes and rhodolith beds in deeper areas towards the shelf edge. For more detail, see the survey report [47]. (Image courtesy of the National Environmental Science Program, Marine Biodiversity Hub and Geoscience Australia; source: Carroll *et al.* 2021, [47])

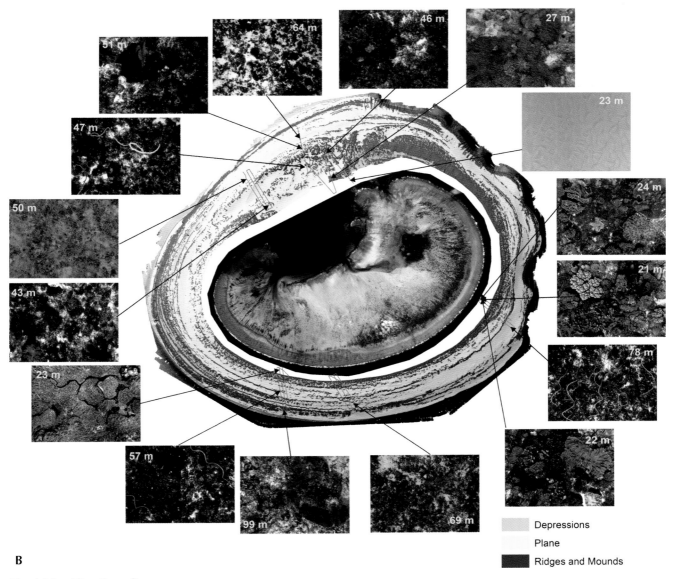

B

Fig. 1.26. (Continued)

Middleton reefs recognised as one of the last remaining strongholds for this species.

High latitude reefs such as Elizabeth and Middleton are extremely vulnerable to acute disturbances that cause extensive coral mortality. This is because the replenishment and recovery of coral populations are limited by both their isolation and environmental constraints on population dynamics. Indeed, recovery of coral populations at Elizabeth and Middleton reefs was slow and protracted in the aftermath of an outbreak of crown-of-thorns starfish in the mid- to late 1980s, with the cover of coral increasing from ~6 per cent cover in 1994 to only ~20 per cent in 2014 [46].

Lord Howe Island

Michelle Linklater

Lord Howe Island is a mid-oceanic, subtropical island in the south-west Pacific Ocean, situated 600 km off the mainland east coast of Australia. The island forms part of an island–seamount chain that includes Balls Pyramid to the south, and Elizabeth Reef and Middleton Reef to the north (see the previous section). These islands were created by hotspot volcanism ~6 million years ago and collectively span a distance of 250 km from the north to the south. Over time, the Lord Howe Island volcano was eroded to become a small island remnant, 10 km in length, with a 6 km fringing coral reef and lagoon on the western coast. Encircling the island is a broad shelf, typically 30–60 m deep, which then plummets steeply into abyssal depths several thousand metres deep. Above water, the island is now an impressive mountain landscape, rich in terrestrial and marine life and colonised by a vast array of endemic land species.

Positioned at the latitudinal threshold of reef-forming seas, Lord Howe Island supports the southernmost true coral reef in the Pacific Ocean and an extensive fossil reef. Coral reef growth is possible at this high-latitude location due to the influx of the EAC, which travels south from the Coral Sea along the Australian mainland and delivers warm, tropical waters to the region via an eastward-flowing eddy (see Fig. 1.1). The influence of the EAC is critical to the distinctive mix of regional tropical and temperate marine life, with some tropical corals surviving at their southernmost limits and some algae living at their northernmost limits [48, 49]. Its isolation from the mainland coast results in high endemism, and the region is recognised as a global biodiversity hotspot. These attributes, together with its natural beauty and the fact that this is one of the world's southernmost coral reefs, have earned the region World Heritage protection status, as well as the designation of a New South Wales state and Commonwealth marine park, the Lord Howe Island Marine Park.

Coral growth at the limits

The island is considered a 'marginal' setting for coral growth as it occurs at the lower limits of temperature tolerance and aragonite saturation, a component of seawater chemistry that determines how organisms like corals build their carbonate skeletons (see 'Marginal reefs: distinct ecosystems of extraordinarily high conservation value' in Chapter 4). Despite this marginal setting, hard corals are abundant, and reef growth has been prolific over the past 10 000 years.

The island's fringing reef formed 6500 years ago following the last ice age, and the reef continued to grow upwards as the sea level rose to its present level [50]. Coral cover greatly varies at different sites around the island, with the highest coral cover occurring within the sheltered lagoon [48, 49]. The coral fauna of Lord Howe Island is further discussed in Chapter 4, 'Marginal reefs: distinct ecosystems of extraordinarily high conservation value'.

High-resolution seabed mapping of the deeper shelf surrounding Lord Howe Island has revealed a substantial fossil reef on the mid-shelf – a remnant of a large reef that has now died, but once would have thrived (Fig. 1.27, [51]). The large, 155 km² fossil barrier reef of Lord Howe Island dominates the mid-shelf in 25–50 m water depth and measures more than 20 times the area of the modern fringing reef [52]. Dated corals from the fossil reef surface indicate that the fossil reef flourished 9000 to 2000 years ago [51]. A similar, smaller fossil reef surrounds the neighbouring volcanic peak of Balls Pyramid [52]. These deeper reefs are likely to have formed atop older fossil corals and other material, and drowned because they were unable to keep up with the fast pace of rising seas following the last ice age.

As technology permits exploration of these deeper waters, the marine communities living on these reefs are increasingly discovered and described. These deeper reef communities are termed 'mesophotic' communities, defined as typically occurring within 30–150 m water depth. High-resolution photographs and video from a towed underwater video system around Lord

Howe Island and Balls Pyramid have revealed abundant and diverse modern mesophotic communities [53]. Communities comprise a mix of hard corals, soft corals, black corals and algae [53]. Hard corals encrust the fossil coral reef surfaces at depths of 30–60 m (Fig. 1.27), and are seen on even deeper reef ridges down to a maximum observed depth of 94 m. The great depth of water in which these corals survive is remarkable, given the high-latitude location and the marginal nature of coral growth in this region. Such deep coral growth is possible at this location due to the exceptional water clarity, which allows light to penetrate to these extraordinary depths such that light-dependent species like hard corals can survive and thrive.

Crossing the threshold into reef-forming seas

The Lord Howe Island chain, which includes Balls Pyramid, Elizabeth Reef and Middleton Reef, presents an interesting example of geological evolution for an island chain that has crossed the threshold

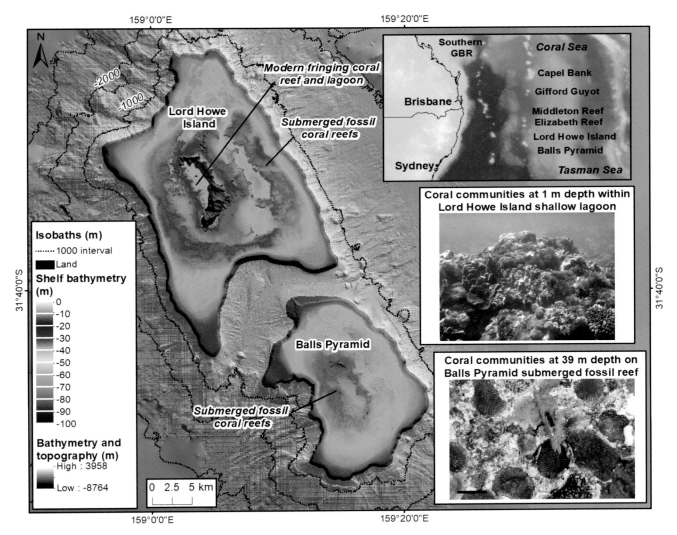

Fig. 1.27. High-resolution seabed mapping of the Lord Howe Island and Balls Pyramid shelves, showing detailed bathymetry of large, submerged fossil reefs. Coral and algae co-exist in the shallows, and corals are abundant in deeper waters. (Image credit: Michelle Linklater)

Fig. 1.28. View of Lord Howe Island taken from the Mount Eliza in the north, looking towards the southern peaks of Mount Lidgbird and Mount Gower, with a sheltered, shallow coral reef lagoon to the west. (Image credit: Matt Curnock)

into reef-forming seas (Fig. 1.29). In a classic Darwinian sequence of atoll evolution, a reef develops around a volcano, and as the island subsides and the reef grows vertically, an atoll forms (see 'Charles Darwin's theory of coral reef formation' in this chapter). If this same process occurs outside reef-forming seas, the volcano may be eroded by wave action to form planated shelves. In the case of the Lord Howe Island chain, the volcanoes originally formed further south in the Tasman Sea, outside reef-forming seas. Here, wave action and subaerial

exposure are believed to have eroded the volcanoes to a fraction of their original size to form broad, near-horizontal shelves around Lord Howe Island and Balls Pyramid [54]. Over time, the islands transported into reef-forming seas by movement of the earth's crust from northward tectonic plate motion, and reefs formed to varying degrees along the island chain. These reefs probably formed successively during the warm interglacial periods between ice ages, and eventually became the sequence of reef evolution we see today [54, 55].

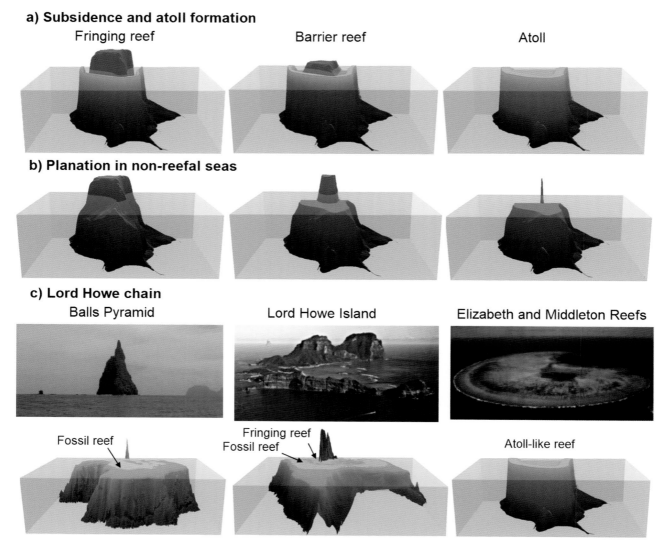

Fig. 1.29. Sequences of island-reef evolution: (a) the classic Darwinian sequence in reef-forming seas; (b) the sequence of erosion outside of reef-forming seas; and (c) the hybrid example presented in the Lord Howe chain where volcanoes were carried into reef-forming seas. (Image credits: Michelle Linklater, modified from [54 and 55])

The southernmost island in the chain, Balls Pyramid, is a dramatic rock pinnacle surrounded by a wide underwater shelf that has seemingly undergone heavy erosion before entering reef-forming seas (Fig. 1.30). A substantial fossil reef is now known to have formed on the shelf, but it did not grow to form a modern fringing reef. This is likely to be due to the lack of shallower substrate available for corals to grow on as sea levels rose. The Balls Pyramid pinnacle plunges steeply beneath the surface, leaving little by way of a shallow platform available for corals to colonise. Lord Howe Island has a similarly flat submerged shelf, but possesses a larger, exposed remnant volcanic island with a more gradual steepening of substrate beneath the water's surface. Here, the greater availability of shallower substrate allowed a modern fringing reef to form on the extensive fossil reef. To the north, Elizabeth and Middleton reefs appear to have kept pace with rising sea levels, forming atoll-like reefs, which encircle a central lagoon or passage with no volcanic island evident at the surface, likely to have kept pace with rising sea levels. Together, this chain of islands requires the Darwinian model of atoll evolution to be broadened to incorporate the movement of tectonic plates across a threshold in environmental conditions necessary for reef formation. Here, instead of coral reefs forming around a subsiding volcano, the volcanoes, initially outside of reef-forming seas, are first heavily eroded to form wide shelves, and are then carried into warmer waters where coral reefs have formed to varying extents around and over them [54, 55].

Coral growth in this region provides key insights into 'marginal' high latitude and deeper, mesophotic reefs, as well as reef evolution over millennia. In turn, this provides important context for global studies of coral reef development. The diverse and distinctive marine life and the interaction of tropical and temperate influences reinforce the high conservation value of Lord Howe Island and its importance as a globally significant hotspot for biodiversity.

The Great Barrier Reef
Mel Cowlishaw and Scott Smithers

The GBR is renowned for its superlative natural beauty and significant biodiversity. As the world's largest coral reef ecosystem, it is made up of a network of reefs and islands that cover an area of 344 400 km² along the Queensland coast (Figs 1.31 and 1.32). The GBR is of cultural, social and economic significance to many Australians. Aboriginal and Torres Strait Islander peoples are the

Fig. 1.30. Balls Pyramid, a pinnacle at the south end of the Lord Howe Island chain that sits on a wide underwater shelf that supports diverse coral communities. (Image credit: Matt Curnock)

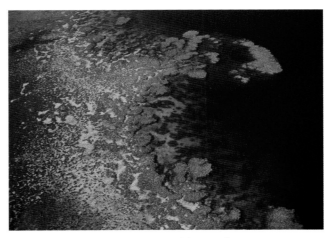

Fig. 1.31. Aerial photograph of far northern Great Barrier Reef in 1981. (Image credit: Len Zell © Great Barrier Reef Marine Park Authority, supplied with kind permission)

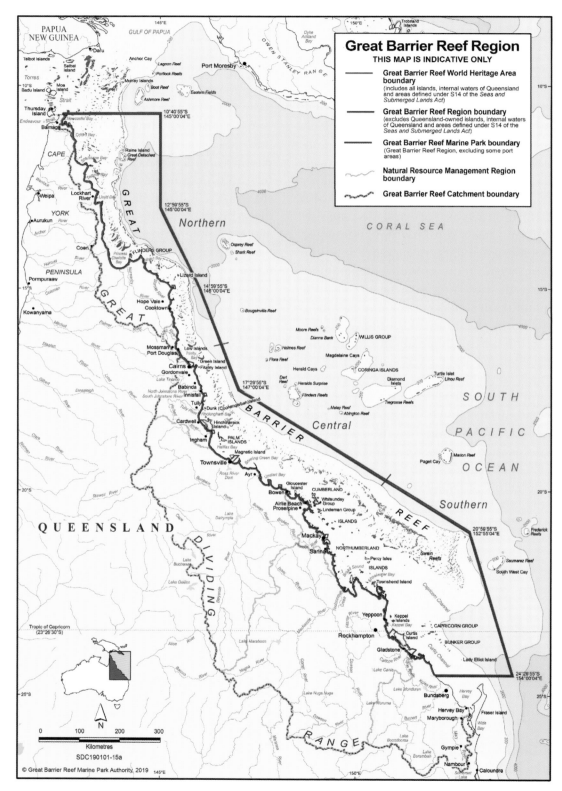

Fig. 1.32.   Great Barrier Reef and Great Barrier Reef Marine Park Boundaries. (Image credit: Great Barrier Reef Marine Park Authority)

Traditional Owners of the GBR and have been linked to it since Australia was first occupied by humans, ~70 000 years ago.

There are over 70 Aboriginal and Torres Strait Islander Traditional Owner groups with cultural authority for sea country management along the GBR and, despite events of the past, they have retained their connection to their land and sea country [56].

Great Barrier Reef: development and diversity

The fundamental existence of the GBR is underpinned by the presence of a large and shallow continental shelf located in tropical waters at a location likely to receive coral larvae transported by major currents. This shelf is most narrow and shallow (mostly less than 30 m deep) north of Cairns, and widest and deepest (mostly ranging between 30–60 m deep) towards the south. The outer barrier is just over 21 km offshore at Cape Melville on the northern GBR, but it is more than 250 km from the outer Swains Reefs to the mainland coast. Now spanning 14° of latitude between 9°30′ S and 24°30′ S, this large shallow area provides an excellent setting for the development of a large reef system.

Although the northern coast of Australia had drifted into tropical waters as early as 24 million years ago [57], cores drilled through the GBR suggest that its foundations did not develop until much later. The ephemeral reefs are analogous to the inshore turbid zone reefs appearing around 700 000 years ago, and the framework reef growth is comparable with today's reefs that began to form later than 450 000 years ago [58]. Since this time, reef growth has been punctuated by glacial-interglacial oscillations in sea levels, with the most recent phase of reef growth only occurring since approximately 10 000 years ago [3].

Perhaps surprisingly, some fundamental attributes of the GBR remain imprecisely known. For example, estimates of its size (constrained by the marine park boundaries to the north and south, and by the shelf edge and coastline to the east and west), vary between around 230 000 km^2 and almost 239 000 km^2. Similarly, attempts to quantify the exact number of individual reefs within the GBR (mostly derived from satellite imagery) yield numbers between around 2900 and 3700 reefs, with differences in observation scale or the classification of some features as single or multiple reefs being responsible for the variation [59]. The earliest attempt to quantify the number of reefs on the GBR used maps produced from LANDSAT satellite imagery with a resolution of around 80 m in the 1980s. From these counts, the number of reefs by reef types include 758 fringing reefs, 566 submerged reefs, 446 small 'patches' of reef at the sea surface lacking consolidated reef flats, 524 crescentic and lagoonal reefs, 544 planar reefs and 66 ribbon reefs. These sum to a total of 2904 named reefs covering a combined area of 20 055 km^2, or around 8 per cent of the GBR shelf area [59]. More recent assessments suggest a larger number of individual reefs and a larger total reef area (24 081 km^2) [60]. It is important to note that these estimates are based on reefs that can be seen from the sea surface. Technology now allows much better observations of the sea floor, including deeper and/or more turbid parts of the GBR. Recently, such technology has revealed more than 14 000 km^2 of mesophotic coral reef within the GBR; this significantly increases the total estimated reef area [61]. Application of the same technology has similarly revealed the distribution, abundance and complexity of *Halimeda* bioherms, and pointed to their importance in supporting overall biodiversity [62].

The GBR includes many important habitats in addition to coral reefs, including more than 6000 km^2 of seagrass, more than 2000 km^2 of mangroves, and around 1000 reef islands. These and other habitats included within the GBR support an astounding diversity of marine life, including more than 600 hard and soft coral species, 3000 species of molluscs, 1625 species of fish, over 30 different species of whales and dolphins, and six of the seven global species of marine turtle. In addition, there are more than 200 species of birds, including 22 species of seabirds and

Fig. 1.33.   Great Barrier Reef Utopia. (Image credit: Melanie Hava)

32 species of shorebirds, and countless species of bacteria, fungi and other organisms living on the reef. The size, complexity, biodiversity and cultural significance of the GBR are among the many values that make it a very special place, to be carefully managed and protected. Although all reefs worldwide are facing some serious challenges [63], the GBR and its associated ecosystem is protected and is considered by many to be one of the best managed coral reef ecosystems in the world (see 'Zoning the Great Barrier Reef' in Chapter 5).

The Great Barrier Reef Marine Park

The Great Barrier Reef Marine Park was established in 1975 under the *Great Barrier Reef Marine*

Park Act 1975, which aims to provide for the long-term protection and conservation of the environment, biodiversity and heritage values of the GBR. In 1981, the GBR was inscribed in the UNESCO Convention Concerning the Protection of the World Cultural and Natural Heritage, on the basis of its natural outstanding universal value, including 'man's interaction with his natural environment'

[64]. The GBR was the first coral reef ecosystem to be declared a World Heritage site and is also formally listed as a national heritage place under the Commonwealth *Environment Protection and Biodiversity Conservation Act 1999*. As it is home to many listed migratory species, threatened species and ecological communities, its wetlands are of international importance. The Great Barrier Reef

Marine Park is a multiple-use area, with the most recent zoning arrangements established in 2003 to protect and conserve its biodiversity while also providing opportunities for its ecologically sustainable use (see 'Zoning the Great Barrier Reef' in Chapter 5).

In 2004, the Queensland Government established the Great Barrier Reef Coast Marine Park to allow for management of the reef within the state jurisdiction under the Queensland *Marine Parks Act 2004*. The Great Barrier Reef Coast Marine Park covers ~63 000 km², is contiguous with the Great Barrier Reef Marine Park, and covers the area between the low- and high-water marks and many State of Queensland waters (refer to Fig. 1.32). Since 1979, the Commonwealth and Queensland governments have worked in partnership to protect and manage the Great Barrier Reef. In 2015, the Commonwealth and Queensland governments renewed their commitment to protecting the Great Barrier Reef World Heritage site, through the release of the Reef 2050 Long-Term Sustainability Plan, which is their overarching framework for protecting and managing the GBR through to 2050.

Managing for resilience

The GBR is an inherently complex social-ecological system, and its management is underpinned by the best available science and knowledge. Understanding how different pressures influence the GBR's values and underpinning ecosystem processes is critical to the design and implementation of effective management strategies. Despite being one of the best-managed marine protected areas in the world, the GBR continues to face mounting pressures from the cumulative impacts associated with climate change, coastal development, land-based runoff and direct use [63]. These pressures have intensified the frequency and intensity of extreme weather events, and the outbreaks of disease and pest species (i.e. the crown-of-thorns starfish). The GBR Marine Park Authority, Australia's lead management agency for the GBR, downgraded its long-term outlook for the GBR to 'very poor' in their 2019 Outlook Report and highlighted the urgency for continued and accelerated action to address threats, both local (i.e. declining water quality and over-fishing) and global (i.e. storms, ocean warming and acidification).

Over the past 45 years, the management approach for the GBR has shifted from actions designed to support its natural recovery to deliberate and proactive measures that use innovative strategies to enhance the GBR's resilience to the effects of climate-driven impacts, poor inshore water quality and other cumulative pressures (see Chapter 5). While global action to mitigate climate change is crucial, effective partnerships across governments, Traditional Owners and stakeholders are also necessary to ensure its long-term protection and conservation.

Under the bunggu: the inspiration of sea country
Melanie Hava

I am a Mamu Aboriginal woman from the Dryribal speaking Dugul-barra and Warii-barra family groups of the Johnstone River catchment in the Wet Tropics of Far North Queensland. This area sits in the middle of the GBR, with sea country and rainforest that is an important source of inspiration for me.

Mamu people are one of ~70 Aboriginal groups on the east Australian coast whose country immediately adjoins the Reef sea country. Our traditional estates are now internationally recognised through the Wet Tropics of Queensland and the GBR both being listed as World Heritage sites for their universal values. The Aboriginal cultural stories of this country are an important part of the outstanding universal value of this World Heritage site.

Family stories passed from generation to generation show that our ancestors travelled a lot around Mamu country, up and down to the coast according to the changes in the seasons for food. Our freshwaters flow out to the sea waters and come back to the land as it rains, so we are all connected. Under the Traditional Use of Marine Resources Agreement (TUMRA), one clan of the Mamu people manage our sea country that stretches to the outer edges of the GBR and covers many reefs between the land and the open ocean. They are working together to increase the numbers of turtle and dugongs for a sustainable future.

Growing up in Tully we used to go down to Mission Beach every Saturday for fish and chips. Here I discovered a love for the mysteries and beautiful visuals of the reef. The first time I realised that I could paint the reef was when I moved to Cairns and took my first trip out to Green Island and saw firsthand the beautiful reef and fishes from the glass bottom boat. On the surface the GBR sea country looks like beautiful plain water, but underneath the 'bunggu' (waves), a flurry of activity and sea life abounds. The rest was left to my imagination and my paintbrush (Fig. 1.33).

I now live in the beautiful and inspiring city of Cairns in Far North Queensland, where I feel close to the spirit of rainforest and reef animals of my mother's country. As an Aboriginal woman, my work highlights our culture, heritage and my environmental values. Like the ancestors who came before me, I am here for the long haul – without our beautiful reef and rainforest, we will not exist.

References

1. Spalding M, Spalding MD, Ravilious C, Green EP (2001) *World Atlas of Coral Reefs*. University of California Press, Berkeley and Los Angeles.

2. Fairbridge RW (1950) Recent and Pleistocene coral reefs of Australia. *The Journal of Geology* **58**, 330–401. doi:10.1086/625751

3. Hopley D (1982) *The Geomorphology of the Great Barrier Reef*. Wiley, New York.

4. Veron JEN (1995) *Corals in Space and Time: The Biogeography and Evolution of the Scleractinia*. University of New South Wales Press, Sydney.

5. Pecl GT, Araújo MB, Bell JD, Blanchard J, Bonebrake TC, Chen IC, Clark TD, Colwell RK, Danielsen F, Evengård B, Falconi L (2017) Biodiversity redistribution under climate change: impacts on ecosystems and human well-being. *Science* **355**, 6332–6338. doi:10.1126/science.aai9214

6. Heyward AJ, Stower M, Wakeford M, Colquhoun J, Spagnol S, Radford BT, Wahab MAA, Richards ZT (2018) 'Shallow coral habitat distributions across the offshore Kimberley region'. Subproject 1.1. 1.8 report prepared for the Kimberley Marine Research Program. Western Australian Marine Science Institution, Perth.

7. Moustaka M, Mohring MB, Holmes T, Evans RD, Thomson D, Nutt C, Stoddart J, Wilson SK (2019) Cross-shelf heterogeneity of coral assemblages in Northwest Australia. *Diversity (Basel)* **11**, 15. doi:10.3390/d11020015

8. Wilson SK, Depczynski M, Holmes TH, Noble MM, Radford BT, Tinkler P, Fulton CJ (2017) Climatic conditions and nursery habitat quality provide indicators of reef fish recruitment strength. *Limnology and Oceanography* **62**, 1868–1880. doi:10.1002/lno.10540

9. Thomson DP, Babcock RC, Haywood MDE, Vanderklift MA, Pillans RD, Bessey C, Cresswell AK, Orr M, Boschetti F, Wilson SK (2020) Zone specific trends in coral cover, genera and growth-forms in the World-Heritage listed Ningaloo Reef. *Marine Environmental Research* **160**, 105020. doi:10.1016/j.marenvres.2020.105020

10. Smith LD, Gilmour JP, Heyward AJ (2008) Resilience of coral communities on an isolated system of reefs following catastrophic mass-bleaching. *Coral Reefs* **27**, 197–205. doi:10.1007/s00338-007-0311-1.

11. Moore JA, Bellchambers LM, Depczynski MR, Evans RD, Evans SN, Field SN, Friedman KJ, Gilmour JP, Holmes TH, Middlebrook R (2012) Unprecedented mass bleaching and loss of coral across 12° of latitude in Western Australia in 2010–11. *PLoS One* **7**, e51807. doi:10.1371/journal.pone.0051807

12. Evans RD, Wilson SK, Fisher R, Ryan NM, Babcock R, Blakeway D, Bond T, Dorji P, Dufois F, Fearns P (2020) Early recovery dynamics of turbid coral reefs after recurring bleaching events. *Journal of Environmental Management* **268**, 110666. doi:10.1016/j.jenvman.2020.110666

13. Gilmour JP, Cook KL, Ryan NM, Puotinen ML, Green RH, Shedrawi G, Hobbs J-PA, Thomson DP, Babcock RC, Buckee J, *et al.* (2019) The state of Western Australia's coral reefs. *Coral Reefs* **38**, 651–667. doi:10.1007/s00338-019-01795-8

14. Collins L (2011) Geological setting, marine geomorphology, sediments and oceanic shoals growth history of the Kimberley region. *Journal of the Royal Society of Western Australia* **94**, 89–105.

15. Richards Z, Bryce M, Bryce C (2018) The composition and structure of shallow benthic reef communities in the Kimberley, north-west Australia. *Records of the Western Australian Museum* **85**(Supplement), 75–103.

16. Heyward A, Radford B (2019) Northwest Australia. In *Mesophotic Coral Ecosystems Coral Reefs of the World*. Volume 12. (Eds Y Loya, KA Puglise and T Bridge) pp. 337–349. Springer, Cham, Switzerland.

17. Cooper TF, Ulstrup KE, Dandan SS, Heyward AJ, Kuhl M, Muirhead A, O'Leary RA, Ziersen BEF, Van Oppen MJH (2011) Niche specialization of reef-building corals in the mesophotic zone: metabolic trade-offs between divergent *Symbiodinium* types. *Proceedings, Biological Sciences* **278**, 1840–1850. doi:10.1098/rspb.2010.2321

18. Underwood JN, Smith LD, van Oppen MJH, Gilmour JP (2009) Ecologically relevant dispersal of corals on isolated reefs: implications for managing resilience. *Ecological Applications* **19**, 18–29. doi:10.1890/07-1461.1

19. Gilmour JP, Smith LD, Heyward AJ, Baird AH, Pratchett MS (2013) Recovery of an isolated coral reef system following severe disturbance. *Science* **340**, 69–71. doi:10.1126/science.1232310

20. Wilson B (2013) *The Biogeography of the Australian North West Shelf: Environmental Change and Life's Response*. Newnes.

21. Solihuddin T, O'Leary MJ, Blakeway D, Parnum I, Kordi M, Collins LB (2016) Holocene reef evolution in a macrotidal setting: Buccaneer Archipelago, Kimberley Bioregion, Northwest Australia. *Coral Reefs* **35**, 783–794. doi:10.1007/s00338-016-1424-1

22. Richards Z, Bryce M, Bryce C (2018) The composition and structure of shallow benthic reef communities in the Kimberley, northwest Australia. *Records of the Western Australian Museum* **85**(Supplement), 75–103. doi:10.18195/issn.0313-122x.85.2018.075-103

23. Richards Z, Sampey A, Marsh L (2014) Kimberley marine biota. Historical data: scleractinian corals. *Records of the Western Australian Museum* **84**(Supplement), 111–132. doi:10.18195/issn.0313-122x.84.2014.111-132

24. Veron J, Stafford-Smith M, DeVantier L, Turak E (2015) Overview of distribution patterns of zooxanthellate Scleractinia. *Frontiers in Marine Science* **1**, 81. doi:10.3389/fmars.2014.00081

25. Richards ZT, Garcia RA, Wallace CC, Rosser NL, Muir PR (2015) A diverse assemblage of reef corals thriving in a dynamic intertidal reef setting (Bonaparte Archipelago, Kimberley, Australia). *PLoS One* **10**, e0117791. doi:10.1371/journal.pone.0117791

26. Richards ZT, Garcia R, Moore G, Fromont J, Kirkendale L, Bryce M, Bryce C, Hara A, Ritchie J, Gomez O, Whisson C (2019) A tropical Australian refuge for photosymbiotic benthic fauna. *Coral Reefs* **38**, 669–676. doi:10.1007/s00338-019-01809-5

27. Williams DG (1994) Marine habitats of the Cocos (Keeling) Islands. *Atoll Research Bulletin* **406**, 1–10. doi:10.5479/si.00775630.406.1

28. Darwin C (1842) *On the Structure and Distribution of Coral Reefs: Being the First Part of the Geology of the Voyage of the Beagle Under the Command of Captain Fitzroy, RN During the Years 1832 to 1836*. Smith, Elder & Co., London.

29. Beeton R, Burbidge A, Grigg G, Harrison P, How R, Humphreys W, McKenzie N, Woinarski J (2010) 'Final report of the Christmas Island expert working group to the Minister for Environment Protection, Heritage and the Arts', <https://www.environment.gov.au/system/files/resources/f8b7f521-0c69-4093-bf22-22baa4de495a/files/final-report.pdf>.

30. Allen GR, Steene RC, Orchard M (2007) *Fishes of Christmas Island*. Christmas Island Natural History Association, Flying Fish Cove.

31. Meekan MG, Jarman SN, McLean C, Schultz MB (2009) DNA evidence of whale sharks (*Rhincodon typus*) feeding on red crab (*Gecarcoidea natalis*) larvae at Christmas Island, Australia. *Marine and Freshwater Research* **60**, 607–609. doi:10.1071/MF08254

32. Richards ZT, Hobbs JP (2014) The status of hard coral diversity at Christmas Island and Cocos (Keeling) Islands. *The Raffles Bulletin of Zoology* **30**, 376–398.

33. Hicks J (1985) The breeding behaviour and migrations of the terrestrial crab *Gecarcoidea natalis* (Decapoda: Brachyura). *Australian Journal of Zoology* **33**, 127–142. doi:10.1071/ZO9850127

34. Burstyn HL (1975) Science pays off: Sir John Murray and the Christmas Island phosphate industry, 1886–1914. *Social Studies of Science* **5**, 5–34. doi:10.1177/030631277500500102

35. Williams M, Macdonald B (1985) *The Phosphateers: A History of the British Phosphate Commissioners and the Christmas Island Phosphate Commission.* Melbourne University Press, Melbourne.

36. Martinez-Escobar DF, Mallela J (2019) Assessing the impacts of phosphate mining on coral reef communities and reef development. *The Science of the Total Environment* **692**, 1257–1266. doi:10.1016/j.scitotenv.2019.07.139

37. Leon JX, Woodroffe CD (2013) Morphological characterisation of reef types in Torres Strait and an assessment of their carbonate production. *Marine Geology* **338**, 64–75. doi:10.1016/j.margeo.2012.12.009

38. Rowland MJ (2018) 65,000 years of isolation in Aboriginal Australia or continuity and external contacts? An assessment of the evidence with an emphasis on the Queensland coast. *Journal of the Anthropological Society of South Australia* **42**, 211–240.

39. Woodroffe CD, Kennedy DM, Hopley D, Rasmussen CE, Smithers SG (2000) Holocene reef growth in Torres Strait. *Marine Geology* **170**, 331–346. doi:10.1016/S0025-3227(00)00094-3

40. McNiven IJ (2011) Torres Strait Islanders: the 9,000-year history of a maritime people. In *The Torres Strait Islands.* pp. 210–219. Queensland Museum, Brisbane.

41. Morgan K (2015) From Cook to Flinders: the navigation of Torres Strait. *International Journal of Maritime History* **27**, 41–60. doi:10.1177/0843871414567075

42. McInnes A (1979) Dangers and difficulties of the Torres Strait and Inner Route. *Journal of the Royal Historical Society of Queensland* **10**, 47–73.

43. Rainbird J (2016) 'Adapting to sea-level rise in the Torres Strait'. Case Study for CoastAdapt, National Climate Change Adaptation Research Facility, Gold Coast, <https://coastadapt.com.au/sites/default/files/case_studies/CS011_Adaptation_in_the_Torres_Strait.pdf>.

44. Ceccarelli DM, McKinnon AD, Andrefouet S, Allain V, Young J, Gledhill DC, Flynn A, Bax NJ, Beaman R, Borsa P, Brinkman R (2013) The Coral Sea: physical environment, ecosystem status and biodiversity assets. *Advances in Marine Biology* **66**, 213–290. doi:10.1016/B978-0-12-408096-6.00004-3

45. Hoey AS, Pratchett MS, Harrison HB (2020) 'Coral reef Health in the Coral Sea Marine Park. Report on reef surveys April 2018 – March 2020'. Report produced for Parks Australia, <https://parksAustralia.gov.au/marine/management/resources/scientific-publications/coral-sea-coral-reef-health-report-2020/>.

46. Hoey AS, Pratchett MS, Sambrook K, Gudge S, Pratchett DJ (2018) 'Status and trends for shallow reef habitats and assemblages at Elizabeth and Middleton Reefs, Lord Howe Marine Park'. Report produced for Parks Australia. ARC Centre of Excellence for Coral Reef Studies, James Cook University, Townsville.

47. Carroll AG, Monk J, Barrett N, Nichol S, Dalton SJ, Dando N, Siwabessy J, Leplastrier A, Evans H, Huang Z, *et al.* (2021) 'Elizabeth and Middleton Reefs, Lord Howe Marine Park, post survey report'. Report to the National Environmental Science Program, Marine Biodiversity Hub. Geoscience Australia, University of Tasmania, Hobart.

48. Veron JEN, Done T (1979) Corals and coral communities of Lord Howe Island. *Australian Journal of Marine and Freshwater Research* **30**, 203–236. doi:10.1071/MF9790203

49. Edgar GJ, Davey A, Kelly G, Mawbey RB, Parsons K (2010) Biogeographical and ecological context for managing threats to coral and rocky reef communities in the Lord Howe Island Marine Park, southwestern Pacific. *Aquatic Conservation* **20**, 378–396. doi:10.1002/aqc.1075

50. Kennedy DM, Woodroffe CD (2000) Holocene lagoonal sedimentation at the latitudinal limits of reef growth, Lord Howe Island, Tasman Sea. *Marine Geology* **169**, 287–304. doi:10.1016/S0025-3227(00)00093-1

51. Woodroffe CD, Brooke BP, Linklater M, Kennedy DM, Jones BG, Buchanan C, Mleczko R, Hua Q, Zhao J (2010) Response of coral reefs to climate change: expansion and demise of the southernmost Pacific coral reef. *Geophysical Research Letters* **37**.1–6. doi:10.1029/2010GL044067

52. Linklater M, Hamylton SM, Brooke BP, Nichol SL, Jordan AR, Woodroffe CD (2018) Development of a seamless, high-resolution bathymetric model to compare reef morphology around the subtropical island shelves of Lord Howe Island and Balls Pyramid, southwest Pacific Ocean. *Geosciences* **8**, 11. doi:10.3390/geosciences8010011

53. Linklater M, Jordan AR, Carroll AG, Neilson J, Gudge S, Brooke BP, Nichol SL, Hamylton SM, Woodroffe CD (2019) Mesophotic corals on the subtropical shelves of Lord Howe Island and Balls Pyramid, south-western Pacific Ocean. *Marine and Freshwater Research* **70**, 43–61. doi:10.1071/MF18151

54. Woodroffe CD, Kennedy DM, Brooke BP, Dickson ME (2006) Geomorphological evolution of Lord Howe Island and carbonate production at the latitudinal limit to reef growth. *Journal of Coastal Research* **221**, 188–201. doi:10.2112/05A-0014.1

55. Linklater M, Brooke BP, Hamylton SM, Nichol SL, Woodroffe CD (2015) Submerged fossil reefs discovered beyond the limit of modern reef growth in the Pacific Ocean. *Geomorphology* **246**, 579–588. doi:10.1016/j.geomorph.2015.07.011

56. Great Barrier Reef Marine Park Authority (2019) *Aboriginal and Torres Strait Islander Heritage Strategy for the Great Barrier Reef Marine Park*. Great Barrier Reef Marine Park Authority, Townsville.

57. Isern AR, McKenzie JA, Feary DA (1996) The role of sea-surface temperature as a control on carbonate platform development in the Western Coral Sea. *Palaeogeography, Palaeoclimatology, Palaeoecology* **124**, 247–272. doi:10.1016/0031-0182(96)80502-5

58. International Consortium for Great Barrier Reef Drilling (2001) New constraints on the origin of the Australian Great Barrier Reef: Results from an international project of deep coring. *Geology* **29**, 483–486. doi:10.1130/0091-7613(2001)029<0483:NCOTOO>2.0.CO;2

59. Hopley D, Parnell KE, Isdale PJ (1989) The Great Barrier Reef Marine Park: dimensions and regional patterns. *Australian Geographical Studies* **27**, 47–66. doi:10.1111/j.1467-8470.1989.tb00591.x

60. Lewis A (2001) *Great Barrier Reef Depth and Elevation Model: GBRDEM*. CRC Reef Research Centre Ltd, Townsville.

61. Harris PT, Bridge TCL, Beaman RJ, Webster JM, Nichol SL, Brooke BP (2013) Submerged banks in the Great Barrier Reef, Australia, greatly increase available coral reef habitat. *ICES Journal of Marine Science* **70**, 284–293. doi:10.1093/icesjms/fss165

62. McNeil MA, Webster JM, Beaman RJ, Graham TL (2016) New constraints on the spatial distribution and morphology of the *Halimeda* bioherms of the Great Barrier Reef, Australia. *Coral Reefs* **35**, 1343–1355. doi:10.1007/s00338-016-1492-2

63. Great Barrier Reef Marine Park Authority (2019) 'Great Barrier Reef Outlook Report 2019'. Great Barrier Reef Marine Park Authority, Townsville.

64. Great Barrier Reef Marine Park Authority (1981) *Nomination of the Great Barrier Reef by the Commonwealth of Australia for inclusion in the World Heritage List*. Great Barrier Reef Marine Park Authority, Townsville.

Imperilled icons and shifting sentiments

From the late 1990s, public awareness of the vulnerability of Australia's coral reefs, and the GBR, grew in response to repeated outbreaks of crown-of-thorns starfish, mass coral bleaching and mortality events (1998 and 2002) and rapid coastal development, particularly around ports associated with a boom in mining exports. The prospect of the GBR being added to its List of World Heritage in Danger elevated both public concerns and international scrutiny, prompting a coordinated government response that eventually led to the development of the Reef 2050 Long-Term Sustainability Plan. This set out investments over 5 years to fund local environmental management initiatives, such as improving water quality, controlling infestations of crown-of-thorns starfish, and reef restoration projects.

Social research into public perceptions of the GBR established its foremost status among Australia's natural and built icons [4], with strong place values (e.g. for biodiversity, aesthetics, recreation and other ecosystem services) shared by local, national and international stakeholders alike [5]. Based on data from 2015 to 2016, the total economic value of the GBR, encompassing its direct contributions via GBR-dependent industries as well as the indirect social and cultural values, was estimated to exceed $56 billion [6], in a report commissioned by the GBRF.

When consecutive summers of mass coral bleaching and mortality occurred on the GBR in 2016 and 2017, an outpouring of public sentiment, including grief and empathy for the stricken GBR, accompanied a major shift in public perceptions of the climate change threat [2, 7]. The GBR was labelled the 'canary in the coal mine' of climate change, as media narratives began to personify the Reef as sick and dying – notably, in 2016 an American outdoor adventure magazine published an obituary for the GBR. This had flow-on effects to the GBR's extensive tourism industry, motivating tourists to 'see the Reef before it's gone'.

In 2017, the Great Barrier Reef Marine Park Authority (GBRMPA) led key stakeholders, managers and Traditional Owners to develop the Great Barrier Reef Blueprint for Resilience, an action plan to strengthen the Reef's resilience to the threats of climate change and other stressors [8]. The Reef Blueprint marked another pivotal period in Australia's coral reef management by proposing novel interventions to protect fundamental ecological processes in the face of a changing climate.

Reefs beyond 2020

Understanding public perceptions of and values surrounding Australia's coral reefs, and the events that change them, plays a key role in shaping effective management responses to the growing threats reefs face. Yet the diversity of human perceptions relating to the vast areas of Australia's reefs remain only partly understood.

Australia's coral reefs continue to be perceived as valuable for their economic contributions, as well as their intrinsic natural values, with a high regard for the marine biodiversity they support. Alongside global awareness of humanity's impact on the Earth's biosphere, there is a growing narrative of 'reefs needing care' with a rise in community stewardship and citizen science initiatives (see Chapter 5). Reef tourism operators strive to be environmental stewards, while also promoting the aesthetic qualities and biodiversity of Australia's coral reefs, on which their livelihoods depend. The evolution of these narratives shapes public perceptions and support for coral reef management at a critical time when the persistence of coral reefs as we know them is at stake.

Three coral reefs in Yanyuwa country, meaningful and powerful

John Bradley with Yanyuwa families

Warrawarra, that's the name for coral, name for coral reef too.

On 3 June 1992, the High Court of Australia ruled that Indigenous titles to land be recognised, thus throwing out forever the legal fiction that Australia was, as defined in 1788, *terra nullius*: an empty land. For many Indigenous people, the sea is also their country and in 1992 the myth of *marae nullius*, an empty sea, was also silenced. The Yanyuwa people of the south-west Gulf of Carpentaria proudly proclaim themselves as li-Anthawirriyarra, or 'Saltwater People', a people whose spiritual and cultural heritage comes from the sea. They are not bush, scrub or dwellers of the inland regions. They are a people who know and can demonstrate what the sea means for them.

Indigenous groups such as the Yanyuwa have claimed and won back ancestral homelands on the mainland and islands, namely the Sir Edward Pellew Group of Islands under the *Aboriginal Land Rights (Northern Territory) Act 1976*. Nevertheless, rights to the sea and intertidal zone are misunderstood, and at times heavily contested and seen by some government agencies and resource managers as problematic. For some, it is too hard to grasp that something as fluid as the sea and the submerged reefs, sandbars and seagrass beds can have Law and be classed as country and home, managed according to Yanyuwa understandings of their own Law, and linked to people via dense connections of human and non-human kin. The Dreaming is one of the fundamental concepts that must be grasped to better understand Yanyuwa Law [9].

The Dreaming – a process of relatedness

The Yanyuwa people use the Aboriginal English term 'Dreaming' to refer to the relationships between people, their country and the Law that set out the realm of Yanyuwa experience. In Yanyuwa, this Law, called *narnu-Yuwa*, embodies beliefs that derive from the Dreaming. This is a term that has popular currency among both Indigenous and non-Indigenous people, although care must be taken with this term because it can carry connotations of an imaginary or unreal time.

For Yanyuwa people, Dreaming refers to a body of moral, jural and social rules and practices. These derive from the actions performed by the original creator beings that gave form and substance to sea country by naming them and leaving parts of their bodies, or objects they carried, in the land and sea. Thus, the coral reefs discussed here are associated with the journeys of groups of Ancestral Women, Ancestral Sea Turtles and a Tiger Shark and they carry meaning for Yanyuwa men and women. The stories that are still told about these reefs constitute a very particular knowledge associated with places. This knowledge is respected and observed in everyday practices, as well as re-enacted in ritual.

Reefs have clan association; any enquiry into ownership of sea and reefs will almost inevitably yield an answer that identifies one of four patrilineal clans that own or, as it is said in Yanyuwa, 'hold' the country. All people belonging to one clan are broadly perceived as being owners of a particular tract of sea country. While all clan categories can be used to express general understandings of land and sea ownership, there are more detailed ways of understanding this system.

As the Dreaming beings travelled, they transformed their bodies, or moved their bodies in certain ways, to create sand bars, reefs, tide and tidal currents. Images such as these dominate the cosmogonies and cosmologies of Yanyuwa people. Such images are in fact central to the relatedness that is at the basis of the Law. This Law provides an understanding of how names and naming are crucial to its activation, transferal and negotiation. People have names given to them from their sea country; they also know the names of the different parts of the sea, the tides and waves as well as knowing the specific names of reefs, sandbars, channels, seagrass beds and beaches.

Each action of the Ancestral Beings or Dreamings established a relationship between the Ancestral Being, a place and a group of people who identify with the sea and reefs under question and own it. It is the image of the journey that is the mechanism that orders, distributes and differentiates a group's rights to the ownership of particular tracts of sea or countries. These are important issues: the image of the journey sometimes crosses many hundreds of kilometres. Law and knowledge associated with these Ancestral Beings becomes the property of various Yanyuwa families. For the Dreamings discussed below, travel links to other groups, such as the Garrwa and Gangalida people to the east, or the Marra people to the west. While there may be no known 'blood' links between these various owning groups, the people who share these Dreamings are seen to also share a substance derived from this common, non-human Ancestor. The groups sharing kinship with the Dreamings see themselves as kin; thus, a regional network is established that translates into daily duties of obligation. Journeys and transformations of the Ancestral Beings become important, pervasive images that convey different levels of relatedness.

Three reefs: Kurrkumala, Aburri-Babalungku and Wurlma

My own travels over the islands by boat and helicopter have identified three reefs that hold particular significance to Yanyuwa people (Fig. 2.2). The site Kurrkumala lies ~5 km from the Rosie Creek mouth. Historically, this site was not visited because of associations with important Ancestral Beings. Modern forms of boat travel, helicopters and aeroplanes have allowed people to view and interact with these sites in ways that were not previously considered

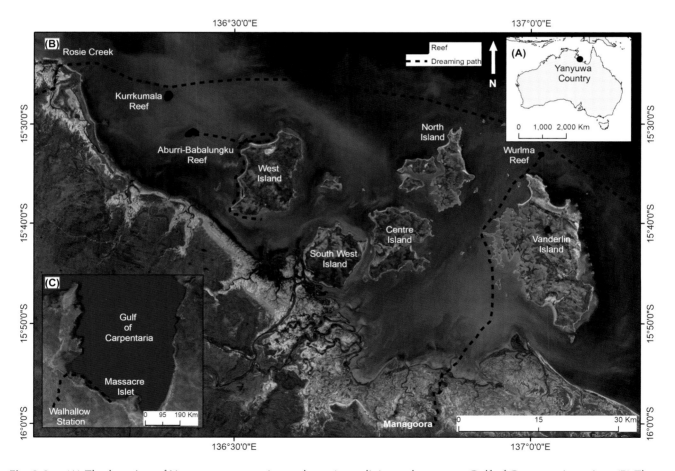

Fig. 2.2. (A) The location of Yanyuwa country in northern Australia's south-western Gulf of Carpentaria region. (B) The three named coral reefs in the area of the Sir Edward Pellew Islands along with the paths of the Dreamings described. (C) The path of the Women's Dreaming associated with Kurrkumala Reef, which is sung as part of the sacred Kunabibi ceremony. (Image credits: Sarah Hamylton)

possible. Kurrkumala is the place where the Dreaming Women broke through the surface of the sea, having travelled under the sea from Massacre Inlet, which lies over 270 km further to the east near the Northern Territory–Queensland border. Kurrkumala is the starting point of two important song lines that belong to the highly sacred Kunabibi ceremony, which concerns elements of the fecundity of country and is in many ways a profound demonstration of how knowledge of land and sea ownership is demonstrated through ritual. During the Kunabibi ceremony, one of these song lines is sung by men on the men's ceremony ground and the other is sung by women on the women's ceremony ground. Another version of this song line is sung by men or women as part of the all-night dancing by women for the a-Marndiwa initiation ceremonies for young boys. The Women Dreaming travels through the sea and comes ashore at the mouth of the Rosie Creek or Warrawarda. From there they travel around 250 km inland, where their journey and the line of the song finishes at a place called Kalabirrangani on Walhallow Station on the Barkly Tablelands.

The reef Aburri-Babalungku is associated with the travels of a group of Marine Turtle Dreamings, these are predominately green turtles (*malurrba*) and flat back turtles (*wirndiwirndi*); however, in the song line that begins at this reef, hawksbill turtles (*karrubu*) and olive-ridley turtles (*liyarnbi*) are also sung. The song line circles around this reef, singing species of reef fish and seagrasses. Babalungku is a specific name given to the southern edge of the reef and it is associated with the beaked sea snake (*a-rikarika*). It too is included within the song line that circles this reef before the song moves across the sea to the north-west coast of West Island.

The reef called Wurlma, known in English as Vanderlin Rocks, is primarily associated with the journey of a Tiger Shark (*ngurdungurdu*) that had travelled from Ganggalida country in the southern part of the Gulf of Carpentaria. This shark carried cycad fruit in a paper bark bundle on his head, and many of the rocks that can be seen exposed at low tide on this reef are bundles of cycad food he left there, because he was tired from his journey (see [9]). Eventually the Tiger Shark Dreaming travelled south up the Wearyan River to a place called Managoora (Manankurra), where he left the rest of the cycad food. There is a very substantial forest of cycads (*Cycas angulata*) at this place today. Wurlma is also the starting point of an important song line that travels south through the sea and then onto Vanderlin Island; this song is associated with both initiation rituals and, in the past, hollow log burial rituals. This song line is the only one that actually has a verse for corals; in particular large, flat 'shelf' corals and brain corals are sung in this song line.

The Dreaming as Law, as philosophy

There is an unfortunate tendency to romanticise and make exotic Indigenous ways of knowing, particularly to emphasise the 'spiritual', as opposed to the deeply philosophical. Yanyuwa philosophy contends that all beings have Law, have culture, are autonomous but at the same time are enmeshed in relations of interdependence, while being self-organising and sentient. Such understandings reiterate a place-based way of being. This, and the emotional geography that comes with the full recognition of kinship, is vital to decolonising the way that people think about Australian places, particularly sea country. Such kinship is not symbolic or rarefied – it is actual and understood, just as relation to human kin.

Aboriginal Law is imbedded in a deeply held philosophical positioning that we in the West might best describe as a kin-centric ecology, whereby both human and non-human kin are bound in relationships. Such a term is defined by epistemologies of place where the world, both above and underwater, is not one of wonder, but rather of great familiarity. With such a positioning, life in any environment is viable only when humans view their surroundings as kin, their mutual roles as being essential for survival. The knowledge of how living beings, including the reefs, turtles, reef fish, seagrasses, sea snakes, corals and tiger sharks and humans, fit together is not just a body of information, but rather this guides processes of action, interaction and connection that encompass the personal, social, environmental, cosmic and spiritual.

Encountering and charting the hazardous reefs of Australia, 1622–1864
Tom Spencer

> *ye 25th daye at 11 of ye clocke in ye night, faire weather and smoth ye ship strocke, I ran to ye poope and hove ye leade and found but 3 faudoms water … ye rock being sharpe ye shipp was presentlie full of water, for ye most part of these rocks lie 2 fadom under watter, it strucke my men in such a mayze when I said ye shipp strocke and they could se neyther breach, land, rocks, change of watter nor signe of danger…* [10]

So wrote Captain John Brookes to the English East India Co. on the wrecking of *The Tryall* on the rocks that now bear the ship's name, 100 km from the Pilbara coast of north-west Australia, on the night of 25 May 1622, thus gaining the grim distinction of being the first recorded shipwreck in Australian waters. Brookes's report graphically chronicles all those components that make coral reefs so dangerous to mariners, particularly to those in slow-moving sailing ships of limited manoeuvrability. This is just one example from the litany of ship losses along the west and north-west coast of Australia in the 17th, 18th and early 19th centuries. Some well-known reef wrecks include the *Batavia* (1629) and the *Zeewijk* (1727) on the Houtman Abrolhos reefs, and the *Rapid* (1810) and the *Correjo da Azia* (1816) both of which perished on Ningaloo Reef. But the wreck of *The Tryall* also demonstrates the deficiencies of 17th-century charting.

Many ships' captains carried charts of dubious value, often composites from earlier mapping, and were faced with the severe difficulty of accurately fixing longitudinal position. Brookes and *The Tryall* were attempting to follow the route between the Cape of Good Hope and the East Indies devised by the Dutch explorer Hendrick Brouwer in 1611. This involved heading eastwards in the 'roaring forties', a subtropical area of gale-force westerly winds often used for sailing between Europe and Australasia, for as long as possible before turning north towards Java and the commercial hub of Batavia (Jakarta). This route was profitable, halving the voyage time compared to the Portuguese route around the eastern margin of Africa and the northern Indian Ocean, but risky, given the lack of positional clues from the few islands in the expanse of the Southern Ocean, which might be missed or misidentified. Brookes grossly underestimated his longitude – being 1000 km further east than the route demanded – and was thus forced to navigate along the Australian coast, with the catastrophic loss of 93 lives.

From 'the very jaws of destruction': Captain Cook and the Great Barrier Reef
One hundred years after Brookes, James Cook in HMS *Endeavour* passed Sandy Cape (24°42′ S) on the night of 22 May 1770, to begin his engagement with the GBR. Having entered the southern lagoon, sheltered by the Swain Reefs over 200 km to windward, the *Endeavour* became increasingly squeezed between the mainland coast and the shelf edge reefs closing in from the east. At a location known ever since as Endeavour Reef (15°45′ S), matters came to a head at 11 p.m. on 11 June 1770 when the depth was:

> *17 [fathoms (31 m)] and before the Man at the [sounding] lead could heave another cast the Ship Struck and stuck fast … upon the SE edge of a reef of Coral rocks.* [11, p. 344]

Cook saved the ship and made the necessary repairs, but was then faced with the difficulty of escaping what Joseph Banks arrestingly termed 'the labyrinth of shoals'. On 12 August, from the summit of Lizard Island, Cook provided the first analytical description of the ribbon reefs of the Outer Barrier:

> *a Reef of Rocks laying about 2 or 3 Leagues without the Island, extending in a line NW and SE farther than I could see on which the Sea*

broke very high. This however gave me great hopes that they were the outermost shoals, as I did not doubt but what I should be able to get without them for there appear'd to be several breaks or Partitions in the reef and deep water between it and the Islands. [11, p. 373]

Exiting the lagoon through one of these 'partitions', now Cook's Passage, on 13 August, they were driven back towards the seaward margin of the reef 3 days later:

All the dangers we had escaped were little in comparison of being thrown upon this Reef where the ship must be dashed to pieces in a Moment. A Reef such as is here spoke is scarcely known in Europe, it is a wall of Coral Rock rising all most perpendicular out of the unfathomable Ocean, always overflown at high-water generally 7 or 8 feet and dry in places at low-water; the large waves of the vast Ocean meeting with so sudden a resistance makes a most terrible surf breaking mountains high. [11, p. 378]

Fortuitously reaching the sanctuary of the inner reef lagoon, through what is now known as Providence Channel, they then sailed onwards to round Cape York (see Chapter 1) and finally escaped the GBR.

Extending and synthesising: Flinders and Baudin

Forty years after Cook, and now aided by multiple chronometers for more accurate position fixing, Matthew Flinders extended and re-surveyed the reefs of northern Australia in HMS *Investigator*, part of a circumnavigation of the entire continent in the years 1801–03. Contemporary with Flinder's surveys was the French expedition led by Nicholas Baudin in the ships *Géographe* and *Naturaliste*; indeed, Flinders named Encounter Bay, south-east of Adelaide, as a result of a meeting between the *Investigator* and the *Géographe* on 8 April 1802. Baudin's expedition surveyed much of the southern, western and north-western coasts of Australia between 1801 and 1803, including their reefs, and

reached as far as Timor. But it was Flinders who favoured the descriptor 'Australia' on his charts ('more agreeable to the ear'), named the 'Corallian Sea' and first published the term 'barrier reefs'.

Perhaps most remarkably, he was the first person to grasp the enormous connectivity of the reefs of the east coast. In his journal, *A Voyage to Terra Australis*, published in 1814 on the day before he died (although he had seen the final charts), he named 'The Great Barrier Reefs', stating:

I therefore assume it as a great probability, that with the exception of this [the Flinders Passage at 10°33'], and perhaps several small openings, our Barrier Reefs are connected with the Labyrinth of Captain Cook; and that they reach to Torres' Strait and to New Guinea … through 14° of latitude and 9° longitude; which is not to be equalled in any other known part of the world.' [12, p. 102]

Understanding reef growth: Beaufort's route to maritime safety

Although by the 1820s, the French naval surgeons Jean-René Quoy and Paul Gaimard, on the *Uranie* circumnavigation of 1817–20 under Louis de Freycinet, had argued convincingly from studies of modern reefs and the Timor raised reefs that corals are shallow-water organisms, the fact that they are near-surface living structures continued to give those charged with charting the reef seas concern: might not coral growth produce new hazardous reefs in areas already charted? How soon would it be before existing charts became not only obsolete but also downright dangerous to use? It is not surprising therefore that as hydrography became systematised in naval administrations, ships' captains were asked not just to chart coral reefs but also to elucidate where and how reefs grow. Perhaps the greatest exponent of this approach was Francis Beaufort who became Hydrographer of the Admiralty in 1829. Prior to Beaufort's appointment, instructions were restricted to purely navigational issues. Thus, the Lord Commissioners of the

Admiralty simply instructed Frederick William Beechey, Captain of HMS *Blossom*'s voyage of 1825–28, to check on the positions of certain islands and to ensure that 'none of them are passed by you unobserved' (a task that Beechey achieved in spectacular fashion, mapping 32 atolls); no mention was made of engagement with theories of reef formation. By contrast, Beaufort's orders of 11 November 1931 to Robert Fitzroy, Captain of the *Beagle*, were on a completely different level:

An exact geological map of the whole island should be constructed, showing its form, the greatest height to which the solid coral has risen, as well as that to which the fragments appear to have been forced. The slope of its sides should be carefully measured in different places, and particularly on the external face, by a series of soundings, at very short distances from each other, and carried out to the greatest possible depths, at times when no tide or current can affect the perpendicularity of the line. A modern and very plausible theory has been put forward, that these wonderful formations, instead of ascending from the bottom of the sea, have been raised from the summits of extinct volcanoes; and therefore the nature of the bottom at each of these soundings should be noted, and every means exerted that ingenuity can devise of discovering at what depth the coral formation begins, and of what materials the substratum on which it rests is composed. [13, p. 38]

Fitzroy chose South Keeling (now part of the Australian territory of Cocos (Keeling)) for these observations (see Chapter 1), but of course it was another member of the ship's company, Charles Darwin, who exploited the island survey to support a new explanation for the formation of low islands and barrier reefs, one which in the process debunked Quoy and Gaimard's 'modern and very plausible theory'. But before the *Beagle* had reached New Zealand, Beaufort was already issuing his next set of instructions for a further voyage, this time by HMS *Sulphur*, to both follow the Fitzroy

template and extend it with a reef boring (which turned out to be a catastrophic failure). And even after Darwin's announcement of his subsidence theory at the Geological Society of London on 31 May 1837, Beaufort was still writing to ships' captains. Thus, in his directions to Francis Blackwood, captain of HMS *Fly*, surveying the GBR and completing two circumnavigations of Australia, 1842–46 (and through its geologist, Joseph Beete Jukes, theorising on the Reef's gross structure through, for the first time, a Darwinian lens), Beaufort advised:

do not hurry over the hidden dangers which lurk and even grow in that part of the world [the Torres Strait]. [14, 16]

And to Owen Stanley, Captain of HMS *Rattlesnake*, cruising along the Reef in 1848–49, Beaufort spelled out:

there would be much discouragement attached to such surveys if changes should be constantly & rapidly at work in[coral] seas' and 'direct [his officers'] attention more particularly to the formation and growth of coral reefs. [15, 16]

Beaufort was, in effect, adding another explanatory layer to the survey process. That effort did not stop; not only Blackwood and Stanley but also Charles Jeffreys, Phillip Parker King (most notably), J. C. Wickham, John Lort Stokes and Henry Denham progressively re-surveyed, filled in the gaps and geographically extended the pioneer charting [17]. Thus by 1864, the 6th Hydrographer of the Navy, Sir George Richards, could confidently state:

twenty years ago this inner passage from Cape York to Moreton Bay was the most intricate in the world; now it was as easy to navigate as the English Channel. Silently but steadily this great work has been progressing during these long years … undeniably one of the most gigantic and splendid undertakings. [18, p. 119]

Fig. 2.3. The wreck of the trawler *Pacific Lady* on Boult Reef. (Image credit: M. Osmond © Great Barrier Reef Marine Park Authority, supplied with kind permission)

Frank Hurley's aquarium and the art of coral reef science

Ann Elias

> *Flowers turned to stone! Not all the botany*
>
> *Of Joseph Banks, hung pensive in a porthole,*
>
> *Could find the Latin for this loveliness,*
>
> *Could put the Barrier Reef in a glass box*
>
> *Tagged by the horrid Gorgon squint*
>
> *Of horticulture* [19]

As the GBR became better known through tourist advertising, so its visual beauty became more prominent in Australian art and literature of the 1920s and 1930s. In *Five Visions of Captain Cook* (1931) quoted above, the poet Kenneth Slessor (1901–71), delighted in the wondrous, life-affirming, first-hand encounter of corals on the GBR, which he contrasted with the deadening effects of captive corals in scientific aquariums. His generation, though, was practised with aquariums. Nearly a decade earlier, Slessor's fellow writer and artist – the photographer, filmmaker and explorer, Frank Hurley (1885–1962) – had set out for the GBR on a scientific and picture-making expedition equipped with a glass aquarium. It was far from Hurley's intention to reduce the living reef to dull or lifeless specimens. Rather, his idea was to combine the creative potential of optical devices including an aquarium and camera with artistic tricks, such as montage and hand-colouring, to improve upon nature and present the GBR as a vivid, modern spectacle.

Elucidated here are the circumstances surrounding Hurley's expeditions to the GBR, including the Torres Strait and Papua. The account begins by explaining Hurley's attraction to tropical coral reefs as a subject for photography and adventure, and the expeditions he made to the GBR, especially in 1922 with Allan McCulloch (1885–1925), an ichthyologist at the Australian Museum. Three angles on the history of Frank Hurley and coral reefs are explored: the part played by an aquarium in the production of Hurley's photographic images; the impact of Hurley's images on public perceptions of coral reefs and how they ushered in a new awareness of coral reef environments; and the entanglement of science and art in the production of Hurley's underwater photographs of the GBR.

Hurley's attraction to tropical reefs

In 1920 Frank Hurley set out for the GBR seeking 'Nature's wonders … from a newer viewpoint' [20]. He planned to become the first photographer to capture the hallucinatory colours and forms of coral reefs from an underwater perspective. Following two expeditions to the freezing climate of Antarctica, including as photographer with the Ernest Shackleton Imperial Trans-Antarctic Expedition from 1914 to 1917, he began to think intensely about the warmth of a tropical undersea and how Oceania promised a new geographical frontier for exploration and photography.

Hurley sensed that waiting at the GBR were dramatic pictures of immense commercial and aesthetic value that other photographers of the GBR before him, including William Saville-Kent (1845–1908), had failed to exploit through the modern mass media channels of cinema, illustrated public lecture events and the syndicated press. What attracted Hurley to the tropical underwater? He knew in 1920 that underwater photography was a largely unchartered field. However, Hurley did not possess the right technologies to get below the surface of the water to photograph the reef from below. It was some time before the widespread availability of SCUBA and portable underwater cameras. The

question for Hurley was how to reach the underwater and photograph it.

At completion of his trip to the reef in 1920–21, Hurley returned with footage for a feature film that was released with the title *Pearls and Savages* (1924). He also returned with a large collection of glass plate negatives including black and white views of corals at low tide. Figure 2.4 shows the living coral reef. But as an image photographed through air looking down on the surface of the water, it lacked the detail that Hurley sought.

In 1922 Hurley returned to the GBR with Allan McCulloch to reshoot footage for *Pearls and Savages*. On the trip he took an aquarium. He was determined to photograph the beauty of a coral reef at eye level with fish and corals, rather than have the creatures of the underwater obscured by the sea's surface. Figure 2.5 is a publicity photograph showing Hurley and McCulloch on a beach at Dauko Island, Papua, surrounded by a team of Papuan helpers who were their reef navigators and who also collected fish and corals for Hurley's aquarium. The aquarium can be seen in the background.

Aquarium images

With an aquarium, Hurley could stage and design his own mental picture of an ideal coral reef by arranging fish and corals at will to control the

Fig. 2.4. *Exposed Coral*, 1921, black and white glass plate negative. (Image supplied with kind permission by the National Library of Australia; image credit: Frank Hurley)

Fig. 2.5. *Hurley, McCulloch Dauko*, 1922. (Image supplied with kind permission by the Australian Museum, photographer unknown)

Fig. 2.6. Fish underwater. (Image supplied with kind permission by the Australian Museum; image credit: Frank Hurley)

overall aesthetic. Furthermore, a glass aquarium enabled him to produce the illusion of having a camera immersed in tropical water. An example of a GBR photograph by Frank Hurley that was choreographed inside an aquarium is shown in Fig. 2.6. The density and profusion of life are evoked not only by stage-managing the arrangement of objects, but also through photographic montage. More than one negative was combined for the printing of the photograph, and the final black and white image was then hand-coloured. The aquarium allowed Hurley to invent an exotic extravagance of dense patterns and by touching up with hand-painting he was able to evoke iridescence.

Being modern in the 1920s was associated with bright colours and bold expression – both were signs of newness and change from tradition. Australia's vividly coloured landscapes and natural wonders, including the red interior of the outback and the kaleidoscopic oceanic edge of the GBR, symbolised a nation that stood out from the rest of the world as a bold, modern, colourful place. It was a colonialist fantasy, and it denied the ancient occupation of Australia by Aboriginal and Torres Strait Islander peoples. Yet in this fantasy of the newness of modernity, the riot of colour and vivid patterns of corals and tropical fish contributed to a distinct direction in Australian culture that Laurie Duggan has described as the 'techno-pastoral' frontier of modern art, the cultural field to which Fig. 2.6 belongs [21].

Like a frame for a painting, an aquarium frames objects for people to consume visually. The aquarium vision that Hurley brought to coral reefs also drew, as Judith Hamera explains about aquariums, 'on the perceptual logics and plots of the shop window' [22]. It enabled Hurley to sell an edited version of the GBR as an aesthetic spectacle for viewers to consume as if they were window shopping. But what viewers of Fig. 2.6 were unable to see was that in the capture, handling, transportation and arrangement of the aquarium, the reef fish and corals suffocated and died. The aquarium was overcrowded; it silted up, and became hot and depleted of oxygen. It produced the conditions for biological degradation that in the 21st century is known as coral bleaching. To get a successful photograph required constant replenishment of the aquarium contents [23]. Hurley lived in an era rather different to our own when it was common to assume that the oceans' riches were virtually inexhaustible and so he did not question the ethics of collecting additional specimens and subjecting them to the same lethal conditions.

The impact of Hurley's images on perceptions of coral reefs

Hurley's aim was to integrate the Reef into a national story of Australia's unique and spectacular place in the world. It was an ambition he would achieve through the expanding mass media that spread his images to Western audiences. In turn, the same photographs stimulated a desire among viewers and readers to travel and experience coral reefs first-hand.

Through a form of photojournalism that Hurley called 'picture-stories', he took advantage of a new era of illustrated magazines and newspapers to take the GBR to the world. The fantasies of corals and fish that Hurley photographed and filmed were published in magazines, including the *Illustrated London News*, and in national newspapers including the Sydney *Sun*. Among the images he selected for audiences in the United States of America and the United Kingdom was a hand-coloured photograph of a clam slightly open (Fig. 2.7). By 1920, the giant clam had become a mythical object. Coral reefs were known as beautiful but they were also feared, especially for human-eating sharks and giant clams that were supposedly able to trap a human foot in a vice-like grip and never let go. Hurley's clam photograph therefore made an exciting accompaniment for adventure stories about the GBR. By applying hand-colouring to a black and white image, he was better able to emphasise the clam's seductive but menacing gape.

The boundaries between science and art

In the process of using the power of the expanding mass media of the 1920s to excite the public

Fig. 2.7. Clam and corals, Great Barrier Reef, *c.* 1921. (Image supplied with kind permission by the National Library of Australia; image credit: Frank Hurley)

imagination about the natural world of coral reefs, Hurley also straddled the boundary between science and art. His images embody dual interests of sensation and knowledge. While the Australian Museum expedition with Allan McCulloch was intended to serve discovery and scientific inquiry, Hurley's image-making relied on artifice, illusion and manipulation. He actively constructed his own visual worlds. Indeed, in 1922 Hurley was celebrated by Australian society not just as a photographer, filmmaker and explorer but also, as one enthusiast put it, as 'an artist. He paints with the camera … [and] in the hands of Captain Hurley the strip of celluloid becomes a magic carpet along which the feet of the audience wander to lands and among people hitherto undreamed of' [24]. While art and science need not be incompatible, Hurley often pushed the limits of scientific practice based in truth to present an edited view of the reef as reality. In the coral reef photographs we see the liberal use of cropping and hand-colouring to produce the effect of an abundant and profuse coral reef, and a reef of latent dangers.

Great Barrier Reef as battlefield

Hurley's coral reef expeditions marked the beginning of the GBR's transformation into a mediated spectacle defined primarily through photographic images. Those who viewed his images in the mass media learned to see the GBR as a collection of vivid colours, teeming with marine life in transparent water. It was the very image that 20th-century society would embody as typical of coral reefs, and the aesthetic experience that visitors expected to have when they travelled there. In 1922, Frank Hurley wrote in the popular press about the aesthetic richness that evolution had brought to the GBR: 'My coral garden is a furious battlefield. It is heavily overpopulated and the struggle for existence is a desperate one' [25]. He could not have known that in the context of a looming environmental crisis in the 21st century, the word 'battlefield' would take on a new meaning as the term 'battleground' was later coined by the poet Judith Wright in reference to the struggle to save the reef from destruction [26].

Coral reefs of Zenadth Kes (Torres Strait)
Leah Lui-Chivizhe

The extensive coral reefs of Zenadth Kes have evolved over thousands of years. Viewing these reefs as outlines or colourful shades on maps does little to prepare you for their ubiquity and grandeur when flying over them. The orientation of the fringing reefs, their growth predominating on the south-east side of islands, skewed by the winds and currents, is noticeable from above. The variation of colour points to deeper water at the edge of reefs in the east and the presence of sand banks or shallow water and turbidity in the central and western parts of the Strait (Fig. 2.8).

For the uninitiated travelling through the narrow body of water, the reefs can be dangerous. Early European visitors who failed to successfully navigate the labyrinth of reefs that stretch from Garboi (Anchor Cay) into the south-west of Zenadth Kes met with tragic ends. Islanders, in contrast, were highly competent navigators. Their skills were honed by several thousand years of experience, observation and the intergenerational sharing of stories of the region's climatic and ecological complexities. Being adept at reading and responding to changing weather conditions, Islanders used the

Fig. 2.8. Long Reef, Mer (Murray) Island, eastern Torres Strait. (Image credit: Kim Wirth)

flow of the current and winds to guide their canoes over, through and around the maze of sometimes barely submerged coral platforms, columns and barrier reefs. The people of Zenadth Kes knew and lived with their reefs and seas in ways vastly different to the European outsiders who ventured to the region throughout the 19th century. Where Europeans saw empty space, Islanders knew, named and owned their seas and what lay beneath. In 1986, Meriam Ie researcher George Passi wrote: 'in Torres Strait every bit of soil, rock, beach, sea, reef is either owned communally, or by clans, families or individuals' [27].

Here, I present aspects of Islander ways of knowing and engaging with reef ecosystems before sustained European contact. I draw on the observations of outsiders and the stories they collected from our ancestors about Islander lifeways *bipotaim* – that is, before colonial occupation paved the way for commercial fishing and the introduction of Christianity upended Islanders' lives.

Knowledges and place
Islander' knowledges of the complex interactions between reef habitats, marine species and climate are acquired in localised contexts and understood through significant ancestral or mythical notions. These knowledges are also rooted to place in historical experiences and have been transmitted and relearned through subsistence and ritual practices

fishing, which may not involve a direct payment, but that require substantial market expenditure (purchase of boats and gear, travel to fishing spots and so on) and which may be compared directly with marketed forms of recreation.

Deloitte estimates the value added in these activities (i.e. output net of purchased inputs) at $6.4 billion a year on the GBR [6]. That's comparable in magnitude to the value added in agriculture (not including forestry and fisheries).

This comparison is useful for considering policies that aim to limit sediment, pesticides and fertiliser runoff from agriculture. Given that, until the introduction of controls, there was no economic cost to farmers from discharging sediment and fertiliser into rivers (though bad soil management and excess fertiliser use might generate on-farm costs), it is likely that significant reductions in discharges could be achieved at low costs. That, in turn, implies a small percentage reduction in value added in agriculture.

On the other hand, there is plenty of evidence to show that runoff is significantly damaging the Reef, and therefore significantly reducing its economic value. A broad finding of this kind does not obviate the need for a careful assessment of proposed interventions, but it suggests that we should be looking for ways to reduce runoff pollution and that governments should be willing to fund them.

In the case of overfishing, the first step is to reduce the total level of fishing where it exceeds the maximum sustainable yield. In Australia, this is typically done by creating a system of individual transferable quotas for commercial fishers, and imposing catch limits on recreational fishers. Beginning with the southern bluefin tuna fishery in the 1980s, most Australian fisheries are now regulated in this way. In general, the socially optimal catch for a commercial fishery will be less than the maximum sustainable yield. In addition, it is necessary to strike a balance between commercial and recreational fishing. Economic valuation is an important tool in this respect.

Runoff pollution and overfishing are major problems, but the existential problem facing the GBR is the combination of rising water temperatures and acidification caused by the burning of carbon-based fuels. This is a global problem that affects human and natural environments throughout the world, and to which everyone in the world contributes to some degree or other. It requires a global solution, in which countries cooperate to achieve an outcome beneficial to all.

Decarbonisation and the threat to reefs

Given the importance to the Australian economy of coal and gas exports, it is necessary to ask whether Australia would benefit from the rapid decarbonisation of global energy production needed to meet the Paris Agreement goal of holding global temperature rises to well below 2°C this century. It might be argued that a slower path to decarbonisation, with a likely outcome of 3°C of warming would yield greater benefits to Australia, even if the world would be worse off.

Among the consequences of 3°C of warming, the loss of nearly all the world's coral reefs is one of the most clearly established [36]. Substantial damage is already occurring, and further damage is inevitable, but, with rapid decarbonisation, the complete loss of Australia's reefs can be avoided. On the other hand, rapid decarbonisation would mean an end to mining of thermal coal (used in electricity generation) by 2030 at the latest, and a rapid transition away from metallurgical coal, used in steel production.

The annual value of coal exports from Australia is around $60 billion, of which about $20 billion is derived from thermal coal and $40 billion from metallurgical coal. Only part of this sum represents a net gain to Australia; a large proportion is paid out to international owners of capital or used in the purchase of imported equipment. Nevertheless, the economic value of coal is significant.

However, the value of domestic and international tourism is substantially greater than that of coal. Deloitte estimates the net benefit of the GBR to domestic tourists at $30 billion a year [6].

The loss of the Reef will affect vastly more people than those who use it to produce or consume market services each year. Placing a monetary value on the survival of the GBR is more difficult, conceptually. Unlike the case of tourist visits, the survival of the GBR is a public good. Either the GBR survives for everyone, or it survives for no-one. Hence, it is problematic to consider individual valuations of the survival of the Reef.

Nevertheless, in considering Australia's position in climate change negotiations, it is necessary to weigh the environmental costs of inaction against the market income that will be lost in an energy transition. There are a variety of ways to estimate this value. My own preference is for a hypothetical referendum in which survey respondents choose between alternative policy programs with different mixes of environmental and employment outcomes and tax changes, an approach that has been successfully used to model optimal land use in GBR catchments [37].

The central point is not the details of estimation, but the fact that estimates of the value of the survival of the GBR are comparable to the economic value of the coal industry. At the upper end, Deloitte estimates this value at $30 billion a year, which is greater than the gross value of thermal coal exports [6]. This comparison suggests that many Australians would rather see an end to the coal industry than lose the GBR.

Economics cannot provide a final and objective answer to the choices we face in relation to the GBR. Nevertheless, even a partial analysis shows that we should be willing to take costly action to save Australia's reefs from the many threats they face.

Bêche-de-mer: the cornerstone of Australian fisheries
Kennedy Wolfe

Sea cucumbers are important seafloor dwellers that are found in all major oceans of the world, from the deep abyssal sea to tropical coral reefs. They are large and abundant animals that are closely related to sea stars and sea urchins. Akin to earthworms in your garden bed, sea cucumbers move slowly along the sea floor where they play important roles in sediment filtering, so are often considered 'vacuum cleaners of the sea'.

Unbeknown to many, sea cucumbers are a widely fished animal. They are harvested as trepang or bêche-de-mer, and provide millions of coastal fishers with a source of income and more than 1 billion people with a source of nutrition, predominantly throughout South-East Asia. Around 10 000 tonnes of dried sea cucumber are traded internationally per year (Fig. 2.12) [38], which roughly corresponds to 200 million animals extracted from global oceans each year. At least 70 species of sea cucumber are commercially exploited in this multi-billion-dollar industry [39], with few farmed successfully through aquaculture.

The way people fish for sea cucumbers has changed rapidly, from traditional gleaning methods in shallow waters at low tide to the use of large boats, bottom dredges and SCUBA. These advances have encouraged widespread overfishing of slow-moving sea cucumbers, even in remote locations and deep-water habitats where they had previously been out of reach. Over recent decades, increased demand in the Asian market has driven exorbitant prices for bêche-de-mer, up to US$3000 per kilogram. This has exacerbated overfishing and illegal fishing, particularly of high-valued species [40].

One of Australia's oldest export industries
From the 1700s onwards, fishing fleets of 'trepangers' sailed from Indonesia to the Arnhem Land and Kimberley coastlines to collect sea cucumbers [41]. They traded other goods with Indigenous communities before returning home to sell their catch with traders from China (Fig. 2.13). This became so lucrative that overfishing and the consequent introduction of licence fees had halted this historical trade in the early 1900s.

In Australia today, sea cucumbers are fished from tropical waters in Western Australia, the Northern Territory, Torres Strait, the GBR and Coral Sea, each area with its own legislations. Many bêche-de-mer fisheries in Australia have suffered from unsustainable practice, including on the GBR [42]. In some cases, severe stock depletion has

Fig. 2.12. (A) Black teatfish. (B) Curryfish. (C and D) Bêche-de-mer in the market. (Image credits: Kennedy Wolfe)

triggered fishery closures, most notably for the high-value black teatfish, *Holothuria whitmaei*, which was protected for two decades (1999–2018) on the GBR. Catch limits, fishery closures and rotational zoning plans have been introduced to manage sea cucumber fishing efforts over space and time. However, increased monitoring of sea cucumber populations is required to understand the successes (or failures) of these management operations.

Impressively, the Western Australia Sea Cucumber fishery, which focuses its catch on just two species (the deep-water redfish *Actinopyga echinites* and sandfish *Holothuria scabra*), was the world's first sea cucumber fishery to be awarded the Marine Stewardship Council certification. This assures buyers that the fishery meets international standards of best practice for sustainable fishing.

However, illegal fishing is undermining sustainable practices and local management efforts in Australia's sea cucumber fisheries.

In the past decade, illegal fishing boats have been found with sea cucumbers in all tropical waters of Australia, from Western Australia to the GBR. In 2017, a foreign fishing vessel caught in the Coral Sea had almost 20 tonnes of live sea cucumber estimated to be worth $250 000 [43]. This vessel alone represented 77–133 per cent of the total allowable catch permitted for particular species within the Coral Sea Fishery. Moreover, up to 90 per cent of the sea cucumbers found were below the minimum size limit [43]. Illegal fishing therefore poses significant risk to the sustainability of sea cucumber fisheries in Australia, perhaps most critically for high-valued species.

Fig. 2.13. *Macassans at Victoria, Port Essington, 1845.* Indonesian traders from Makassar visited Arnhem Land and the Kimberley coastlines between the 1700 and 1900s to trade for trepang, which was smoked and transported back to Indonesia to be sold on to China. (Image credit: Hardern Melville)

Sea cucumbers experience similar serial exploitation to archetypal hunted animals such as rhinos, tigers and sharks. Their endangered status has been internationally recognised and measures have been taken to restrict their export and trade. In 2013, significant declines in sea cucumber numbers, globally, resulted in the listing of 16 species on the IUCN Red List for endangered species. This sent a stern message to resource managers for the conservation of at-risk bêche-de-mer targets [45]. However, global overfishing of key bêche-de-mer species continues. In 2020, three species known as 'teatfish' (the black teatfish *Holothuria whitmaei*, white teatfish *H. fuscogilva* and one not present in Australian waters, *H. nobilis*) were placed on

CITES Appendix II to further restrict their international trade and export [44]. Teatfish are among the most valuable sea cucumbers [40] and have consequently experienced population declines up to 90 per cent [45].

As one of Australia's oldest export industries, sea cucumbers hold great cultural and economic value. They also carry out a key ecological role on the seabed. Given the extent of exploitation of sea cucumbers globally, and the current vulnerability of their populations, bêche-de-mer fisheries operating in Australia must aim to protect vulnerable species, avoiding the classic 'boom and bust' cycles of populations that are harvested. In doing so, they can demonstrate the value of sustainable fishing practices to this nationally important and treasured species, both in Australia and elsewhere.

Western rock lobster and the Houtman Abrolhos Islands
Howard Gray

For many people, the Houtman Abrolhos Islands are synonymous with western rock lobster. The islands provide the shore base for a successful fishery targeting the delectable *Panulirus cygnus* (herein referred to as 'rock lobster') that thrive in surrounding waters.

The three shallow reef platforms of the Abrolhos, separated and deeply cut by channels, lie 60–80 km offshore from Geraldton (Western Australia), near the edge of the continental shelf and stretch 100 km from north to south. The two hundred or so islands are mostly small, many just banks of shingle and sand on reef tops, others exposed reef platforms from past sea-level highs, and just three of any great elevation or extent, topped with loose and lithified dunes. The latter large islands, with their cargo of mainland species and reef-built foundations underlying each of the Abrolhos platforms, reflect the dramatic sea-level changes of the last half-million years.

Coral reef building occurs here at high latitude under the influence of the warm south-flowing Leeuwin Current, opposed by the temperate algae-dominated ecosystems shaped by cold currents from the south. Seabirds have found ideal nesting habitat on these rare shelf-edge platforms on the eastern edge of the vast Indian Ocean, their immense colonies astounding early ornithologists. More notoriously, these suddenly emerging reefs snared at least two Dutch East India Company ships – the *Batavia* in 1629 and the *Zeewijk* in 1727 – and many more vessels, both large and small, over the last two centuries. The richness of the natural and human history of this archipelago is extraordinary [46].

Western rock lobster
Rock lobster are suitably adapted in form and behaviour to the shelf-edge Houtman Abrolhos Islands' habitat: crevassed ledges and coral hideaways providing daytime shelter; an abundant and diverse range of food items; favourable current systems that carry the phyllosoma larvae out into the Indian Ocean where they can circulate in eddies for a year while developing to the puerulus stage, ready to return to the nooks and crannies of the reefs (Fig. 2.14, [47]).

Fig. 2.14. (A) A western rock lobster, *Panulirus cygnus*, occupying its typical daytime shelter (image credit: Graeme Gunness). (B) Western rock lobster close up. (Image credit: Karl Monaghan)

Their success is evident in the prodigious numbers caught by fishers, for whom they have been the primary target species since World War II, perfectly meeting the requirements for commercial viability: they grow to a reasonable size, congregate in significant numbers, live at accessible depths in navigable waters, can be trapped easily, are robust so survive capture and handling, and they meet the culinary, aesthetic and cultural requirements of worldwide marketplaces.

In the early 1900s rock lobster were a sideline for a few fishers, either sold to mainland markets, used as finfish bait or, for a few years during the 1930s, canned at a rudimentary factory on West Wallabi Island. During World War II Abrolhos Island waters became the major source of rock lobster for 'Sea Coast Canneries' in Geraldton, filling contracts to supply the armed forces [48].

Post-World War II when a lucrative market opened in the United States of America for frozen rock lobster tails, the abundant almost virgin stock in Abrolhos waters enabled inexperienced fishers to make a profit while learning their distribution and habits. Indeed, so prolific were the rock lobster that pioneers only half-jokingly recall that they set their 'pots' only where they could see antennae protruding from under coral ledges. Baited traps were set by day to capture the nocturnally wandering omnivores when fishers returned the next morning to harvest the catch (Fig. 2.15). Catchability varies with moon phases (rock lobster don't venture far from their dens on moonlit nights), moulting periods (once the exoskeleton hardens after moulting, which occurs around March and November, they are very hungry), migratory activity (newly mature adults undertake mass migrations offshore and northwards along the shelf) and according to the sea conditions of the shelf-edge environment (rock lobster are more catchable when disturbed from their haunts by ocean swells).

Early post-World War II fishers concentrated on the shallow inner reefs of the Island platforms that suited small sailing craft and hand-pulled pots. Nearly 1 million kg of legal market size, or 1.5–2 million lobsters, were produced by these

Fig. 2.15. A fisherman retrieves rock lobster from a pot on the deck of his boat as the Houtman Abrolhos fishing industry was rapidly expanding, 1969. (Image supplied with kind permission by the National Archives of Australia, photographer unknown)

reefs at their peak in 1958/59. As the virgin stock was fished out and the catch from the inner reef systems declined, fishers turned their attention to the outside of deeper reefs, aided by larger vessels with pot pullers and echo sounders. In 1961/62 the inner and outer grounds together produced nearly 1.8 million kg or 3 million or more individual lobsters [49].

An expanding fishing base in the Abrolhos

Before World War II there had been only sporadic occupation of the Abrolhos: wet-line fishers from as early as the 1840s sleeping on their boats or in simple shacks ashore during the few days required to fill the live 'well' or icebox before a dash to the markets of Fremantle or Geraldton; guano miners from the mid-1880s to the early 1920s gathering the phosphate-rich rookery accumulations of several thousand years; young men at a short-lived crayfish cannery during the 1930s; a few holiday-makers taken by local fishers on shooting and fishing adventures; and the occasional scientist, most often an ornithologist enraptured by the millions of shearwaters and terns nesting on the islands.

In places where the islands were too far from the mainland to make a return crossing each day, fishers established bases on islands where sheltered anchorages were close to fishing grounds. Although rivalry at sea was intense, they worked cooperatively ashore, so several islands became crowded with 'camps'. Each camp grew to consist of a main dwelling, one or two shacks for deck crew, and generator and equipment sheds. Each fisher usually built their own jetty out to deep water and island occupation steadily increased (Fig. 2.16).

The Abrolhos season (mid-March to mid-August, later shortened to the end of June) would begin with a frantic fortnight of getting ready, all requirements for the months ahead prepared and carted to the islands by carrier boat. Opening day signalled a month of intense make-or-break fishing of the accumulated stocks. As catches dwindled away to the end of the season, just a few stragglers would be left keeping company with the resident ospreys, eagles and gulls.

Fig. 2.16. Rat Island in the Easter Group. The tramlines of the late 1800s guano mining operations are still visible. (Image credit: Karl Monaghan)

Some women soon joined as independent fishers or worked alongside their partners, others staying shore-bound with young children. Carrier boats arrived each day with supplies and mail, returning to the mainland with the catch. On the larger islands of the Easter and Wallabi groups and the isolated North Island, each with 20–30 camps, community halls and schools were built and airstrips constructed, and the population swelled to ~4000 people on the island during the peak fishing season. In the southern Pelsaert Group, a dozen small islands were each occupied by one or a very few camps and so communal facilities did not eventuate. Such distinctive seasonal fishing communities as were found at the Abrolhos are rare in Australia. It was a way of life that by the 21st century was, for many into third, even fourth or fifth generations, embedded in family identity and folklore with the tales of pioneering days and the characters such isolation breeds, the lifestyle as valuable as the monetary return from fishing. It must be remembered that not all succeeded at the fishing game and that, in the treacherous waters of the island's reef systems, tragedy stalked many who ventured to sea, with boats wrecked and lives lost.

Boom and bust: understanding and managing rock lobster populations

The course of a fishery such as that for rock lobster is not just the result of the skills and work of fishermen and -women. It is also subject to an array of factors, including technological innovation and adoption, economic conditions, entrepreneurship, social changes, government management and political decision-making. However, all is ultimately dependent on maintaining the stocks of the target species. For that, a deep knowledge of the biology of both the species and the ecosystem of which they are part is essential [50].

The unique coral reef ecosystems of the Abrolhos caught the attention of two well-known pioneering marine biologists in the late 1800s and early 1900s. William Saville-Kent visited the Abrolhos in 1894 as Commissioner of Fisheries in

Western Australia to investigate the fishery and pearl-shell potential, endorsing both. He was excited to find luxuriant coral reef communities so far south, identifying the unusual warm-water influence. William Dakin, Professor of Biology at the University of Western Australia, undertook expeditions in 1913 and 1915, collecting many specimens and describing the reef structures.

Keith Sheard from the Australian Government's Commonwealth Scientific and Industrial Research Organisation (CSIRO) began systematic studies of the rock lobster populations and the emerging fishery in the post-World War II period, which were continued by Ray George in the early 1950s. Sheard's compilation of detailed catch data for the period 1945–61 provided valuable insights into the impact of the rapidly escalating number of fishers drawn to the lucrative fishery. George went on to study the basic biological attributes of *P. cygnus*, and to discover the whereabouts of the planktonic phyllosoma larvae during their 'lost year', trawl studies finding them as much as 1500 km to the west of the Abrolhos [51].

By the early 1960s a classic 'tragedy of the commons' scenario was unfolding. The work of Sheard, George and Bernard Bowen, a research officer and later Director of the WA Department of Fisheries, led to the 1963 closing of the rock lobster fishery to new entries and a limit on the number of pots each fisher was entitled to work. This approach, with many tweaks and addendums, became the basis for the management of the fishery over the next half century.

Rock lobster phyllosoma transform to shrimp-like puerulus, which return to settle on the reefs of the continental shelf. CSIRO research in the late 1960s found that the puerulus would cling to frames of artificial seaweed and that the number caught gave a good indication of the potential catch 4 years later, when they had grown to commercial size. This provided a means for regulators to monitor the breeding success, while for fishers it was an important guide for future investment and fishing strategies.

The WA Fisheries Department research division, CSIRO's coastal ecosystem and oceanography teams and the WA Museum conducted investigative and monitoring work through the 1970s and 1980s to better understand the creature, its habitat and design appropriate management strategies. During these decades many economic and marketing factors (such as exchange rates, tax breaks, more lucrative Asian markets and live export) and the continued high catches transformed the rock lobster fishery. Increased profitability financed larger craft and sophisticated equipment, adding to the ever-increasing skill and knowledge levels of the fishers (Fig. 2.17). Sustained high catches hid the inexorable creep in effort, more pots being set on more days across even the most remote areas. The introduction of GPS units increased efficiency even further and the regulators imposed ever-increasing restrictions. For the previous 50 years, over 80 per cent of the rock lobster caught each year were those just moulted to legal size (76 mm carapace length), and few remained as breeding

Fig. 2.17.   (A) Fishing craft and equipment at Basile Island in the Pelsaert Group of the Houtman Abrolhos Islands *c.* 1950. (B) Some of the nearly 100 vessels that fish the Abrolhos for rock lobster, now equipped with high levels of fishery-specific technology. (Image credits: Neil McLaughlan (A), Howard Gray (B))

stock in the wild. Despite the warning signs, in 2000 the Western Rock Lobster Fishery, after intensive scrutiny by the Marine Stewardship Council (MSC), became the first fishery in the world to achieve MSC's endorsement as a sustainably managed fishery [50].

It wasn't to last. Ever-increasing fishing intensity with wide-ranging craft using the latest technology saw catches dwindle in the early 2000s and, alarmingly, the puerulus count declined to almost zero, foreshadowing even lower catches in coming years. Exacerbated by the global financial crisis, the industry was on the point of collapse. A long-promoted quota management system was finally accepted by the fishers in 2010, initially cutting the catch to less than 50% of long-term averages. Many fishers sold out to retain some personal financial security, while others bought quota entitlements at what would later be seen as bargain prices.

Rock lobster stocks recovered over the following decade. The shift to a quota fishery reduced the number of fishers and changed the social fabric of the Abrolhos. Those fishers that remained could come and go as they pleased across the year, often fishing only when demand and prices maximised their income. Families no longer made the annual 4-month pilgrimage and the community schools and social clubs and events on the islands were no longer viable. Some camps were abandoned, others rarely used and then often just to retain an island foothold, keeping alive family connections and traditions (Fig. 2.18).

Despite the sudden, unexplained decline of the rock lobster stocks the MSC status has been retained, management by a flexible quota ensuring the fishery is able to respond to changes that may become apparent. For the western rock lobster, the reduction in fishing pressure allowed populations to recover. Fishers found that their profit margins increased, making more but working less. By early 2020, profitability exceeded any previous period, especially for those unencumbered by debt and with large quota entitlements.

The live lobster export market has been dominated by China in recent decades, with over 90 per cent of the catch reaching that destination. In early 2020, however, diplomatic tensions led to a complete ban on the importation of western rock lobster into China. Overnight, fishing for rock

Fig. 2.18.   Fishing huts and jetties at the Pelsaert Group, Houtman Abrolhos. (Image credit: Matt Curnock)

lobster ceased, as the industry plummeted from one of its most lucrative phases to one of loss. It is gradually clawing its way back through domestic and alternative overseas sales, and through so-called 'grey trade' into mainland China, via Hong Kong.

With minimal fishing pressure, the rock lobster populations on the reefs of the Abrolhos and in surrounding waters have begun to resemble those of the untapped fishery, with implications for the overall ecosystems of the reefs. Although the Houtman Abrolhos Islands are already home to one of the most intensively studied crustaceans and fisheries, there is still much to be investigated.

William Saville-Kent

Howard Gray

William Saville-Kent was born in 1845 at Sidmouth, Devon, England (Fig. 2.19). Saville-Kent came to Australia in 1883 to work as Superintendent and Inspector of Fisheries for the Tasmanian Government, which was seeking to restore depleted oyster beds. He introduced many management measures to ensure fisheries sustainability.

After joining a survey cruise along the north-western coast of Australia in 1888, Saville-Kent accepted an appointment as Queensland Commissioner of Fisheries from 1889 to 1892 to oversee the pearl oyster, fish, bêche-de-mer, corals, sponges, dugong and turtle fisheries.

Saville-Kent immersed himself in collecting, photographing, painting and drawing the colourful life forms he encountered. His magnificent book, *The Great Barrier Reef*, was published in 1893 and featured 49 of his

Fig. 2.19. (A) William Saville-Kent. (B) Book cover of Saville-Kent's *The Great Barrier Reef of Australia: Its Products and Potentialities*, an illustrated account of the corals and coral reefs, pearl and pearl-shell, bêche-de-mer, other fishing industries, and the marine fauna of the Australian Great Barrier Region [33]. (Image supplied with kind permission by the Biodiversity Heritage Library, Harvard University)

photographs (pioneering the camera as a scientific recording device) and 16 beautiful lithographs based on his original watercolour sketches. While in Queensland he found a way to artificially induce pearl shell to produce hemispherical or 'blister' pearls.

Given the declining fortunes of the once-prosperous pearling industry in Western Australia, that state's first Premier, John Forrest, offered Saville-Kent the position of Commissioner of Fisheries for Western Australia. In his 2 years in this role (1893–95), Saville-Kent surveyed fish stocks and investigated fishery laws, the economics of the industry and the marketing of the catch. He visited the Houtman Abrolhos Islands twice and was intrigued by the interblending of tropical and temperate marine life at such a southern latitude, concluding that an ocean current from the equatorial area of the Indian Ocean must penetrate this far south (now known as the Leeuwin Current). The Abrolhos so captivated Saville-Kent that he devoted a whole chapter of his 1897 book, *The Naturalist in Australia*, to its terrestrial and marine ecology; other chapters reflected his diverse fascination with Australia's unique reptiles, amphibians, ants, termites, marsupials and monotremes.

Seeing the potential for pearl shell culture at the Abrolhos, he organised for *Pinctada margaritifera*, the black-lipped pearl, to be introduced from the north-west into Pelsaert lagoon. They survived for some years before they were lost. Further plans to cultivate mother-of-pearl shell at the Abrolhos, Sharks Bay and the Lacepede Islands on a large scale, buy a pearling fleet and establish a depot at Broome for a business in turtle, bêche-de-mer and general fishing never eventuated.

After a 9 year spell back in England, Saville-Kent returned to Australia in 1904 to experiment with pearl culture for Levers Pacific Plantations, a British company pursuing copra and phosphate interests. A generous bonus applied if he succeeded, but after 18 months he retired. While passing through Brisbane on his way to England he found support for a venture off Cape York, floated as The Natural Pearl Shell Cultivation Co. of London. By 1908 he had produced blister and spherical pearls of commercial quality.

In mid-1908 Saville-Kent returned to England and, after surgery, succumbed to an attack of pneumonia, aged 63. His grave in All Saints' churchyard, Milford-on-Sea, Hampshire, was decorated with corals he had collected in Australia.

William Saville-Kent's scientific approach to fisheries laid the foundation for Australia's sustainable fishing industries. His enthusiasm for Australia's coral reefs, matched by his pioneering techniques and extraordinary output, were appropriately acknowledged when a reef off Townsville and an island in the Houtman Abrolhos archipelago were named in his honour.

Coral reef tourism

Graeme S. Cumming, Matthew I. Curnock and Michelle Dyer

A recreational paradise

Australia's diverse and accessible coral reefs have long been a premier tourist destination. They offer visitors the opportunity to enjoy sun, sand and warm water with stunning snorkelling and diving in secure and well-serviced locations (Fig. 2.20). The large number and extent of reefs, and the wide variety of different budget options and activities on offer, have been both a cause and consequence of rapid growth in the tourism industry.

Reef tourism worldwide is economically important. In Australia, total income from tourism to reefs and other marine habitats was just under $15 billion in 2015–16 (note that this figure represents total tourism revenue across 2 years, whereas the aforementioned $6.4 billion income is a net annual figure encompassing all reef-related activities). In 2017, the economic, social and iconic value of the GBR was calculated as $56 billion [6]. Coral reef tourism also generates 'reef-adjacent' revenue through related industries, such as accommodation and transport. The tourism industry associated with the GBR creates nearly 59 000 jobs, of which 38 000 are directly reef-dependent [6]. On the West Coast, Ningaloo Reef contributed around $110 million per annum to the Western Australian economy in 2018–19, with 90 per cent of this value deriving from tourism [51].

Fig. 2.20. Tourists snorkelling and feeding fish on the Great Adventures pontoon, Norman Reef, 1991. (Image credit: J. Oliverv © Great Barrier Reef Marine Park Authority, supplied with kind permission)

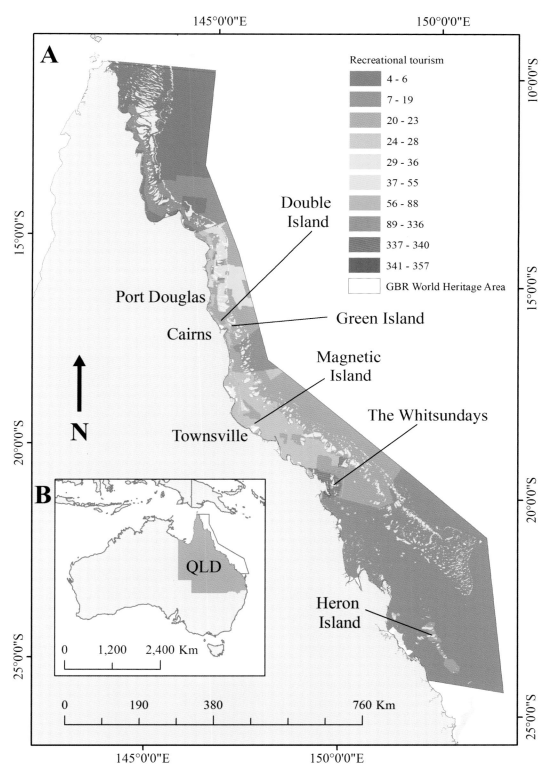

Fig. 2.23. (A) Numbers of recreational tourism permits by 2 × 2 km grid cell across the Great Barrier Reef Marine Park. (B) Location of the Great Barrier Reef Marine Park, along the Queensland (QLD) coastline, Australia (see [55] for further details). (Image credit: Sarah Hamylton)

Ningaloo, Shark Bay and within the Lord Howe Group.

A rocky, turf-dominated or biodiverse future?

Mass coral bleaching events have raised significant concerns for the future of the coral reef tourism industry. Some researchers have argued that the GBR is perceived as a 'last chance' tourism destination; ecological grief (the feeling of personal loss resulting from the degradation of a cherished ecosystem) and empathy were also reported from Australian and international tourists [1]. Support within the tourism industry for urgent climate action and interventions to protect coral reefs have been galvanised, and the tourism industry has become a proactive and willing partner in changing the focus of GBR tourism activities. Commercial tourism is now an important vehicle for encouraging positive interactions with the environment and increasing environmental stewardship. At present, there are ~80 GBRMPA-certified, high-standard tourism operators spreading the length of the Reef and over 60 Master Reef Guides (i.e. graduates of the GBRMPA's 10-module 'Reef Discovery Course' who are intensively trained in reef interpretations, master storytelling and experience delivery) working along the breadth of the GBR. Citizen science is also a growing area of interaction for tourists with the Reef (see Chapter 5).

Tourism to Australian coral reefs is changing again. At the time of writing, with the industry heavily impacted by travel restrictions following the COVID-19 pandemic, it seems that without international visitors, recovery is likely to be slow despite the provision (in mid-2021) by the Australian Government, through the GBRMPA, of a $3.2 million dollar relief initiative for reef tourism operators impacted by the pandemic.

Coral reef tourism also faces high uncertainty about the condition and future of reefs. It remains to be seen whether coral reefs, and dependent reef tourism, will successfully adapt to the challenges of global warming and other anthropogenic impacts. The current expert consensus is that coral reefs will persist under low to mid-range warming trajectories, but with different dominant species and a different three-dimensional structure and function (e.g. potential shifts to rubble, algal domination or turfs) from today's reefs. For Australian tourism specifically, it is unclear whether Australian reefs will remain globally competitive tourist destinations, or will gradually become degraded, poorquality reef destinations relative to alternative tourist reefs in other countries. Under a global outlook of coral reef decline, Australia's ability to invest resources in restoring and maintaining its coral reef assets may become a competitive advantage. If coral declines make it difficult for operators to find highquality tourism locations, will a quality over quantity model (i.e. focusing on upper-end tourism with associated luxury accommodation and educational experiences) be more effective than a bulk tourism model? Will this drive increased competition between operators in the future, and what steps can be taken to encourage cooperation?

The drastic reduction of tourist numbers that followed the onset of the COVID-19 pandemic seems to have led to cooperation rather than competition among Cairns GBR tourism operators. In 2020 and 2021, many companies cooperated to run tours on different days and cross-deck crew to share the domestic-only tourist market. Will it be possible for Australian coral reef-related tourism to continue to grow, or even return to pre-pandemic levels? We do not yet have answers to these questions, but it seems that in the short term the Australian coral reef tourism industry may be on the cusp of another transformation. While travel difficulties may make Australia less accessible to international visitors, one thing that seems certain is that the future of coral reef tourism in Australia will be defined by innovation and calculated risk-taking.

References

1. Curnock M, Marshall NA, Thiault L, Heron SF, Hoey J, Williams G, Taylor B, Pert PL, Goldberg J (2019) Shifts in tourists' sentiments and climate risk perceptions following mass coral bleaching of the Great Barrier Reef. *Nature Climate Change* **9**, 535–541. doi:10.1038/s41558-019-0504-y

2. Thiault L, Curnock M, Gurney G, Heron S, Marshall N, Bohensky E, Nakamura N, Pert P, Claudet J (2021) Convergence of stakeholders' environmental threat perceptions following mass coral bleaching of the Great Barrier Reef. *Conservation Biology* **35**(2), 598–609. doi:10.1111/cobi.13591

3. Lloyd RJ (2016) Fathoming the reef: a history of European perspectives on the Great Barrier Reef from James Cook to GBRMPA. PhD thesis. James Cook University, Australia, <https://researchonline.jcu.edu.au/49776/>.

4. Goldberg J, Marshall N, Birtles A, Case P, Bohensky E, Curnock M, Gooch M, Parry-Husbands H, Pert P, Villani C, *et al.* (2016) Climate change, the Great Barrier Reef, and the response of Australians. *Palgrave Communications* **2**, 15046. doi:10.1057/palcomms.2015.46

5. Gurney GG, Blythe J, Adams H, Adger WN, Curnock M, Faulkner L, James T, Marshall NA (2017) Redefining community based on place attachment in a connected world. *Proceedings of the National Academy of Sciences of the United States of America* **114**, 10077–10082. doi:10.1073/pnas.1712125114

6. Deloitte Access Economics (2017) 'At what price? The economic, social and icon value of the Great Barrier Reef'. The Great Barrier Reef Foundation, Brisbane, <https://www2.deloitte.com/content/dam/Deloitte/au/Documents/Economics/deloitte-au-economics-great-barrier-reef-230617.pdf>.

7. Marshall N, Adger WN, Benham C, Brown K, Curnock MI, Gurney GG, Marshall P, Pert PL, Thiault L (2019) Reef grief: investigating the relationship between place meanings and place change on the Great Barrier Reef, Australia. *Sustainability Science* **14**, 579–587. doi:10.1007/s11625-019-00666-z

8. Great Barrier Reef Marine Park Authority (2017) Great Barrier Reef blueprint for Resilience. Great Barrier Reef Marine Park Authority, Townsville, <https://elibrary.gbrmpa.gov.au/jspui/handle/11017/3287>.

9. Bradley J, Yanyuwa families (1988) *Yanyuwa Country: The Yanyuwa People of Borroloola tell the History of Their Land*. Greenhouse Publications, Richmond.

10. John Brookes to English East India Company, 25 August 1622, India Office Original Correspondence Vol. 9, no: 1072.

11. Beaglehole JC (Ed.) (1955) *The Journals of Captain James Cook on his Voyages of Discovery: The Voyage of the Endeavour 1768–1771*. Cambridge University Press for the Hakluyt Society, Cambridge.

12. Flinders M (1814) *A Voyage to Terra Australis: Undertaken for the Purpose of Completing the Discovery of that Vast Country, and Prosecuted in the Years 1801, 1802, and 1803, in His Majesty's Ship the Investigator*. Volumes 1 and 2. G and W Nicol, London.

13. Fitzroy R (1839) *Narrative of the Surveying Voyages of His Majesty's Ships Adventure and Beagle between the years 1826 and 1836, describing their examination of the southern shores of South America, and the Beagle's circumnavigation of the globe. Proceedings of the Second Expedition, 1831–36, Under the Command of Captain Robert Fitz-Roy, R.N.* H Colburn, London.

14. Proposed orders for Captain Blackwood, UK Hydrographic Office (UKHO) MB 3: September 1837–May 1842, 409–416. Beaufort to Blackwood, UKHO LB 12.

15. Beaufort to Stanley (1846) UKHO LB 14, 214–215.

16. Sponsel A (2015) From Cook to Cousteau: the many lives of coral reefs. In *Fluid Frontiers: Exploring Oceans, Islands, and Coastal Environments*. (Eds J Gillis and F Torma) pp. 139–161. White Horse Press, Cambridge, UK.

17. Ingleton GC (1979) Flinders as cartographer. In *Matthew Flinders: The Ifs of History*. (Ed. RW Russell) pp. 63–80. University Relations Unit, Flinders University, Adelaide.

18. Richards G (1863–64) Discussion on survey within the Great Barrier Reef, Royal Geographical Society London, 11 April 1864. *Proceedings of The Royal Geographical Society of London* **8**, 114–121.

19. Kenneth Slessor *Five Visions of Captain Cook* (1931), <https://www.poetryfoundation.org/poems/47089/five-visions-of-captain-cook>.

20. Anon (1920) Secrets of deep. *The Sun,* Sydney, 1 December, p. 8.

21. Duggan L (2001) *Ghost Nation: Imagined Space and Australian Visual Culture 1901–1939.* pp. 100–109. University of Queensland Press, St Lucia.

22. Hamera J (2012) *Parlor Ponds: The Cultural Work of the American Home Aquarium 1850–1970.* pp. 1–2. University of Michigan Press, Ann Arbor MI.

23. Elias A (2019) *Coral Empire: Underwater oceans, colonial tropics, visual modernity.* Duke University Press, Durham NC.

24. Anon (1922) Pearls and savages, *Daily Herald,* Adelaide, 2 January, p. 4.

25. Hurley F (1922) My coral garden: wonders of the ocean bed. Riot of color. *The Daily Mercury,* Mackay Qld, 4 November, p. 7.

26. Wright J (1977) *The Coral Battleground.* Thomas Nelson Australia, West Melbourne.

27. Passi G (1986) Traditional resource knowledge, western education and self-management autonomy of Torres Strait. MA thesis. p. 10. University of Queensland, Australia.

28. Rowland M, Ulm S (2011) Indigenous fish traps and weirs of Queensland. *Queensland Archaeological Research* **14**, 1–58. doi:10.25120/qar.14.2011.219

29. Neitschmann B (1989) Traditional sea territories, resources and rights in Torres Strait. In *A Sea of Small Boats.* (Ed. John Cordel) pp. 62–93. Cambridge, Mass. Cultural Survival, c1989,.

30. *Akiba on behalf of the Torres Strait Islanders of the Regional Seas Claim Group v State of Queensland (No 2)* [2010] FCA 643, 606.

31. *Akiba v State of Queensland,* ibid., 612.

32. Lawrie M (1970) *Myths and Legends of Torres Strait.* University of Queensland Press, St Lucia.

33. Saville-Kent W (1893) *The Great Barrier Reef of Australia: Its Products and Potentialities.* pp. 334–335. Allen, London, UK.

34. Le Roux G (2016) Transforming representations of marine pollution. For a new understanding of the artistic qualities and social values of ghost nets. Anthrovision [Online], 4.1, 1–20. GhostNets Australia, <https://www.ghostnets.com.au/>.

35. Oxford Economics (2009) *Valuing the Effects of the GBR Bleaching.* The Great Barrier Reef Foundation, Brisbane.

36. Australian Academy of Science (2021) 'The risks to Australia of a 3°C warmer world', <https://www.science.org.au/files/userfiles/support/reports-and-plans/2021/risks-Australia-three-deg-warmer-world-report.pdf>.

37. Mallawaarachchi T, Quiggin J (2001) Modelling socially optimal land allocations for sugar cane growing in North Queensland: a linked mathematical programming and choice modelling study. *The Australian Journal of Agricultural and Resource Economics* **45**, 383–409. doi:10.1111/1467-8489.00149

38. Purcell SW, Mercier A, Conand C, Hamel JF, Toral-Granda MV, Lovatelli A, Uthicke S (2013) Sea cucumber fisheries: global analysis of stocks, management measures and drivers of overfishing. *Fish and Fisheries* **14**, 34–59. doi:10.1111/j.1467-2979.2011.00443.x

39. Purcell SW, Conand C, Uthicke S, Byrne M (2016) Ecological roles of exploited sea cucumbers. *Oceanography and Marine Biology – an Annual Review* **54**, 367–386. doi:10.1201/9781315368597-8

40. Purcell SW, Polidoro BA, Hamel JF, Gamboa RU, Mercier A (2014) The cost of being valuable: predictors of extinction risk in marine invertebrates exploited as luxury seafood. *Proceedings, Biological Sciences* **281**, 20133296. doi:10.1098/rspb.2013.3296

41. Macknight C (2011) The view from Marege: Australian knowledge of Makassar and the impact of the trepang industry across two centuries. *Aboriginal History* **35**, 121–143. doi:10.22459/AH.35.2011.06

42. Eriksson H, Byrne M (2015) The sea cucumber fishery in Australia's Great Barrier Reef Marine Park follows global patterns of serial exploitation. *Fish and Fisheries* **16**, 329–341. doi:10.1111/faf.12059

43. Skewes TD (2017) *Coral Sea Sea Cucumber Catch Sampling. Catch of two foreign fishing vessels, Coral Sea, February 2017.* Tim Skewes Consulting. Brisbane.

44. Food and Agricultural Organisation (2019) FAO expert advisory panel assessment report: COP18 proposal 45. In *Report of the Sixth FAO Expert Advisory Panel for the Assessment of Proposals to Amend Appendices I and II of CITES. Concerning Commercially Exploited Aquatic Species.* FAO Fisheries and Aquaculture Report No. 1255. pp. 62–105. Rome.

45. Conand C, Polidoro B, Mercier A, Gamboa R, Hamel JF, Purcell S (2014) The IUCN Red List assessment of

aspidochirotid sea cucumbers and its implications. *SPC Beche-de-mer Information Bulletin* **34**, 3–7.

46. Gray HS (in press) *Abrolhos – The Natural and Human History of the Houtman Abrolhos Islands*. Westralian Books, Geraldton.

47. Gray HS (1992) *The Western Rock Lobster* Panulirus cygnus *Book 1: A Natural History*. Westralian Books, Geraldton.

48. Gray HS (1999) *The Western Rock Lobster* Panulirus cygnus *Book 2: A History of the Fishery*. Westralian Books, Geraldton.

49. Sheard K (1962) *The Western Australian Crayfishery, 1944–1961*. Paterson Brokensha, Perth.

50. Gray HS (2000) Skinnin the pots – a history of the western rock lobster fishery. PhD thesis. Murdoch University, Australia.

51. Department of Biodiversity Conservation and Attractions (2020) 'Economic contribution of Ningaloo: one of Australia's best-kept secrets'. Commissioned report provided by Deloitte Access Economics, Department of Biodiversity, Conservation and Attractions, Perth.

52. Lloyd R (2015) 'Wealth of the reef': the entanglement of economic and environmental values in early twentieth century representations of the Great Barrier Reef. *Melbourne Historical Journal* **43**(1),40–62.

53. Daley B (2014) *The Great Barrier Reef: An Environmental History*. Routledge, Oxford, UK.

54. Day JC, Dobbs KA (2013) Effective governance of a large and complex cross-jurisdictional marine protected area: Australia's Great Barrier Reef. *Marine Policy* **41**, 14–24. doi:10.1016/j.marpol.2012.12.020

55. Cumming GS, Dobbs KA (2019) Understanding regulatory frameworks for large marine protected areas: permits of the Great Barrier Reef Marine Park. *Biological Conservation* **237**, 3–11. doi:10.1016/j.biocon.2019.06.007

56. Great Barrier Reef Marine Park Authority (2019) 'Great Barrier Reef Outlook Report 2019'. Great Barrier Reef Marine Park Authority, Townsville.

3

The evolution of science
on the Great Barrier Reef

The Great Barrier Reef (GBR) is recognised as one of the major coral reefs in the world and it is a place where many important scientific discoveries have been made. European scientific research on Australia's coral reefs dates to the observations of early explorers who navigated through the different reef regions. Notable voyages on the GBR were made by James Cook (HMS *Endeavour*, 1770), Matthew Flinders (HMS *Investigator*, 1802) and Joseph Jukes (HMS *Fly*, 1843).

The Great Barrier Reef Committee was established in 1922 by the University of Queensland and the Queensland State Governor to initiate and support reef studies by individuals or groups, and to drill through the GBR to better understand its origins. Subsequently, working with the British Great Barrier Reef Committee, a more biological focus was adopted, leading to the 1928–29 Great Barrier Reef Expedition, based at Low Isles, northern GBR, for a little over a year. The 1928 expedition resulted in many pioneering discoveries about coral reef ecology, physiology, and geomorphology, leaving behind a legacy that lingers on today in coral reef science.

This chapter provides an overview of geological and biological research undertaken on the GBR in the last 100 years. Establishing the age of the reef has demanded both a long-term perspective of when the underlying reef platform was shaped by continental movements hundreds of thousands, even millions of years ago, and a more geologically 'modern' understanding of how the reef islands and platforms formed in relation to sea-level changes over the last 10 000 years. Tracking changes in coral communities and associated marine life over the last century has meant that researchers have had to learn to work underwater, initially under challenging circumstances.

The establishment of research stations at Heron Island, One Tree Island and Lizard Island in the 1960s and 1970s supported scientists to carry out applied research towards a better understanding of how coral reefs function. By the end of the 1970s, an informal network of six research stations plus AIMS had been established to support scientists and students to learn about the GBR across its entire length. The capacity to make observations and carry out analytical experiments *in situ* allowed a wealth of applied research projects to answer fundamental questions about coral reefs, with an increasing focus on the impacts of climate change following the 1998 mass coral bleaching event on the GBR.

The Great Barrier Reef Expedition of 1928–29
Barbara Brown, Tom Spencer, Sarah Hamylton and Roger McLean

The 1928–29 Great Barrier Reef Expedition was an important turning point at which modern coral reef science evolved from a 19th-century reliance on theoretical deduction, as typified by Charles Darwin's subsidence theory of oceanic reefs, to a 20th-century focus on empirical and analytical studies. The expedition differed fundamentally from its shipborne predecessors, being housed on a single reef and sand cay of the Low Isles on the northern GBR for 13 months, although small trips were made to other parts of the northern GBR (Fig. 3.1). Led by marine zoologist Dr Maurice Yonge, the interdisciplinary work program brought together English and Australian scientists with expertise in zooplankton, phytoplankton, zoology, geography and botany. They carried out meticulous microscopic work and made painstaking laboratory and field observations, measurements, and experiments, as well as surveying life on the shallow, intertidal reef flat and subtidal reef slope. Over the following three decades, results from this work were published in scientific reports and journals, forming a body of work that had a profound effect on global coral reef science.

Fig. 3.1.   The sand cay at Low Isles at the time of the Great Barrier Reef Expedition (1928–29). (Image supplied with kind permission by the Royal Geographical Society, photographer unknown)

Novel features of the 1928–29 Great Barrier Reef Expedition

The expedition was ground-breaking in several respects. The interdisciplinary nature of the work had a clearly articulated philosophy, from the planning stages right through joint fieldwork and publications. Locating the work at Low Isles enabled the local reef environment to be used as a 'natural aquarium' in which relatively sophisticated instruments were fashioned for research that took both field and laboratory observation and experimentation to a new level. In an early version of SCUBA technology, a diving helmet into which air was pumped from a boat was used for underwater observations. Other inventive equipment included a 'clock tower' for growing corals on the reef and a coffin-shaped, light-proof box for testing the effects of darkness on corals.

The expedition was also noteworthy for the participation of female scientists, who actively investigated phytoplankton production, coral reproduction and growth, sedimentation impacts on corals, as well as undertaking detailed ecological and reef surveys. The inclusion of women in the research party, widely commented upon in newspaper and other popularist accounts at the time, served as a catalyst for greater involvement of women in Australian science – an involvement that continues to be a feature of coral reef research in Australia today (see the box, 'Sidnie Manton: pioneer of coral reef community surveys' later in this chapter).

The scientific findings from the Great Barrier Reef Expedition

The expedition yielded scientific advances of particular significance to coral reef biology and geology. These included work on coral physiology, reproduction, and growth rates, the effects of sediments on corals, and detailed surveys and mapping of ecological attributes and coral reef landforms (see 'Revisiting the corals of the Low Isles 90 years later' later in this chapter).

Coral physiology: feeding, metabolism and bleaching

C. M. Yonge's work on coral physiology investigated feeding mechanisms, determining what corals fed on, how they digested food, and the function and significance of their symbiotic algae. Working in outdoor aquaria, Yonge determined that corals are carnivores that feed in a highly specialised manner [1, 2]. By running an extensive set of experiments on corals in a light-tight box on the reef flat, Yonge found that massive corals could survive in highly shaded conditions without zooxanthellae [2]. He also revealed the significance of oxygen production and consumption in the functioning of reef-building corals, finding them to be capable of surviving in water of extremely variable oxygen content [3]. Through measuring corals with and without zooxanthellae in both dark and light conditions, Yonge revealed the role of zooxanthellae in excreting waste products such as phosphorus and nitrogen from the coral host [2]. While much later work revealed the importance of zooxanthellae in coral nutrition, these early experiments paved the way for our appreciation of zooxanthellae as a critical factor in the overall success of corals.

On 29 February 1929, the expedition scientists observed widespread coral bleaching on the reef flats surrounding Low Isles, noting that many corals were killed during a calm spell of weather with seawater temperatures of 35.1°C in coral pools, although some corals recovered up to 3 months later [2]. This represents the first published account of thermally induced whitening or coral 'bleaching' in the field, which led to temperature experiments and early descriptions of possible bleaching mechanisms.

Coral growth, reproduction and effects of sedimentation on reef corals

Work on coral growth and reproduction took place in the 'aquarium-like' natural environment of the Low Isles reef flat and anchorage. A diving helmet was used to collect and observe marked corals underwater at depths of 4–9 m. The growth rates of young corals were measured using a hollow framed 'clocktower' structure that was secured into the reef on iron legs. Planulae collected from corals were fixed in 'fingerbowls' where they could be photographed as they grew. One-hundred-and-sixty-nine corals of various species were monitored over a 6-month period, yielding some of the earliest observations of coral tentacle budding and the regeneration of broken branches in *Acropora* corals. For the first time, the effect of habitat on the growth form of coral colonies was noted in several species, as well as variations in growth rate between massive and branching corals.

The work on the sexual reproduction of coral evaluated whether this took place all year round, or at a particular time of year. The gonads and production of planula larvae were examined on a monthly basis in 10 species of corals for over 13 months. This work yielded the first observations of the lunar periodicity of coral planula production, with the species *Pocillopora damicornis* planulating on the new moon between December and April and with the full moon during July and August [4].

The production of sediments and its effects upon corals was monitored by deploying a network of sediment traps across the Low Isles reef over 7 months. This work demonstrated the role of waves and tides in sorting sediments on the reef flat and found that corals can survive in turbid waters, being able to withstand sediment cover with help from wind-driven currents, tidal water movement and the coral's ciliary action [5, p. 131].

Coral reef ecological survey and reef island mapping

Detailed ecological surveys were made along traverses ~100–400 m in length using a rectangular wooden frame that was subdivided into a series of smaller quadrats. Counts of corals and algae were made inside the frame to examine the ecological character of different reef zones, including the reef flat, the moat, the boulder tract and the seaward slope to a depth of ~5.6 m, the beach sand, beach sandstone, the inshore reef and seaward slope to a depth of ~1.5 m, and the outer rampart and windward reef slope to a depth of ~20 m. Underwater sections were examined either from a boat in deep areas or with the help of a diving helmet in the shallows (Fig. 3.2). This enabled comparisons to be made of different reef areas and, importantly, the detail of these surveys and their accurate mapping has provided an exceptional baseline that has allowed the vertical

Fig. 3.2. Expedition member A. P. Orr modelling the diving helmet from which the first underwater surveys of the reef were made during the 1928 Great Barrier Reef Expedition. The helmet was made in Plymouth Marine Laboratory based on the designs of one used by A. G. Mayor in Samoa. It was fed by hand-pumped air from a boat at the water surface. (Image supplied with kind permission by the National Library of Australia, photographer unknown)

distribution of reef communities to be related to a tidal datum and, later, further studies to elucidate ecological changes over time.

Low Isles has the longest history of mapping on the GBR, with a collection of maps produced to varying degrees of resolution and by different methods since 1928, including plane-table sketches, compass and measuring-tape survey, reversible level and theodolite tachometry and, most recently, drone and ground survey (see 'Mapping Australia's coral reefs' in Chapter 5). These exceptionally detailed maps feature prominent landforms of the Low Isles complex, including two shingle ridges on the windward side of the reef; a small sand cay on the leeward side of the reef; and an extensive reef-top flat, part of which is occupied by mangroves (Fig. 3.3). The expedition also took the innovative step of checking ground observations against aerial photography flown on 24 September 1928 by the Royal Australian Air Force that was later used to assess the impacts of the tropical cyclone of 1934 [6].

The long cartographic record at Low Isles has allowed the evaluation of changes to both reef-top landforms, including sand cays and mangrove forests [8] and finer scale ecological changes in the coral reef communities. The 90-year evaluation (see 'Revisiting the corals of the Low Isles 90 years later' later in this chapter) represents a considerable timespan within the global record that has allowed consideration of the fundamental processes determining the nature of such changes. Changes revealed by the maps provide insights into two different models of reef responses to environmental forcing, particularly the question of whether or not reef-top landforms such as islands and mangrove forests should be viewed as existing within a dynamic equilibrium, or as part of a longer-term geomorphic evolutionary sequence. In the first equilibrium model, the form of reef landscapes at any one time represents a comparatively stable balance that has developed over time with local environmental and weather conditions. The alternative model of the reef top as existing within an evolutionary sequence

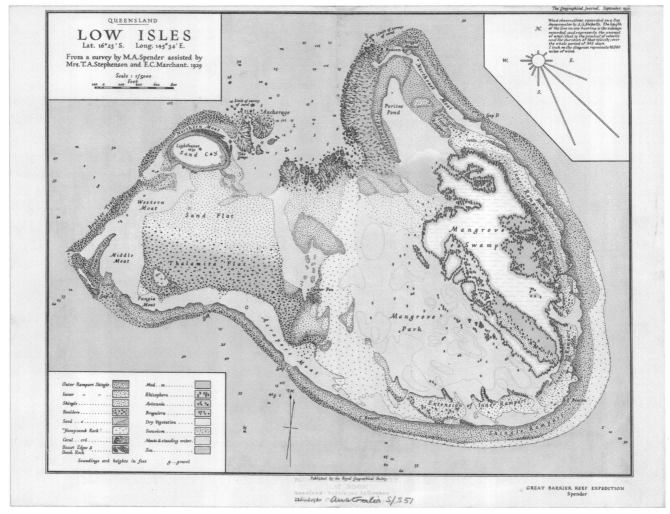

Fig. 3.3. The detailed map of Low Isles reef made using a combination of theodolite triangulation and plane-table sketching to triangulated networks of features with the first aerial photographs of Low Isles, which were flown in September 1928 at a scale of 1:2400 to fill in details of reef surface cover. Swathes of shingle, boulders, sand and mud were traced on the reef flat, while features such as the outer boundary of the reef perimeter, channels and shingle ramparts were fixed by triangulation. This is the earliest known record of ground observations being cross-referenced against aerial photography for the purpose of reef mapping. (Reproduced from [7] with kind permission from the Royal Geographical Society; image credit: Michael Spender)

suggests that landforms have stabilised behind a protective rim, or 'bassett edge', within which mangroves have spread across the platform surface from the windward margin towards the leeward sand cay, ultimately filling the reef-top accommodation space at the final evolutionary stage. Both views can be reconciled by considering the longer-term evolution of reef landforms in which shorter-term expansion of mangroves, with

changes to substrate topography and protection from shingle ramparts, invariably extends to cover the reef top.

Comparisons against the ecological surveys reveal a long-term decline in coral and invertebrate richness at Low Isles since 1928–29. This is likely due to repeated cyclone damage and coral bleaching, on top of increased influence from regional mainland agricultural activity. These findings echoed the later observations of Yonge himself:

When we there in 28/29 the reef flat when exposed at low tide was literally an aquarium. I was briefly there again for some hours in 1965 and over it again in 1975 in a light aeroplane. But I really saw it properly again in 1978 (50 years on) when I was working at AIMS south of Townsville. All that exposed reef was covered with sediments with only holothurians in their element and flourishing. The sediment had come from the mouth of the Daintree River some 10 miles away. This is the result of replacing the rain forest by sugar cane fields.' (quoted in [9] p. 38)

Despite these sombre results, the value of the ecological surveys carried out by the 1928–29 expedition should not be understated, providing as it does the basis for one of the longest known geological and ecological coral reef records (see the next section: 'Revisiting the corals of the Low Isles 90 years later').

The Great Barrier Reef Expedition's legacy for coral reef science

The agenda of the Australian coral reef science community continues to bear the hallmarks of experimental field research that was undertaken during the expedition. As the influence of environmental change has increasingly been felt on the world's reefs, the coral growth experiments have informed later reef-restoration efforts, while the early observations of coral bleaching informed later experiments to unravel its causes and implications for the GBR. In the fullness of time, the value of early, accurate measurement of reef surface features as a benchmark against which both ecological and geomorphological changes could be ascertained has become clear. On the 45th and 90th anniversaries of its production, the expedition's detailed reef-flat map has served as a comparative record for studies of landform evolution, yielding insights into the dynamic nature of sand cays, shingle ramparts and mangrove forests on the low wooded islands of the GBR.

This expedition set new standards and defined new goals in reef studies, as laid out in more detail elsewhere [9]. This was true not only of Maurice Yonge's work on coral physiology, but of the ecological and geomorphological studies, too. This cooperative work represented the first major advance on the predominately theoretical and deductive mode of work which had long dominated discussions of the 'coral reef problem'.

Sidnie Manton: pioneer of coral reef community surveys

Maoz Fine

Sidnie Manton was a young British scientist who joined the Expedition after it had been running for 7 months (Fig. 3.4). As is clear from her diaries [17], Sidnie was fascinated by the natural world. Before joining the expedition, she travelled to Tasmania to study *Anaspides* shrimps, where she also paid homage to the last living Tasmanian tiger (thylacine) in Hobart Zoo.

On arrival at the Low Isles, Sidnie learned to identify many species of corals within 2 weeks, assisted researchers studying sea cucumbers and crustaceans, and did a lot of what she called 'donkey jobs'. Sidnie threw herself into surveying the coral communities along the three traverses, using a heavy diving helmet that enabled her to dive along a metal line that stretched from the intertidal to the deeper edge of the reef. Significantly, this was one of the earliest known uses of sub-aqua diving in coral reef studies. Using a 2 square yard quadrat, she played a major role in the surveying of the three 100–400 m traverses, identifying coral species, counting the number of colonies, and measuring their size, and recording non-coral invertebrates and algae. Sidnie drew the corals underwater on roughened glass and identified the species underlying the quadrat.

Fig. 3.4. Sidnie Manton holding a brain coral on the beach at Low Isles. Coral samples were stored in wooden boxes that had originally been used for holding kerosene tins. (Image supplied with kind permission by Elizabeth Clifford and Jeanie Teasdale, photographer unknown)

Above her, tenders on the boat above pumped air by hand into her diving helmet.

Executing these surveys was no simple task. The diving helmet glass often fogged, communication with the tender was limited to rope pulling signals, and taking notes was accomplished using floating pencils tied to her helmet. Despite these technical difficulties, Sidnie remained fascinated by marine life. In a letter to her family dated 30 March 1929, she described the underwater landscape: 'The colours of things are marvellous, brilliant corals, clams of huge size – up to 2 feet, they are big bivalves sitting with wide open mouths the inside all frills of the richest and most brilliant tints imaginable and every one differently coloured and patterned' [17].

Sidnie Manton's enthusiasm, love for nature and science, and meticulous work ethic positioned her well to make a pioneering contribution. Sidnie and T. A. Stephenson's early reef community surveys served as the baseline for many follow-up studies, becoming some of the most important accounts of coral communities in the history of coral reef research in Australia and globally.

Revisiting the corals of the Low Isles 90 years later

Maoz Fine, Ove Hoegh-Guldberg, Efrat Meroz-Fine and Sophie Dove

Long-term studies of reefs are rare yet critical for understanding the trajectory of coral reefs under environmental changes such as ocean warming and acidification. The 1928–29 Yonge expedition to Low Isles initiated one of the longest studies on a coral reef globally. This pioneering study focused particularly on the coral communities inhabiting the depression running along the northern reef margin of the horseshoe-shaped reef known as the 'Anchorage'.

The Anchorage is the closest site inhabited by live corals to the camp established by the expedition, and it was therefore readily accessible to ecological surveys and experiments. It was meticulously surveyed across spatial scales that ranged from aerial photography to close up underwater observations over the 12 months of the expedition [10],

yielding a high-quality set of baseline measurements that could be revisited by future scientists and used to evaluate changing ecological dynamics in the Low Isles reef over the intervening 90 years. This is arguably the longest study of coral reefs ever [6, 10–13] making this reef iconic in coral reef science.

Changes to the coral reef community

Yonge's team established three 'traverses' around the Low Isles, along which ecological and physiological studies were undertaken. At the Anchorage, traverse II ran 150 m from the sandy cay to the seaward reef edge. Close-range observations of seafloor communities along this traverse were the responsibility of Sidnie Manton under the direction of Dr T. A. Stephenson (see above box, 'Sidnie Manton: pioneer of coral reef community surveys').

In 1928–29, the shallow section of the Anchorage reef had hard coral cover of 100 per cent with many coral colonies overlapping as they competed for space. The reef was dominated by *Montipora angulata*, *Acropora aspera* and *Acropora pulchra*, the rounded bushes and platforms that were exposed at low tide, with plentiful soft corals (*Lobophytum* sp.), clams and holothurians among them [14]. Further seaward, the subtidal coral community was more diverse than in shallower waters, with an abundance of hard corals belonging to the genus *Acropora*, foliose *Montipora* corals and massive corals (*Favia* and *Goniastrea*) interspersed with soft corals such as *Sarcophyton* and *Sinularia*.

Ninety years of collaboration and perspective

An understanding of long-term change in coral reef communities has emerged as scientific groups have revisited the sites. This included several cycles of cyclone damage followed by recovery. Stephenson *et al.* (1958) observed the Anchorage community in 1928, which had probably recovered from a severe cyclone in 1911 [11]. Similarly, the healthy state of the reef in 1945 was thought to reflect recovery from the 1934 cyclone, suggesting that a period of at least 10–12 years is required for a full recovery of hard-coral communities from cyclones, although increasing levels of siltation from regional agricultural activity since European settlement may have slowed recovery rates since then [13]. Living corals were again damaged by a cyclone in 1954, in many areas that had previously been covered by live hard coral. Living corals became sparse at these sites as dead scleractinian corals were overgrown by abundant soft coral (*Sarcophyton*) [11]. By the early 1990s, soft corals and benthic algae dominated the reef at the Anchorage [12]. Interestingly, Stephenson *et al.* (1958) predicted a prolonged period would elapse before the flourishing communities observed at the Anchorage in 1928–29 returned [11]. Over 60 years later, the Anchorage reef community has not returned to its flourishing state; rather, since the 1950s, the reef communities of the Anchorage have shifted from being dominated by scleractinian hard corals towards soft-coral-dominated reefs [13]. In 2004, 2015 and 2019, the abundance of soft-coral colonies in the Anchorage had increased by fivefold

on 1928 levels (Fig. 3.5). This shift to a community of less complex coral growth forms has led to a 50 per cent reduction in species richness, with likely impacts on the associated diversity of other invertebrates and fish.

Reasons for change on the Low Isles reefs

Ecological changes at the Anchorage over the past 90 years are most likely a result of global disturbances, including warming and acidification, interacting with local eutrophication, sedimentation flux, and storms [13]. The changes in the reef communities at the Low Isles have long-term implications for reef resilience [15, 16]. Stress reduces the capacity of a reef to absorb disturbance without dramatic changes to the seafloor communities that live there. Chronic, ongoing stress may result in a transition towards resistant species over time, which in turn will enable recovery of coral cover after disturbance events. Many inshore reefs on the GBR are already dominated by resistant, slow-growing corals. Observations like those made at the Low Isles highlight the importance of long-term studies for our understanding of the dynamics of coral reefs into the future. Indeed, the Australian Institute of Marine Science has a well-established long-term monitoring program for much of the GBR which is proving to be similarly important at broad scales.

The 1973 expedition to the northern Great Barrier Reef

Roger McLean and David Hopley

The 1973 expedition to the northern GBR was enthusiastically championed by two leaders of the 1928–29 expedition: biologist Maurice Yonge (later to become Sir Maurice) and, less surprisingly, geographer Alfred Steers. It was Steers who realised the need for further geomorphological information about how reef landscapes are structured following from his field experiences in the region in 1928 and 1936. In 1971, Steers proposed an expedition with a geological and geomorphological focus to

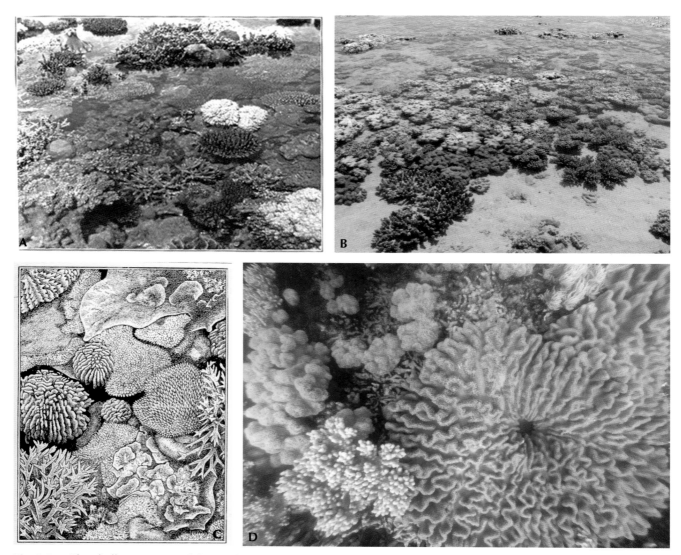

Fig. 3.5. The shallow section of the Anchorage at Low Isles in (A) 1928 and (B) 2019 highlighting the shift from an *Acropora*-dominated reef to a high cover of soft corals. Similarly, a seaward section 8 feet long and 6 feet wide of the Anchorage, drawn by T. A. Stephenson in 1928 (C) is now (D) dominated by *Lobophytum*, *Sinularia* and *Sarcophyton* with some *Montipora* rubble underneath. Note the dramatic shift from a highly diverse and complex reef in 1928 to one that is more homogenous today. (Image credits: Alan Stephenson (A and C), Maoz Fine (B and D))

the Southern Zone Research Committee of the Royal Society whose Chair was Sir Maurice Yonge. Fortuitously, after the Pacific Science Congress in Sydney in 1971, both Yonge and Steers visited the University of Queensland in Brisbane and James Cook University in Townsville and obtained local commitments for an expedition to the northern regions of the GBR in 1973, together with agreement that David Stoddart (Department of Geography, University of Cambridge) should be the expedition's leader.

Key attributes of the 1973 expedition and of 'team Stoddart'

Many differences are evident between the 1928–29 and 1973 expeditions, notably that the majority of the 1973 members came from Australia and not the United Kingdom, as acknowledged in the expedition's formal name: the 'Royal Society and Universities of Queensland Expedition to the northern Great Barrier Reef'. The all-male expedition was dominated by earth scientists with four marine biologists among its 24 members. The expedition began in mid-July 1973 (shortly after the shipborne Second International Coral Reef Symposium aboard the *Marco Polo*), lasted 4 months and was divided into three phases: phase I in the Howick Group; phase II from Low Isles to the Turtle Group; and phase III in the far north from Cape Melville to Raine Island (see the next section). It was a mobile, boat-based expedition that employed several vessels for the geophysical survey, sediment sampling and drilling. In the first two phases, a shore party camped for a few days at East Hope Island, Three Isles, the Turtle Group and Low Isles, the southernmost site of the 1973 expedition, and home base of 1928–29 expedition (see 'The Great Barrier Reef Expedition of 1928–29' and 'Revisiting the corals of the Low Isles 90 years later' in this chapter).

Significantly, the team assembled by David Stoddart in 1973 had a wealth of prior collective fieldwork experience on reefs in the Indian Ocean (Aldabra, Chagos, Seychelles, Maldives), the Caribbean (Bahamas, Barbados, Belize) in the central and south-west Pacific (Caroline Islands, Cook Islands, Marshall Islands, Solomon Islands, Tuvalu), and on the GBR itself. Critically, the team members had strongly opposed views on the pattern of sea-level change during the Holocene based on their prior field observations. In geological terms, the formation of the reef during the Holocene, (i.e. the last 11 000 years) is viewed as covering a 'recent' timeframe with direct implications for the current configuration of reefs that can be seen today. One group supported a mid-late Holocene high sea level [18, 19], while the other held the view that sea level had not reached above its present position in the last few millennia [20, 21]. At the time, both groups believed there was just one global sea-level pattern. Such differences were resolved during the expedition's collaborative field program, in the post-expedition analyses and through publication of the

results in the *Philosophical Transactions of the Royal Society* in 1978, all of which led sea-level historians to realise that there was no single universal sea-level story, but several regional ones.

Evidence from reef-top landforms, inter-reef sediments and subsurface stratigraphy

The main aim of the 1973 expedition was 'to elucidate the recent history of the reefs especially in response to Holocene sea level change' [22]. This was achieved using evidence from inter-reef geophysical soundings and sediments, carrying out shallow coring on two reefs (Bewick and Stapleton), and making observations of surface geomorphology, including the lithology of exposed limestones and sediment deposits including shingle ramparts, islands, and sand cays on over 30 island reefs and low wooded islands. Field observations and measurements were temporally constrained through radiometric dating of 79 samples by the Australian National University (Fig. 3.6). Over 600 sediment samples were collected, and their texture (grain size and shape) and composition determined from three key environments: reef flats, sand cays and shingle ramparts/islands, and inter-reef shelf sediments.

Interpreting the sea-level history clues: reef growth and reef-top morphology

That the 1973 expedition achieved its main objective of linking sea-level, reef growth and reef-top landforms can be illustrated through three examples of results. First, cross-shelf surficial sediments from the Queensland coast to the outer ribbon reefs in the northern GBR showed a zonal pattern in which terrigenous facies transitioned to impure carbonate and ultimately high carbonate facies, while the subsurface seismic record revealed a sequence of disconformities at depths ranging from 60 to 120 m. Shallower bores on Bewick and Stapleton reefs passed from mid- to late- Holocene accretionary sediments into altered older limestones at depths of only 5.6 m and 14.6 m respectively. Above this discontinuity, material was dated at less than 7000 years BP (before present),

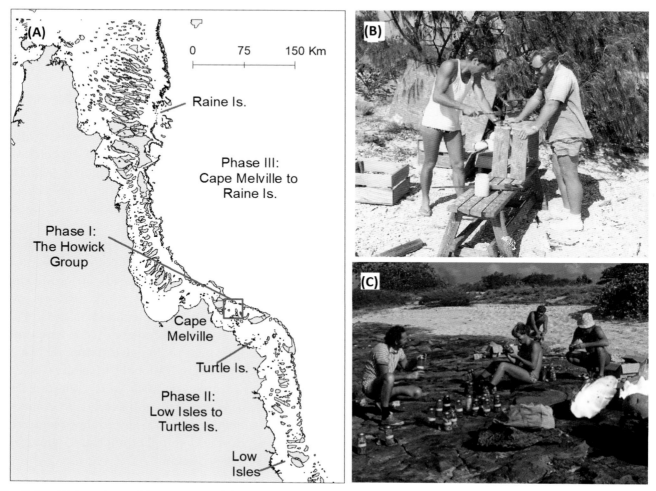

Fig. 3.6. (A) Locations of the sites visited during the 1973 expedition to the northern Great Barrier Reef: the Howick Group (phase 1); Low Isles to the Turtle Group (phase 2); and Cape Melville to Raine Islands (phase 3). (B) Terry Scoffin and David Stoddart making sample boxes at Turtle Island. (C) Labelling samples for analysis and radiometric dating at Howick Island, 9 August 1973. From left: Henry Polach, Radiocarbon Dating Laboratory, Australian National University; John Veron (aka Charlie Veron – see the next section), Biological Sciences, James Cook University; and Roger McLean and Bruce Thom, Biogeography and Geomorphology, Australian National University. (Image credits: Sarah Hamylton (A), Roger McLean (B and C))

while below the discontinuity material was clearly Pleistocene in age and dated greater than 30 000 years BP. Consistent with this finding was the fact that nowhere in the expedition's territory was Pleistocene reef framework found above the present sea level, suggesting that subsidence, rather than stability or uplift, has dominated reef development in this area.

Second, extensive fields of fossil microatolls (dated between 6300 and 4800 years BP) were found on several reef tops, many within the vertical range of modern corals (Fig. 3.7). This suggested that the sea first reached its present level in the region by ~6000 years BP and then passed above that level reaching a high-stand peak at least 1 m above present around 3700 years BP. Analysis of microatoll data during the 1973 expedition was the first demonstration of the nature and significance of coral microatolls as water level recorders [23], using fossil microatolls to constrain the mid–late Holocene sea-level chronology for the northern GBR [24, 25].

Fig. 3.7. The fossil microatoll field immediately east of sand cay on Leggatt Reef in the northern GBR was typical of those associated with mangroves in the Howick Group. This site was dated at 5800 +/– 120 years BP (ANU-1286). (Image credits: Sarah Hamylton)

Third, the expedition identified two other groups of prominent reef-top features that had formed in the period 4000 to 3000 years BP. These were windward storm ramparts cemented into solid conglomerate platforms and the core areas, or upper terraces, of the leeward sand cays. During this time interval, the basic outlines of modern reef-top deposits, including the low wooded islands, were established, suggesting that this was a time when the reefs were at their most productive. Lower, loosely cemented ramparts, cay surfaces and fossil microatolls with younger ages of 2000–3000 years BP were thought to represent a subsequent fall in sea level. Many

sea-level studies on the GBR have been undertaken since publication of the 1973 expedition report, several of which have also used coral microatolls as critical indicators. These studies have agreed with the general pattern of mid–late Holocene sea-level history established during the expedition, although improved calibration of radiocarbon-based ages has modified the specifics of the timeframe.

The team of earth scientists who participated in the 1973 expedition utilised their geographically extensive reef field experience to significantly advance our understanding of the links between sea-level change, reef growth and reef-top landforms in the northern GBR. Notably, they accelerated our understanding of a globally variable sea-level history; elucidated patterns in the age, composition and texture of sediments, both across the shelf and across different sub-reef environments to understand reef formation, uncovered evidence of regional subsidence through the Holocene–Pleistocene discontinuity revealed by shallow boring and seismic work; and established the importance of microatolls as sea-level indicators. There were also several spin-off projects, including resurveys of the surficial features of Low Isles and Three Isles after 45 years since the 1928–29 expedition; recognition that the basement and much of the volume of the northern barrier reefs are Miocene in age; and that the outer barrier reefs north of the ribbon reefs are composed of two distinctly different reef types – deltaic reefs and dissected reefs. Collectively, these contributions revealed important insights on how the reef landscapes of the northern GBR have formed.

Determining the age of the Great Barrier Reef: a voyage of discovery

Charlie Veron

Despite the fame of the 1928 Yonge Expedition, the following 40 years saw a dearth of initiatives to provide even a basic understanding of the biology and geology of the GBR. Thus, there was plenty of fallow ground for another expedition, one that was ship- rather than land-based. As previously stated, in 1973 a ship-based expedition was organised by Professor David Stoddart from the University of Cambridge to the northern GBR. The final, third phase of the Stoddart expedition was on RV *James Kirby*, James Cook University's new research vessel, the first scientific expedition ever undertaken to the far-northern GBR. I was honoured to be a part of this expedition, which brought together bathymetric, geological and bio-geographical evidence to shed new insights on the age of the GBR.

What lies beneath: where the Great Barrier Reef meets the Queensland Trough

Given a free hand to go wherever we chose, we headed for Tijou Reef, because of its close proximity to the abyssal depths of the Queensland Trough (Fig. 3.8A). Admiralty charts being dangerously inaccurate, we depended on World War II aerial photographs for both the ships navigation and planning our dive sites.

We crossed to the eastern side of Tijou (Fig. 3.8B) early on a December morning. These reefs on the outer, oceanic boundary of the immense barrier reef, slope downwards, densely covered in armour-plated coral (Fig. 3.8C) before plunging down at an average slope of 45° to depths beyond sight. The corals form a dense crust, resistant to all but extreme wave action.

As there were no charts with nautical soundings for this area, we made the first observations that connected the GBR to the western edge of the Queensland Trough: clearly the two were the same structure, coming together at the base of this steep reef slope. While this may seem obvious now, these were the first scientific observations to be made on the oceanic side of any ribbon reef. It also seems obvious, then as now, that the GBR lagoon could not possibly have remained intact so close to an abyssal trough without there being a wall – a reef wall – to contain it.

A geologist's domain

At that time, the mystery of how old the underlying reef platform structure of the GBR might be was a big question over which geologists had intellectual ownership. The focus was on the ribbon reefs, stretching many hundreds of kilometres along the outer edge of the northern GBR, rather than the 5000 or so platform reefs distributed over the lagoon surface. Darwin's (1842) famous theory of atoll formation had greatly influenced global thinking on reef formation. There were, however, rivals to Darwin's theory. The Canadian geologist Reginald Daly expounded the now mostly forgotten glacial control theory of reef development in which he proposed that during

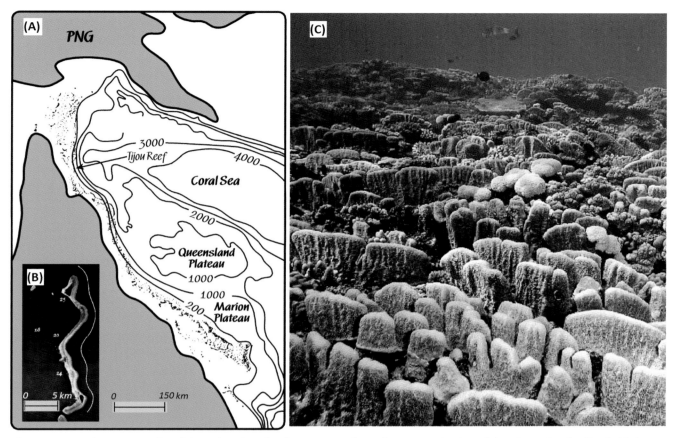

Fig. 3.8. (A) The western Coral Sea, showing depth contours along the seaward edge of the Great Barrier Reef. (B) An aerial photograph of Tijou Reef taken in World War II showing depth in metres to the west and the 1000 m contour to the east. (C) The outer face of Tijou Reef at 8 m before it plunges steeply. (Image credits: Charlie Veron)

the Pleistocene (2.5 million–11 000 years ago), low sea levels enabled exposed reefs to be planed off by wave action [26]. They would then grow afresh when the sea level rose again. No fewer than 20 theories – or rather variations on the themes of Darwin and Daly – were put forward by geologists in the century after Darwin's theory was published [27]. Darwin was finally vindicated when the US Atomic Energy Commission sank two boreholes at the Marshall Islands, reaching a volcanic basement at depths that were close to Darwin's predictions.

Interest in the origins of the GBR never slackened against this global backdrop, the main development being that plate tectonics gave rise to the notion that the GBR formed when Australia drifted into the tropics [28].

It was therefore surprising that, after the most expensive study of its time, the International Drilling Consortium announced that Ribbon Reef 5 on the central GBR started growing 600 000 ± 280 000 years ago, a view that most reef geologists endorsed [29, 30]. A range of other evidence would suggest that the reef was much older.

Enter coral historical biogeography

At this time, biology played virtually no part in the debate over the age of the GBR. However, long before the above-mentioned drilling was being planned, biologists knew that corals were not dispersed by plate tectonics as geologists had once supposed. Rather, they were dispersed by their larvae, which could make long-distance journeys on ocean surface currents before they settled and ultimately built reefs.

Fig. 3.9.   Scenes from research activities at Lizard Island Research Station. (Image credits: Paul Jones)

At the scale of the world's oceans, the distribution of reefs is largely determined by the availability of shallow platforms on which corals can survive. By the close of the Oligocene epoch (24 million years ago), the gross bathymetric and geological characters of the GBR's seafloor were in place. At that time, Australia was nearer to Antarctica than to its present position, so water temperatures may have been low enough (< 18°C) to preclude reef development. However, ever since the Early Miocene (20 million years ago) or earlier, the developing equatorial Indonesian–Philippines island archipelago to the north of Australia has been the global centre of coral diversity and corals have been dispersing from this centre ever since [31]. This includes coral dispersal to northern 'Greater Australia', the name given to the combination of Australia and Papua New Guinea, as they have been united through most of geological time. As the region of the GBR has an extensive rocky coastline not dominated by river discharge, and a stable continental shelf adjacent to deep ocean, it would have been a suitable coastline for reef development throughout this time.

The evidence for an Early Miocene origin

In view of the above-mentioned age of the GBR being in the latter half of the Pleistocene, as determined by drilling, there are pertinent pieces of evidence that clearly indicate a much earlier origin. These relate to rates of seafloor spreading, the diversity of nearby fossil reefs, and the age of surrounding modern reefs and the reefs of Western Australia.

Extensive reef limestone of Miocene age occurs in the northern Coral Sea, including the far-northern GBR, and outcrops of reef limestone of all Cenozoic epochs from the Late Eocene (40 mya) onward occur in Papua New Guinea. It is hard to imagine how reefs, especially those of Papua New Guinea, could have existed – for perhaps 40 million years – without forming in the region of the GBR just to the south (Fig. 3.8A). Furthermore, reefs extended as far south as New Zealand in the Lower Miocene [32], at that time around 45°S, indicating that conditions for reef growth were *better* at higher latitudes than they are today.

Australia has been drifting north at a speed of ~6 cm per year. It would therefore have been ~35 km south of its present position 600 000 years ago or ~150 km south at the beginning of the Pleistocene. This is < 10 per cent of the length of the GBR today, making plate tectonics an irrelevant concept to any proposed Pleistocene origin.

The Early Pleistocene Era beds near Port Moresby, only 300 km east of the northernmost

GBR, have the most diverse coral communities of any fossil reef [33]. The age of comparable modern Australian reefs provides a further clue about the age of the GBR. Elizabeth and Middleton reefs, 1300 km south-east of the southernmost reefs of the GBR, and the extensive carbonate beaches of Lord Howe Island still further south, are likely to have a Pleistocene foundation [34]. Even the southernmost reefs of the GBR had reached the present latitude of these places by the Early Miocene. Moreover, the reefs of Western Australia have a similar history. Today, reefs occur south to the Houtman Abrolhos Islands, 500 km further south than the southernmost reefs of the GBR. They are probably Pleistocene in origin as are reefs limestone outcrops near Perth, 500 km further south [35].

Given the wide capacity for corals to disperse, the idea that they seeded the north-east Queensland coast much earlier, when environmental conditions were amenable to their growth for tens of millions of years, is hard to refute. Drawing from a broader body of evidence suggests that the reef must be much older than the 600 000 years suggested by drilling. When details from geology and biology are pooled, the most parsimonious conclusion to be drawn from this evidence is that the outer reefs of the GBR have looked something like they do today for at least 20 million years [36].

The Australian Museum's quest for a research station

Frank Talbot

In my early thirties, I moved from South Africa to Australia to take up a position as Curator of Fishes at the Australian Museum. Having recently completed a PhD on the ecology of coral reef fishes on the island of Zanzibar, Tanzania, I was keen to work on the 'majestic' GBR. With the naivety of youth, I hoped to visit a small coral reef with a group of interdisciplinary scientists and begin to

unravel some of the ecological complexities of coral reef fish.

As with many museums globally around that time, a large part of the business of the Australian Museum was to document the diversity of life across Australia's deserts, rainforests and coral reefs. Throughout the GBR, there was a focus on mapping the biodiversity of both the coral reefs and the islands. Several expeditions took place to collect specimens both under and above water with a view to describing new species and documenting the biodiversity of reefs. Initially these were short camping trips, both to islands on the GBR (Fig. 3.10) and further afield in Western Australia (Houtman Abrolhos Islands, the Dampier Archipelago and Kimberleys) and off the Eastern coast (Elizabeth and Middleton reefs and Lord Howe), where extensive surveys of crustaceans, polychaetes and molluscs were undertaken by Australian Museum researchers. However, it became apparent that museum staff could not continue to sleep in tents or on ships, or stay in fisheries huts. A more permanent field base was needed for the detailed investigation of the distribution and diversity of coral reef fauna, including fish and other invertebrates.

Fig. 3.10. Australian Museum entomology and fish researchers at an expedition camp on North-west Islet, Capricorn Group, 1925. From left to right: Mr. A. Musgrave, Mr. P. A. Gilbert, Dr W. McGillivray and Mr. G. Whitley. (Image supplied with kind permission by the Australian Museum, photographer unknown)

The first major funding for a research station at Lizard Island came from an American, Henry Loomis. Loomis had a long and distinguished career working for the American Government, first as a naval lieutenant commander, then as Director of the US Office of Research and Intelligence (a post that saw him take many international trips), and later in public broadcasting. Loomis became fascinated by coral reefs when he and his brother leased a Florida island for their holiday home and encouraged university staff to use it for research. When the Florida Island was needed for a national park, he sought a family holiday home that could continue operating as a research centre on Australia's GBR.

I encouraged Henry and his wife to visit One Tree Island, thinking that a venture there might suit Henry while also supporting a research outpost for the museum. The visit helped to clarify where our visions diverged. The two visitors helicoptered in at midday and we spent the afternoon together, wandering over the small coral-rubble island with its central pond, its windblown patch of trees sheltering boobies, and terns breeding on the open flat. As the evening approached, we poured some glasses of wine and began discussing the possibilities of a Loomis–Australian Museum venture. The idea fell rapidly apart. Mrs Loomis felt the island was too small for the sort of grand holiday home she envisaged. Things took a further downward turn when she insisted that she would have to have her dogs with her. Bringing dogs to an island heavily used by breeding birds did not seem a good idea. They helicoptered off the next morning, having clearly decided not to build a research station on One Tree Island!

We began considering sites for a research station further north, and several years later the possibilities changed. I visited Loomis in Virginia to let him know that we had set our sights elsewhere. He showed strong support, offering to hire a boat and join in a search for a suitable more northern site and, if one was found, he agreed to help with some set-up funding for a research station. Henry made his way to Australia and hired a boat out of Cairns with Pat Hutchings, also from the Australian Museum, and me. After visiting several islands,

two sites were considered as feasible: Rocky Isles and Lizard Island. Rocky Isles was an attractive prospect because it was uninhabited, but it had no year-round anchorage and so we discounted this as a possibility. Lizard Island, on the other hand, had a wider variety of habitats, including its lagoon. But there was a small problem. The northern bay, perhaps the best place for a research station and anchorage, was already leased to a group who planned to build a resort there. This meant that the island and its surrounding reef would not be the sort of pristine environment ideal for scientific studies as it could have been impacted by visitors. At the time, the resort itself was not built, and a caretaker family was living in a caravan on the site. Henry Loomis was impressed with Lizard Island, but nothing was finalised.

In July 1973, the International Coral Reef Symposium on board the *Marco Polo* provided a chance to let other reef researchers see the Lizard Island site and discuss the possibility of a research station there. On this trip, I had extensive discussions with Don McMichael from the Department of Environment and Pat Hutchings about establishing a research station on the eastern site of the island. We had the advantage of having several high-profile overseas researchers in our company to offer their opinions. The international group were strongly supportive of the idea and felt the resort in the northern bay would not be damaging, and that the airfield provided an important practical service. So, Lizard Island was chosen. The remaining funds were provided by another American philanthropist, Mrs Topsy Waters, through her Taiping Foundation. Having met Topsy while taking part in the Smithsonian Tektite 2 underwater habitat research program in 1970, I later visited her in Maine and she agreed to match Loomis's donation. The future of Lizard Island Research Station was set.

Things moved fast once funding had been secured. Tents were set up for cooking and accommodation, plus a workshop and dive facility (Fig. 3.11A). Houses for the Director and Maintenance Officer soon followed; the first directors were Steve and Alison Domm, who brought experience

Fig. 3.11. Developments to Lizard Island Research from the 1970s to present. (A) Cooking tents in the 1970s. (B) Talbot House, a researcher facility for cooking and accommodation built in 2005. (C) Aquaria facilities at the station in the 1970s and (D) now. (E) The dive shed in the background (1970s) next to the washing up area (foreground). (F) The dive facility at Lizard Island today. (Image credits: Pat Hutchings (A), Penny Berents (E), Lyle Vail and Anne Hoggett (B, C, D, F))

with them from what had, in the intervening years, become a small research station at One Tree Island. Potable water was collected in tanks to begin with, then pumped from groundwater while power was initially provided by a diesel generator, then by wind and solar power. In the fullness of time, research laboratories and aquaria were established (Fig. 3.11B and D). A fleet of boats was brought in for working around the island and visiting the outer barrier and other island groups (Fig. 3.13).

The first museum collecting trip to Lizard Island was undertaken by the Fish Department in December 1973, and since that time the station has been visited by many thousands of Australian and international researchers. The station is run as a facility of the Australian Museum supported by the Lizard Island Reef Foundation, which has raised funds to develop and maintain the station, as well as to support researchers and sponsor focused workshops on fish, molluscs, amphipods and polychaetes, resulting in many new species being described. The station records environmental data such as rainfall, air and sea temperatures dating back to the 1970s

and acts a node for the Great Barrier Reef Ocean Observing System (GBROOS). Over 2500 scientific publications have been written by Australian and international researchers detailing work undertaken at Lizard Island since 1973 and the facility continues to support 100 research projects annually. Sir David Attenborough visited the research station in 2015 to film a documentary on the GBR. He also undertook an underwater dive in a submersible submarine, which was filmed for production of a virtual reality immersive experience for visitors of many museums worldwide. What began life as a pipedream for staff at the Australian Museum has now become a major, internationally recognised research facility (Figs 3.12 and 3.14).

Early coral reef science at Heron and One Tree islands
Peter Sale

In 1951, the GBR was mostly inaccessible, being a long way from shore with remarkably few islands.

Fig. 3.12.   An aerial photograph of Lizard Island Research Station taken with a drone. (Image credit: Anne Hoggett)

Fig. 3.13.   Research boats in the water at Lizard Island Research Station. (Image credit: Paul Jones)

Fig. 3.14.   Sir David Attenborough is lowered underwater at Lizard Island in the MV *Alucia*'s Triton submersible. (Image reproduced with kind permission from Atlantic Productions; image credit: Freddie Claire)

The southern end of the reef is well north of Brisbane, which was the only significant city in Queensland at the time and home to the most northerly Australian university. If a scientist wanted to be in the field for extended periods, they found a corner in the back of a resort so that they could work on an inshore reef. Then Heron Island Research Station was born. Buoyed by the first 'permanent' research facility on the GBR, Australia took off to become a world leader in reef science.

Memories of a young Heron Island Research Station

At the end of the 1960s, Heron Island was the only site that supported research offshore. It was a primitive place. There were modest houses for the Director and the Maintenance Officer, a kitchen–dining room, toilet–shower complex, ~10 two-bed sleeping huts, the laboratory building, a seawater aquarium and one research vessel. The laboratory was entirely devoid of equipment, the aquarium had only two functional aquaria, the seawater system only operated at high tide, and the research vessel, called the *Dory*, lived up to its name.

The *Dory* was an in-board powered, high-sided, heavy wooden tender ~4.5 m long, with a one-cylinder, gasoline motor that was started by rotating the flywheel as fast as possible with your bare hand before disengaging the clutch. It would putt-putt along at a stately 1 kn. Getting into it wearing dive gear involved remembering before you got in the water to leave a line hanging over the gunnel on which to tie the dive tank (I learned this lesson the hard way). Then, having completed your dive, you took off your tank and tied it to the line, threw your fins high over the gunnel, and clambered up

Fig. 3.15. (A) Researcher measuring a nesting Green turtle, on sandy beach at Heron Island. (B) Research divers photographing *Goniastrea* coral, on reef around Heron Island, 1984. (Image credits: J. Jones (A), S. Parish (B))

onto the rudder, so that you could climb over the stern and fall into the boat. With a bit of luck, you then remembered to reach over and haul in the dive tank.

Immersed in a coral reef environment

Heron had easy access to a rich coral reef where it was possible to do *in situ* field experiments that revealed far more about how reefs functioned than any aquarium experiment ever could (Fig. 3.15). Could field experiments transform the lessons learned by scientists who simply collected organisms to bring back to the laboratory?

On my earliest trips to Heron, I relished the opportunity to really commune with a coral reef. This was something at the forefront of my mind many years later in 1974, when the lease to One Tree Island transferred from the Australian Museum to the University of Sydney and I discovered I was managing a reef field station. I made sure it had ample small boats suitable for travelling about in the lagoon and built a research program based substantially on field experimentation within that natural laboratory. Fundamentally, Heron and One Tree provide the 'living platforms near reef habitat' that field research requires. And a vital opportunity for a young researcher to become immersed in a coral reef environment.

One of my students, long ago, was examining the ecology and behaviour of the common fangblenny, *Meiacanthus lineatus*. He packed a substantial breakfast and lunch, took several full SCUBA tanks, and went out to submerge himself in the One Tree lagoon from before sunrise until after sunset. He did this for several days in a row, watching the reef wake up as the sun rose and then go to sleep as the sun set. That opportunity to spend time on a reef is priceless, and that is what the world's reef field stations can provide.

An outback experience

Getting out to the islands in the early days was a lengthy and often tedious process. With insufficient research funds to fly, I took the overnight train to Brisbane, then spent the whole day in a convenient pub before overnighting to Gladstone. Once I got to Gladstone – now a far, far larger coal, LPG and alumina export terminal than it was then, there was more waiting before boarding the launch to Heron Island. I got to know the main street of downtown Gladstone in exquisite detail, wandering up and down while waiting for the launch. The hardware store with a collection of bicycles that rivalled any museum of technology, and a window display slowly being covered by decades of dust. The newsagent's with postcards depicting the glory of 'the alumina plant at

(Fig. 3.18). As Director, I was shocked when I arrived soon afterwards but realised that we were extremely fortunate not to lose anyone. The station had just said goodbye to a hundred or so school students, and staff were in the middle of the 'switch over' between student groups. A few long-term experiments were lost and there was significant damage to a number of important projects – extremely sad for the researchers and experimental organisms involved. The station building and infrastructure were rebuilt and opened to the public in October 2008. Those who were on the island at the time and exposed to the horror of a catastrophic blaze on a confined island will never forget how terrifying that night was.

To this day, the island research stations continue to provide a critical platform for studying the GBR and coral reefs in general, especially under the growing pressure of climate change. Their role in supporting the *in situ* applied science continues to be crucial to understanding coral reefs in the context of global change.

A world-leading centre for coral reef studies

The ARC Centre of Excellence for Coral Reef Studies commenced operations in 2005 with a mission to: 'Lead the global research effort in the provision of scientific knowledge necessary for sustaining the ecosystem goods and services of the world's coral reefs during a period of unprecedented environmental change.'

Headquartered at James Cook University, the ARC Centre has three additional nodes at the Australian National University, the University of Queensland and the University of Western Australia. National and international partner institutions include the Australian Institute of Marine Science, the Center for Ocean Solutions (Stanford University, USA), Centre National de la Recherche Scientifique (CNRS, France), the Great Barrier Reef Marine Park Authority, the International Union for Conservation of Nature (IUCN, Switzerland), and WorldFish (Malaysia). Terry Hughes was the inaugural Director from 2005 to 2020, with Professor Ove Hoegh-Guldberg (University of Queensland) and Professor Malcolm McCulloch (University of Western Australia) serving as inaugural deputy directors from 2005 to 2021. The current Director is Professor Graeme Cumming. The Centre's Advisory Board is chaired by Professor Hugh Possingham, the Chief Scientist of Queensland.

The centre currently has three major research programs. The first program, 'People and Ecosystems', expands the scope of contemporary coral reef science from its predominantly biological and geological focus to encompass a broader evaluation of the linkages between coral reef ecosystems, the goods and services they provide to people, and the wellbeing of human societies. The second program, 'Ecosystem Dynamics: Past, Present and Future' aims to understand the multiscale dynamics of coral reefs, through the innovative integration of ecology, evolution, genetics, oceanography and palaeontology. The third program, 'Responding to a Changing World', focuses on the responses of coral reef organisms to rapidly changing local and global environments. The goal is to advance fundamental knowledge of physiological, cellular and molecular processes underpinning reef resilience in three vital areas: the dynamics of coral-microbial associations, the integrity of carbonate reef, and the capacity of coral reef organisms to adapt to a challenging future.

The centre hosted the 12th International Coral Reef Symposium (ICRS 2012), in Cairns, Queensland, which was attended by more than 2000 delegates from 80 countries. In the lead-up to mass coral bleaching and mortality in 2016, the centre convened the National Coral Bleaching Taskforce to coordinate the scientific response to bleaching on the GBR, in northern New South Wales and in Western Australia (Fig. 3.19). This work led to a series of high-profile papers on the responses of coral reefs to anthropogenic heating, published in *Science*, *Nature*, *Nature Climate Change*, and elsewhere.

The ARC centre publishes over 400 research articles each year. Roughly half of the centre's publication outputs are focused on the GBR, and on tropical and subtropical reefs elsewhere in Australia. Centre researchers provide public briefings, presentations to governments and agencies throughout Australia and a wealth of media commentary.

Fig. 3.19. (A) Professor Terry Hughes (Director of the Centre of Excellence 2005–20) surveys coral bleaching from an aircraft. (B) The National Coral Bleaching Taskforce plan coral bleaching surveys. (C) Bleached coral reefs, as seen from above on an aerial survey in 2016. (Photos supplied with kind permission from the ARC Centre of Excellence for Coral Reef Studies, photographer unknown)

In 2020, the centre's membership included 65 chief investigators, research fellows, and associates; 26 partner investigators, resident international scholars, and adjunct researchers; and 156 research students. Internationally, centre researchers have ongoing projects and collaborations in > 50 countries.

The ARC centre continues to play a globally significant role in research training and career development, providing support and mentoring to many early and mid-career researchers.

Fig. 3.20. Returning to the boat after a scientific dive. (Image credit: Mikaela Nordborg)

Fig. 3.21. Divers surveying the specimens to be collected for coral spawning observations. (Image credit: Mikaela Nordborg)

References

1. Yonge C (1930) Studies on the physiology of corals. I. feeding mechanisms and food. *Scientific Reports of the Great Barrier Reef Expedition 1928–29 (British Museum)* **1**, 13–56.

2. Yonge C (1931) Studies on the physiology of corals. IV. The structure, distribution and physiology of the zooxanthellae. *Scientific Reports of the Great Barrier Reef Expedition 1928–29 (British Museum)* **1**, 135–176.

3. Yonge C, Yonge M, Nicholls A (1932) Studies on the physiology of corals. VI. The relationship between respiration in corals and the production of oxygen by their zooxanthellæ. *Scientific Reports of the Great Barrier Reef Expedition 1928–29 (British Museum)* **1**, 213–251.

4. Marshall SM, Stephenson TA (1933) The breeding of reef animals. *Scientific Reports of the Great Barrier Reef Expedition 1928–29 (British Museum)* **3**, 219–245.

5. Marshall SM, Orr AP (1931) Sedimentation on Low Isles Reef and its relation to coral growth. *Scientific Reports of the Great Barrier Reef Expedition 1928–29, British Museum* **5**, 94–133.

6. Fairbridge RW, Teichert C (1948) The Low Isles of the Great Barrier Reef: a new analysis. *The Geographical Journal* **111**, 67–88. doi:10.2307/1789287

7. Spender M (1930) Island-reefs of the Queensland Coast. *The Geographical Journal* **76**, 193–214. doi:10.2307/1784796

8. Hamylton S, McLean R, Lowe M, Adnan F (2019) Ninety years of change on a low wooded island, Great Barrier Reef. *Royal Society Open Science* **6**, 181314. doi:10.1098/rsos.181314

9. Spencer T, Brown B, Hamylton S, McLean R (2021) A close and friendly alliance: biology, geology and the Great Barrier Reef Expedition of 1928–29. *Oceanography and Marine Biology – an Annual Review* **59**, 89–138.

10. Hamylton SM (2017) Mapping coral reef environments: a review of historical methods, recent advances and future opportunities. *Progress in Physical Geography* **41**, 803–833. doi:10.1177/0309133317744998

11. Stephenson W, Endean R, Bennett I (1958) An ecological survey of the marine fauna of Low Isles, Queensland. *Marine and Freshwater Research* **9**, 261–318. doi:10.1071/MF9580261

12. Bell P, Elmetri I (1995) Ecological indicators of large-scale eutrophication in the Great Barrier Reef lagoon. *Ambio* **24**, 208–215.

13. Fine M, Hoegh-Guldberg O, Meroz-Fine E, Dove S (2019) Ecological changes over 90 years at Low Isles on the Great Barrier Reef. *Nature Communications* **10**, 4409. doi:10.1038/s41467-019-12431-y

14. Stephenson TA, Stephenson A, Tandy G, Spender MA (1931) The structure and ecology of Low Isles and other reefs. *Great Barrier Reef Expedition 1928–29. Scientific Reports* **3**, 17–112.

15. Lam VY, Chaloupka M, Thompson A, Doropoulos C, Mumby PJ (2018) Acute drivers influence recent inshore Great Barrier Reef dynamics. *Proceedings, Biological Sciences* **285**, 20182063. doi:10.1098/rspb.2018.2063

16. Berumen ML, Pratchett MS (2006) Recovery without resilience: persistent disturbance and long-term shifts in the structure of fish and coral communities at Tiahura Reef, Moorea. *Coral Reefs* **25**, 647–653. doi:10.1007/s00338-006-0145-2

17. Clifford E, Clifford J (Eds) (2020) *Sidnie Manton: Letters and Diaries. Expedition to the Great Barrier Reef 1928–1929*, <https://www.amazon.com.au/Sidnie-Manton-Letters-Expedition-1928–1929-ebook/dp/B08BG3Z52B>.

18. Hopley D (1971) Sea level and environment changes in the Late Pleistocene and Holocene in North Queensland, Australia. *Quarternaria* **14**, 265–276.

19. Gill ED, Hopley D (1972) Holocene sea levels in eastern Australia – a discussion. *Marine Geology* **12**, 233–242. doi:10.1016/0025-3227(72)90041-2

20. Thom BG, Hails JR, Martin ARH (1969) Radiocarbon evidence against higher postglacial sea levels in eastern Australia. *Marine Geology* **7**, 161–168. doi:10.1016/0025-3227(69)90038-3

21. Newell ND, Bloom AL (1970) The reef flat and 'two-meter eustatic terrace' of some Pacific atolls. *Geo-*

logical Society of America Bulletin **81**, 1881–1894. doi:10.1130/0016-7606(1970)81[1881:TRFATE]2.0.CO;2

22. Stoddart DR (1978) The Great Barrier Reef and the Great Barrier Reef Expedition 1973. *Philosophical Transactions of the Royal Society of London. Series A, Mathematical and Physical Sciences* **291**, 5–22. doi:10.1098/rsta.1978.0086

23. Scoffin TP, Stoddart DR (1978) The nature and significance of microatolls. *Philosophical Transactions of the Royal Society of London. Series B, Biological Sciences* **284**, 99–122.

24. McLean RF, Stoddart DR, Hopley D, Polach H (1978) Sea level change in the Holocene on the northern Great Barrier Reef. *Philosophical Transactions of the Royal Society of London. Series A, Mathematical and Physical Sciences* **291**, 167–186. doi:10.1098/rsta.1978.0097

25. Stoddart DR, McLean RF, Scoffin TP, Thom BG, Hopley D (1978) Evolution of reefs and islands, northern Great Barrier Reef: synthesis and interpretation. *Philosophical Transactions of the Royal Society of London. Series B, Biological Sciences* **284**, 149–159.

26. Daly RA (1915) The glacial-control theory of coral reefs. *Proceedings of the American Academy of Arts and Sciences* **51**, 157–251.

27. Harvey N (1984) A century of ideas since Darwin: evolution of the Great Barrier Reef. *Proceedings of the Royal Geographical Society of South Australia* **81**, 1–21.

28. Davies PJ, Symonds PA, Feary DA, Pigram CJ (1989) The evolution of the carbonate platforms of northeast Australia. *Special Publication – Society of Economic Paleontologists and Mineralogists* **44**, 233–258. doi:10.2110/pec.89.44.0233

29. International Consortium for Great Barrier Reef Drilling (2001) New constraints on the origin of the Australian Great Barrier Reef: results from an international project of deep coring. *Geology* **29**, 483–486. doi:10.1130/0091-7613(2001)029<0483:NCOTOO>2.0.CO;2

30. Braithwaite CJ (2004) The Great Barrier Reef: the chronological record from a new borehole. *Journal of Sedimentary Research* **74**, 298–310. doi:10.1306/091603740298

31. Veron JEN, *et al.* (2009) Delineating the Coral Triangle. *Galaxea* **11**, 91–100. doi:10.3755/galaxea.11.91

32. Hayward B (1977) Lower Miocene corals from the Waitakere Ranges, North Auckland, New Zealand. *Journal of the Royal Society of New Zealand* **7**, 99–111. doi:10.1080/03036758.1977.10419340

33. Veron JEN, Kelley R (1988) *Species Stability in Reef Corals of Papua New Guinea and the Indo Pacific.* Association of Australasian Palaeontologists.

34. Woodroffe CD, *et al.* (2004) Geomorphology and late quaternary development of Middleton and Elizabeth Reefs. *Coral Reefs* **23**, 249–262. doi:10.1007/s00338-004-0374-1

35. Kendrick GW, Wyrwoll KH, Szabo K (1991) Pliocene-Pleistocene coastal events and history along the western margin of Australia. *Quaternary Science Reviews* **10**, 419–439. doi:10.1016/0277-3791(91)90005-F

36. Veron JEN (2008) *A Reef in Time: The Great Barrier Reef from Beginning to End.* Harvard University Press, Boston MA.

37. Kinsey DW, Domm A (1974) Effects of fertilization on a coral reef environment – primary production studies. *Proceedings of the Second International Symposium on Coral Reefs* **1**, 49–66.

38. Smith SV (1981) Responses of Kaneohe Bay, Hawaii, to relaxation of sewage stress. In *Estuaries and nutrients.* (Eds BJ Neilson and LE Cronin) pp. 391–410. Humana Press, Totowa NJ.

39. Koop K, Booth D, Broadbent A, Brodie J, Bucher D, Capone D, Coll J, Dennison W, Erdmann M, Harrison P, *et al.* (2001) ENCORE: the effect of nutrient enrichment on coral reefs. Synthesis of results and conclusions. *Marine Pollution Bulletin* **42**, 91–120. doi:10.1016/S0025-326X(00)00181-8

40. Hatcher BG, Larkum AWD (1983) An experimental analysis of factors controlling the standing crop of the epilithic algal community on a coral reef. *Journal of Experimental Marine Biology and Ecology* **69**, 61–84.

41. Hughes TP, Rodrigues MJ, Bellwood DR, Ceccarelli D, Hoegh-Guldberg O, McCook L, Moltschaniwskyj N, Pratchett MS, Steneck RS, Willis B (2007) Phase shifts, herbivory, and the resilience of coral

reefs to climate change. *Current Biology* **17**, 360–365. doi:10.1016/j.cub.2006.12.049

42. Hoegh-Guldberg O, Smith GJ (1989) The effect of sudden changes in temperature, light and salinity on the population density and export of zooxanthellae from the reef corals *Stylophora pistillata* Esper and *Seriatopora hystrix* Dana. *Journal of Experimental Marine Biology and Ecology* **129**, 279–303.

43. Berkelmans R, Oliver JK (1999) Large-scale bleaching of corals on the Great Barrier Reef. *Coral Reefs* **18**, 55–60. doi:10.1007/s003380050154

44. Hoegh-Guldberg O (1999) Climate change, coral bleaching and the future of the world's coral reefs. *Marine and Freshwater Research* **50**, 839–866. doi:10.1071/MF99078

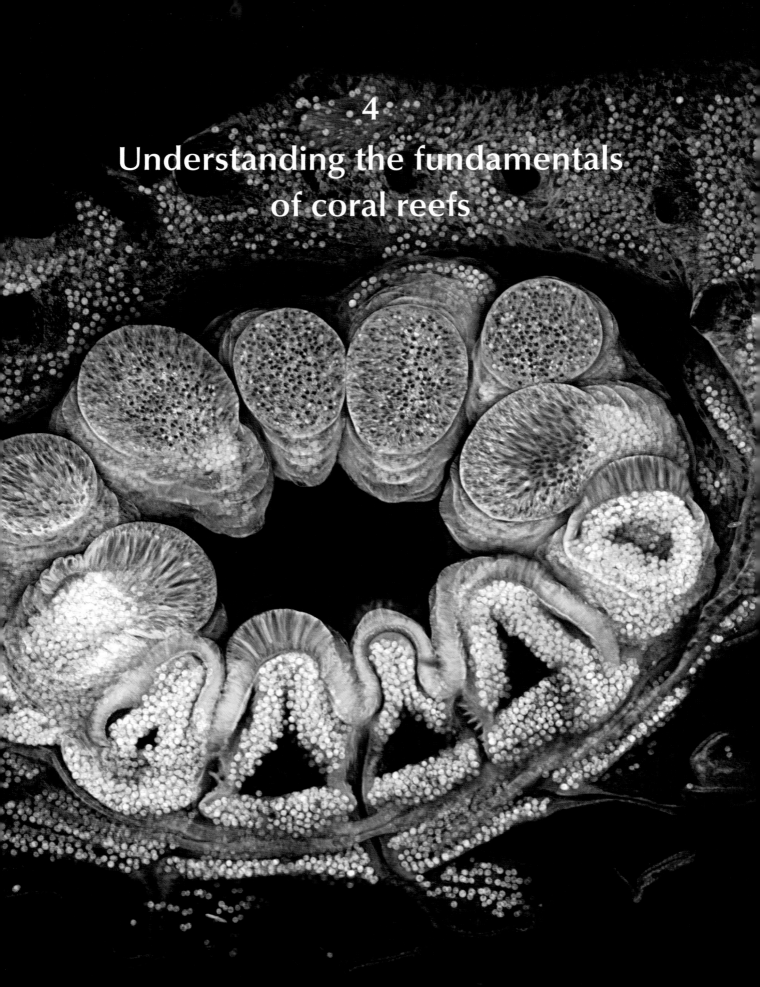

4
Understanding the fundamentals
of coral reefs

Coral reefs are a biological and geological phenomenon. Australia's coral reefs have persisted for hundreds of thousands of years, supporting a wealth of marine biodiversity that, in turn, is responsible for forming the islands we see today around Australia's tropical coastline.

Reefs build up on top of ancient reefs, over layered generations of growth during interglacial periods when conditions are favourable for reef growth. Drilling off Cairns has revealed eight cycles of Great Barrier Reef (GBR) growth corresponding to previous periods of interglacial high sea levels over the last 600 000 years. Modern reefs of the GBR developed around 8500 years ago with the most recent sea level. The islands we see today formed since this time. Reef islands around the Australian coastline serve as turtle rookeries (e.g. Raine Island), provide habitable land for Torres Strait Islanders and a base for fishing communities in the Houtman Abrolhos (Western Australia). As low-lying landforms, these are vulnerable to sea-level rise (SLR), and many are already showing signs of erosion.

Coral reefs and islands are impacted by cyclones around the northern Australian coastline. These intense periods of fierce winds and damaging seas can break corals, erode islands and impact marine life associated with reefs, including fish. Marine communities inhabiting coral reefs are highly specialised to their local environmental conditions and can persist in challenging settings, including at cooler, high latitude locations, or in deeper water depths where ambient levels of sunlight are low (the mesophotic zone), or in turbid inshore waters.

Reef building corals are a biological curiosity. The coral holobiont combines a coral (an animal), an algal symbiont and a community of microbes that work together in a functional unit to build a reef. Corals have multiple ways of reproducing, with species being cued by the moon to spawn, renewing generations of corals across hundreds of kilometres of coastline.

Fish use coral reefs for reproduction, feeding and as habitat to shelter them from predators. Other marine wildlife such as sharks, rays, whales, dugong and sea snakes play a different role within the ecosystem, either residing permanently on reefs or visiting them on ocean migrations. Sharks act as predators, rays oxygenate reef sands, while dwarf minke and humpback whales overwinter in shallow, warm reef waters. Dugong feed on seagrass meadows found on older, larger reefs.

Scientific drilling on the Great Barrier Reef: unlocking the history of the reef
Gregg Webb and Jody Webster

As with Rome, reefs were not built in a day. The modern coral reef communities that inspire us with their beauty, and command our attention, represent only thin veneers of living biology that occur on much larger rocky structures built up by successive, but intermittent, generations of reef builders over many thousands of years. As reefs have been undergoing change for thousands of years, it is critical that we understand the history of the GBR over both longer (geological) and shorter (ecological, historic) timescales. That allows us to understand true rates, trajectories, and drivers of recent changes in reefs within a broader and better-informed context of natural reef behaviour and environmental thresholds.

Many fundamental questions about the GBR relate to its history. How old is the GBR and why did it start forming in the first place? What happened to the GBR during the major ice ages, including the last one ~20 000 years ago when Australia's continental shelves were exposed by sea levels around 120 m lower than today? How resilient have reefs been to previous times of climate change and extreme events? How adaptable are reef communities over differing timeframes? What happens to reefs when the sea level rises and falls, when cyclones are more abundant, when sea surface temperatures are warmer or colder and when rainfall increases or decreases on the mainland?

Why do we still not know the age of the GBR? There are two main issues. First, the oldest parts of the reef are buried under younger parts and hence they are difficult to access. Second, it is technically very difficult to date older reef rock. Radiocarbon (C-14) dating can date ancient corals, but only back to around 40 000 years ago. Corals can be dated back to 500 000 years ago using U-Th techniques, but only if the corals are very well preserved. Unfortunately, the latter is rarely the case. For ages older than that, strontium (Sr) isotope curves can be used because the ratios of Sr isotopes have varied in seawater through time and these ratios are preserved in the skeletons of growing corals. However, the dated corals still must also be very well preserved to use the technique, plus these types of ages are not precise. Regardless, the age of the modern GBR was speculated upon for more than a century before the first and oldest coral samples were obtained by drilling a 210 m hole through Ribbon Reef 5 in the northern GBR (Fig. 4.1A). Decades later, those core samples remain the only physical pieces of the GBR's earliest years, the accurate dating of which remains a fundamental challenge. Surprisingly perhaps, we have dated more rocks from the surface of the moon than from the base of the GBR.

Despite the stigma associated with the word 'drilling', scientific drilling has provided the majority of what we know about the GBR's longer-term history and how reef communities have reacted to natural environmental changes. The earliest scientific drilling on the GBR suggested that individual reefs consist of stacked packages of reef material separated by surfaces of reef death and exposure formed on land during times of lower sea level (Fig. 4.2) [1]. Hence, the reefs are much like layer cakes, with reef growth occurring during times of high sea level (like today), when the continental shelf was flooded, making layers as in a cake. However, the layers are separated by intervals of nongrowth (like icing) formed when the reefs were

eroded while sea levels were lower and the shelf was exposed. The underlying continental shelf subsides through each growth cycle, providing room for the next layer of reef growth.

The GBR that we know today is relatively thin (mostly 10–30 m thick), but is growing on a series of ancient reefs that were exposed and killed during times of low sea level. More sophisticated drilling off Cairns produced a core (RR5 core, Fig. 4.1A) that contains at least eight cycles of reef growth corresponding to previous interglacial sea-level highstands with the remaining time representing exposure and soil development on limestone hills and mesas on land [2]. RR5 drilling also provided the first means of dating actual reef rock. However, owing to the age of the rocks and their preservation, only Sr isotopes could be used and hence many questions remain.

While deep reef drilling confirmed that the reef we know today resembles a 'layer cake' of reef growth that represents only snippets (~10 per cent) of time when the sea level was adequately high over the last 600 000 years or so, the question remained as to what happened to the reef during lower sea levels. That mystery was solved by ship-based ocean drilling when the Integrated Ocean Drilling Program (IODP) Exp 325 recovered cores from fossil reefs offshore of the modern GBR in water depths between around 45 and 130 m [3] (Fig. 4.1B). That drilling showed that when the sea level fell during the last ice age, reefs tracked the shore seaward to lower elevations, and then tried to keep pace with rising sea levels as the glaciers melted and sea levels rose back to modern levels (Fig. 4.3). However, reefs were not able to keep up with rising sea level at certain times, especially when SLR rates were highest and when muddy sediment from the continent increased as the continental shelf was re-flooded. The modern reef was finally re-established on older reefs as the exposed limestone hills were flooded around 8500 years ago [1].

How these youngest reefs of the GBR reestablished and grew to the modern sea level is particularly important for understanding how the modern reef may respond to predicted climate

Fig. 4.1.   Scientific drilling techniques applied on the Great Barrier Reef. (A) Jack-up rig used to recover core from Ribbon Reef 5, north-central GBR. (B) Integrated Ocean Drilling Program drilling ship used to recover low sea-level reef cores from below the sea floor offshore of the modern GBR. (C) Jack-up rig from the University of Queensland's research vessel. (D) Hand-held core drill for recovering shallow cores from reef flat of Heron Reef in 2012. (E) Pleistocene reef core obtained from beneath the Heron Reef reef flat dated to the last interglacial (~125 000 years ago). (Image credits: Peter Davies (A), Graham Lott, Integrated Ocean Drilling Program (B), Gregg Webb (C, E), Luke Nothdurft (D)).

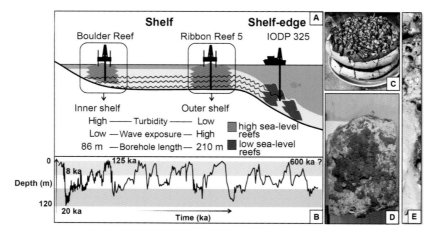

Fig. 4.2.   (A) Schematic diagram showing 'layer cake' reefs on the continental shelf and lowstand reefs offshore (modified after [2]). (B) Approximate sea-level curve spanning the last 600 000 year superimposed onto vertical 'cake' layers. (C) Illustrative layered cake. (D), Fossil soil 'palaeosol' layer from core. (E) Fossil coral cores. (Image credits: Jody Webster)

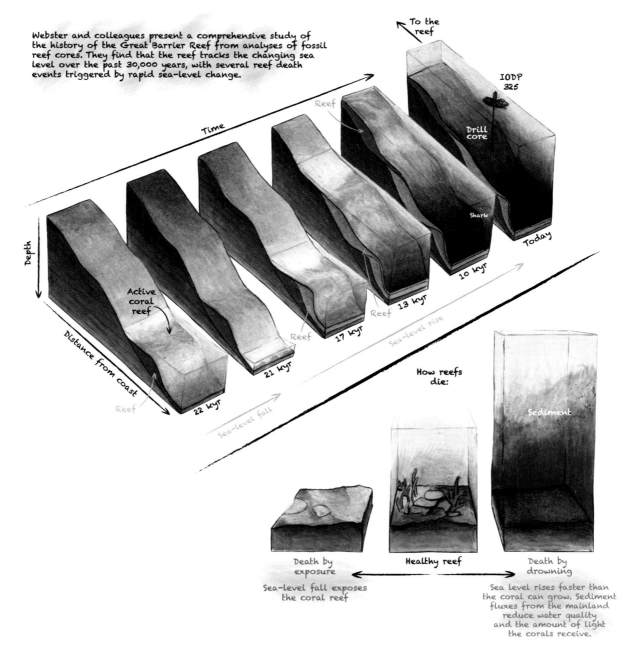

Webster and colleagues present a comprehensive study of the history of the Great Barrier Reef from analyses of fossil reef cores. They find that the reef tracks the changing sea level over the past 30,000 years, with several reef death events triggered by rapid sea-level change.

To the reef

Time

IODP 325

Drill core

Reef

Depth

Shark

Today

Active coral reef

10 kyr

13 kyr

Reef 17 kyr

Distance from coast

21 kyr

Reef

sea-level rise

22 kyr

Reef

Sea-level fall

How reefs die:

Sediment

Death by exposure

Healthy reef

Death by drowning

Sea-level fall exposes the coral reef

Sea level rises faster than the coral can grow. Sediment fluxes from the mainland reduce water quality and the amount of light the corals receive.

Fig. 4.3.   Schematic diagram showing the relationship of lowstand reefs offshore of the Great Barrier Reef. (Image reproduced with permission from [3]; image credit: James Turtle Keane)

change and SLR in the future. Cores recovered from reefs over the last decades showed that the modern reef grew both upwards and laterally at the current sea level, thus providing a basic model for reef development [1, 4]. In this model, the initial highest reef growth was in the high-energy windward part of the reef and then it grew towards the leeward. More recently, improved dating techniques, especially with the introduction of new high-precision U-Th dating [5], and recovery of

Fig. 4.4. Aerial view of Watson Island, Great Barrier Reef. (Image credit: Sarah Hamylton)

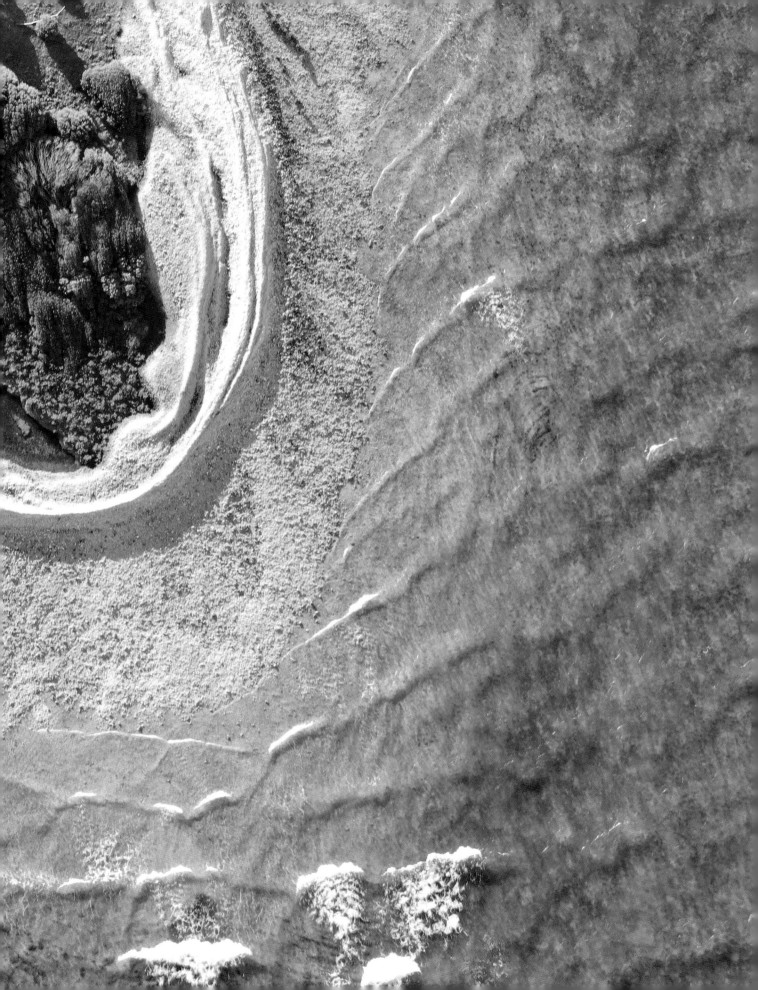

more closely spaced cores along transects have improved our understanding of the internal architecture of reefs considerably. Fringing reefs have been studied using dated soft-sediment cores to understand how they have been affected by minor sea-level fluctuations and local sediment sources [6]. Core transects on offshore reefs have greatly aided our understanding of the relationship between vertical and lateral reef growth through time [e.g. 5, 7] (Fig. 4.1C). These cores have allowed study of the relationships between SLR and reef growth, shape of reef versus underlying topography, community types through time, effects of changing climate, and water quality on the same reefs that persist today, suggesting resilience over the last 8500 years or so.

Dated transects of very short, ~1 m depth cores recovered using handheld core drills (Fig. 4.1D) have allowed us to understand reef growth after it reached the modern sea level. Contrary to expectation, Dechnik and others showed that reefs did not grow out against the highest energy zones [8]. Rather, they grow away from them, towards the lee, filling the back-reef areas. However, on lower energy margins, the reefs grew seaward, actively enlarging the size of the reef. In some cases, minor sea-level fluctuations in the last 5000 years or so caused reef growth to stop for a time on many reef margins. Understanding the dynamics of reef growth and sediment production and distribution in areas of differing wave energy on the same reef is critical for predicting the effects of changing wave parameters, through increased storminess or rising sea level on modern reefs.

As new cores become available through scientific reef drilling and old cores are analysed using new dating and environmental geochemistry approaches, we are continually refining our understanding of the history of the GBR. Scientific drilling in the GBR continues to provide fundamental information about key questions regarding the timing and rates of sea-level change, the temperature of past seawater, the influence of past ENSO behaviour and effects of the Australian-Indonesian summer monsoon on Australian rainfall. As reef cores record those fundamental climate data, they also, and possibly most importantly, provide the records of the responses of individual corals, reef communities and reefs to those changes, allowing us to better understand the environmental thresholds for reef communities at differing timescales. That information will allow us to better manage the GBR in the future.

Australia's reef islands
Scott Smithers and Nicola Browne

Island characteristics and classification
Along Australia's tropical coastline lie some of the world's most well-developed, biodiverse, and well-managed coral reefs. These include the GBR in Queensland and Ningaloo Reef in Western Australia (WA). Both are World Heritage sites because of their outstanding natural and cultural values. Many reef islands occur within these tropical waters. Some formed as outcrops of continental rocks were isolated by rising seas that flooded across continental shelves following the last ice age. The distribution, size and character of these islands are strongly influenced by geological setting. They are usually called continental high islands, in reference to their composition and elevation above sea level [9]. Fringing reefs at least partially surround many high islands. Examples include Hamilton Island in the Whitsundays, Murray Island (Mer) in Torres Strait and Barrow Island on Australia's western coast.

Low-lying reef islands are another broad class of islands found within Australia's tropical waters. As the name suggests, they are typically low, rarely rising more than a few metres above sea level. They form where carbonate sediments produced by reef organisms are transported by reef-top waves and currents to a focal point where they accumulate. The location of this focal point is largely determined by reef shape, prevailing wave direction, sediment size and biological provenance. Sandy islands typically form near leeward reef margins whereas coarse 'shingle' islands usually develop towards the windward reef edge, but the exact position depends on many factors. Multiple islands can also develop on top of a single reef platform. For example, low-wooded islands are a particular class of reef island found on the inner northern GBR that includes a windward shingle and leeward sandy cay on a single reef platform, often with mangrove forest between. Low reef islands can be very dynamic, with large changes in island size and shape occurring in response to shifts in prevailing winds and waves, or because of extreme conditions during tropical cyclones [9].

Low-lying reef islands are relatively young landforms that develop on the reef once it has reached sea level. Most are less than 6000 years old, with many much younger. Some reef islands are in their early stages of development even today. Reef islands are often assumed to be simple landforms because of their youth and small size. However, an assessment of sediment type (sand, shingle or a mixture), island position within the reef platform (windward or leeward), island shape (linear or compact), vegetation cover (vegetated or unvegetated), and the number of islands on a reef reveals significant diversity as captured in most classification schemes. Classifications distinguish a range of reef island types, for example leeward linear vegetated sand cays or windward compact unvegetated shingle cays [9].

Reef islands of the Great Barrier Reef

On the GBR there are 617 high islands with attached reefs, and over 300 low-lying reef islands. Neither high nor low reef islands are distributed evenly across shelf, or latitudinally. The distribution of different reef island types is strongly influenced by spatial patterns in reef size and relative sea-level history [9]. Greater than half of the high islands belong to the Whitsunday, Cumberland and Northumberland Island groups between 20–22°S. More than 70 per cent of the low-lying reef islands sit upon planar reefs concentrated inshore north of Cape Tribulation, or above outer-shelf reefs in the Pompey Complex and Swains

Reefs further south. Unvegetated cays comprise > 70 per cent of low-lying reef islands on the GBR. Many GBR reef islands are highly culturally significant for Indigenous communities. Larger high islands with dependable freshwater were more regularly inhabited, whereas those with intermittent water were visited sporadically. Although unreliable water limited settlement, many low reef islands are culturally significant to Aboriginal and Torres Strait Islander peoples, providing valuable resources, and important linkages to creation stories. European uses of GBR reef islands include guano mining, navigation and military staging posts, pastoral uses, fishing, aquaculture, residential uses and tourism. Critically, many GBR islands are of great ecological value, providing habitats for plants and animals, including threatened species of seabirds and turtles. An example is Raine Island, a 30 ha vegetated cay on the outer northern GBR where the world's largest green turtle population nests (Fig. 4.5). As many as 20 000 female green turtles may visit the beach on a night in an attempt to successfully nest in exceptionally large nesting seasons. It is also a major seabird rookery, a site of great cultural significance for Traditional Owners from both Cape York and Torres Strait, and the location of one of the oldest European built structures in Queensland, the heritage-listed beacon tower built by convicts in 1844 [10].

Reef islands in Torres Strait

Torres Strait is a shallow body of water between the northern tip of Australia and the southern Papua New Guinea coast (see Chapter 1). It covers ~48 000 km^2 and includes around 270 islands, including small unvegetated and vegetated low-lying reef islands, low-lying muddy islands, and high islands dominated by volcanic or granitic rocks (Fig. 4.6). The volcanic high islands are concentrated in the eastern Torres Strait and the low mud-dominated islands towards the north-west. Cays and granitic high islands are more widely distributed. Eighteen islands are permanently inhabited, but uninhabited islands are also used for a

Fig. 4.5.   (A) Raine Island viewed from the north-west. (B) The tower at Raine Island, built by convicts as a beacon in 1844 to assist navigation between the Great Barrier Reef and the Torres Strait, an area that was notorious for shipwrecks before a northern passage was discovered through the Torres Strait. (Image credits: Queensland Government (A), Matt Curnock (B))

Fig. 4.6.   (A) Dauar in the foreground and Wauar behind – two volcanic high islands located close to Mer in Eastern Torres Strait. (B) Poruma – a low, vegetated cay in central Torres Strait viewed from the north-west. Erosion on the north-eastern shoreline of the cay has been treated with the construction of a geotextile wall, but shoreline change is of great concern to island residents. (Image credits: Jacobus Pretorius and Scott Smithers)

variety of purposes. Torres Strait Islanders have occupied the region for the last 2500 years at least, developing a unique and complex culture focused on their relationship with the sea known as Ailan Kastom [11]. Climate change impacts that degrade these systems threaten the cultural identity of Torres Strait Islanders, especially those living on low reef islands who fear they may be Australia's first climate change refugees as rising seas erode and inundate their islands [12]. Similar fears arose following severe tidal flooding of low-lying muddy islands in north-western Torres Strait in the 1940s, resulting in the voluntary relocation of community members to the mainland.

Reef islands in Western Australia

Over 3700 islands lie off the WA coast. The majority (> 2600) are in the Kimberley region, with others in the Pilbara (~320), the central west coastal region (~210), and along WA's southern coast (~260). Most are rocky high islands situated close to shore. Low-lying reef islands are less common, and mostly concentrated along the Kimberley to Western Pilbara coastline, or offshore further south at the Houtman Abrolhos, where tropical waters delivered by the Leeuwin Current enable coral reef growth at higher latitude. Most reef islands are uninhabited. Exceptions include the Houtman Abrolhos, where a small-scale fishing industry has existed since the early 1800s, and several reef islands in the Pilbara, where oil and gas infrastructure are located (e.g. Thevenand Island).

The Houtman Abrolhos islands lie ~60–70 km west of Geraldton and are the highest latitude reef complex in the Indian Ocean (see Fig. 1.3 in Chapter 1). They include 122 islands within three main groups, each on a separate carbonate platform – the southern Pelsaert Group, the central Easter Group, and the northern Wallabi Group [13]. Most islands are low (< 4 m), but some in the Wallabi Group include cemented sand dunes rising 10–20 m above sea level [13]. The majority are composed of coral framestones and sand sheets dominated by coral rubble and coralline algae debris, with bryozoan, foraminiferan and molluscan shells. More than four centuries since the islands were first sighted by Dutch mariner Frederick von Houtman, the Houtman Abrolhos National Park was designated in 2019 to protect their exceptional historic and natural values. Following British occupancy, guano mining and commercial fishing have dominated activities on the islands. Around 56 000 tonnes of guano were mined before ceasing in 1904. The islands currently host WA's most important western rock lobster fishing base, worth between $30 and $50 million annually (see 'Western rock lobster and the Houtman Abrolhos Islands' in Chapter 2) [14].

The Western Pilbara extends south from the Dampier Archipelago to the Exmouth Gulf, and includes 174 small islands (64 per cent are < 50 ha) (Fig. 4.7). Most are carbonate sand dunes deposited since around 6000 years ago over older limestone. They generally comprise an old central stable core made of Pleistocene limestone, 10–15 m high dunes on the eastern shoreline, a sandy spit along the south-eastern margin, and a fringing reef platform (Fig. 4.8) [15]. Since the 1800s, Europeans have used the islands for pearling, guano, fishing, and hunting of marine turtles. Recently the area has experienced extensive oil and gas exploration, with abandoned drill stems, well heads and radio towers remaining on many islands. The islands provide important refuge for terrestrial (e.g. eastern curlew,

Fig. 4.7. Bird life of crested terns (*Thalasseus bergii*) and silver gulls (*Chroicocephalus novaehollandiae*) on Eva Island situated in the Exmouth Gulf, southern Pilbara, Western Australia. Y Island is in the background. (Image credit: Josh Bonesso)

Fig. 4.8. Aerial view of Ashburton island (~33 ha) viewed from the east-south-east showing typical inshore Pilbara island features, including small 'round' island, increased sand deposition along eastern margin and extensive fringing reef flat. (Image credit: Josh Bonesso)

curlew sandpiper) and marine life (e.g. various turtles). Currently, 97 islands are within nature reserves and another 77 are proposed for inclusion in future conservation plans [16].

The Kimberley region includes > 1000 km of coastline. Rocky high islands with fringing coral reefs subjected to very large macro-tides (fluctuating

> 8 m in vertical height) and extreme turbidity occur inshore here. In clear oceanic waters, low-lying islands occur offshore, including 'atolls' such as Seringapatam Reef. Many of these islands are sea-bird breeding sanctuaries. The Lacepedes (situated 30 km offshore) support 1 per cent of the world's population of brown boobies and roseate terns while Adele Island – 100 km from the Kimberley coast – is a globally important lesser frigatebird rookery (see the map in Fig. 1.4 in Chapter 1). A further 175 km offshore, Browse Island was a major seabird rookery into the early 1900s, but by 1949 few seabirds remained, following guano mining and the intro-duction of feral cats and mice to the island [17]. Reef islands within the offshore Oceanic Shoals region include three low-lying vegetated cays on Ashmore Reef, the unvegetated sand cay Cartier Island, Sandy Islet on Scott Reef, and Bedwell Island in the Rowley Shoals. Indonesian fishers have visited these islands for centuries, using them as stop-offs on longer voy-ages to other Kimberley islands. A memorandum of understanding with the Indonesian Government allowing Indonesian fishers access to surrounding fishing grounds and reefs was established in 1974, but landing on islands is prohibited.

Future prospects and challenges

The future of low-lying reef islands and the ecosys-tem and cultural services they provide is uncertain as sea levels rise and the reefs that supply the sedi-ments they are built from degrade (Fig. 4.9). Increas-ingly, interventions will be required if reef islands are to persist and possibly survive climate change impacts. An example of such an intervention is the Raine Island Recovery Project where excavation machinery was used to re-profile turtle nesting beaches on what is perhaps the GBR's most ecologi-cally valuable island, to reduce the areas of nesting beach that were vulnerable to inundation from 79 per cent (in 2009) to just 21 per cent (in 2019). In turn, this has significantly improved hatchling output from the island. However, many reef islands are experiencing severe erosion. In some cases, sea-walls are constructed to address the erosion issue at site, but such treatments often transfer the

Fig. 4.9. A coral rubble rampart, or ridge, on the northern Great Barrier Reef. These deposits from high-energy cyclones play a role in island formation. (Image credit: Paul Jones)

problem alongshore. Moreover, the efficacy of localised engineering solutions is not without limit and unless steps are taken to address more funda-mental drivers of climate change, such localised management actions can only be temporary solu-tions. Carbon dioxide emissions must be urgently and drastically reduced if Australia's low reef islands, and all the ecosystem and cultural func-tions they support, are to survive into the next century.

Tropical cyclones and Australia's coral reefs
Marji Puotinen

Tropical cyclones, known as hurricanes and typhoons elsewhere around the world, are low-pressure weather systems that form in warm tropi-cal waters and can persist for hours to weeks before weakening over land or cool water. They are a dis-tinctive and often destructive feature of Australia's tropical coastline. Cyclone tracks can be very erratic, particularly in north-east Australia, and are notoriously hard to predict ahead of time. Most cyclones around Australia last for about a day and affect a given location at high intensity for a few hours. Notable exceptions include severe cyclone Debbie (2017), which pounded the Whitsunday islands and reefs with extreme waves for more than a day, and severe cyclone Larry (2006), which tracked so quickly that it only affected reefs in its path for less than an hour.

During a cyclone, strong winds, of gale force (63 km h^{-1}) or more, form around a small (~30 km) core of calm at the centre of the storm, known as the eye. In the southern hemisphere, these winds are strongest at the eye boundary in the front left quadrant in the direction towards which the cyclone is moving. Wind speeds gradually lessen with distance from the eye. The size of a cyclone circulation is defined as the distance from the centre of the eye to the point where winds drop to gale force. For cyclones in the Australian region, this distance averages ~200 km, but can be much smaller or larger than this. Cyclone size is not the same as severity (intensity). Cyclone severity is ranked from 1 to 5 based on maximum wind speeds recorded or estimated at the eye boundary. In Australian waters, such wind speeds for severe cyclones (categories 3, 4 and 5) reach or exceed ~120 km h^{-1}. As cyclone-generated winds blow over water, they can build up extreme waves at the water surface given sufficient time and water depth. This is particularly likely when the cyclone is slow moving (forward speed < 11 km h^{-1}), tracks within a small area, and covers a large area with its circulation. Where a source of deep cold water exists below the mixed portions of the water column, cyclone waves can transport the deep cool water to the ocean surface. When this happens during times of thermal stress, the resultant cooling can reduce the severity and duration of mass coral bleaching and mortality [18], but can also cause algal blooms [19] and disrupt organisms in the water column such as fish (see the box, 'What happens during a cyclone? The perspective of a reef fish' later in this chapter).

How does tropical cyclone activity vary around Australia?

There are two main hotspots of consistently frequent cyclone activity around Australia (Fig. 4.10) [18]. The first and greatest hotspot is in the far eastern section of the Coral Sea Marine Park where extreme winds and waves from cyclones occur, on average, up to 5.4 days per year. The second hotspot is around the Pilbara coast of WA where cyclone activity averages almost 4 days per year. In contrast, cyclone activity typically affects the GBR Lagoon and the Gulf of Carpentaria half as frequently, up to 2 days per year [18].

What makes a cyclone damaging to a coral reef?

It has long been observed that wave energy plays a major role in shaping the geomorphology and ecology of coral reefs [20]. Tropical cyclones can generate waves that are considerably more energetic than those typically experienced by reefs and thus often exert forces on reefs beyond what they have adapted to absorb without damage. This damage ranges from minor impacts, such as broken parts of coral colonies, to more extreme impacts, such as sand burial, dislodgement of massive colonies and destruction of entire reef structures. The spatial distribution of this damage within reefs is patchy for two key reasons. First, the vulnerability of individual coral colonies to waves varies depending on many site-specific factors like growth form, colony shape and size [21] and strength of attachment to the reef framework. Second, reefs are very effective at blocking wave energy, especially when clustered together as on the GBR [22]. This means that some reefs bear the brunt of a cyclone while others nearby can be untouched.

Due to higher wind speeds, severe cyclones (intensity categories 3, 4 and 5) cause more damage on land than weak cyclones (intensity categories 1 and 2). Any consideration of the damage that cyclone waves inflict on coral reefs must look beyond intensity to other factors that determine how a given cyclone generates extreme waves. While cyclone intensity reveals the maximum wind speeds that are possible, the extent to which these winds translate into big waves depends on the distance they blow unobstructed across the ocean at that speed and for how long. In turn, this depends on the depth of the water, whether there is land nearby, and on the overall size of the cyclone

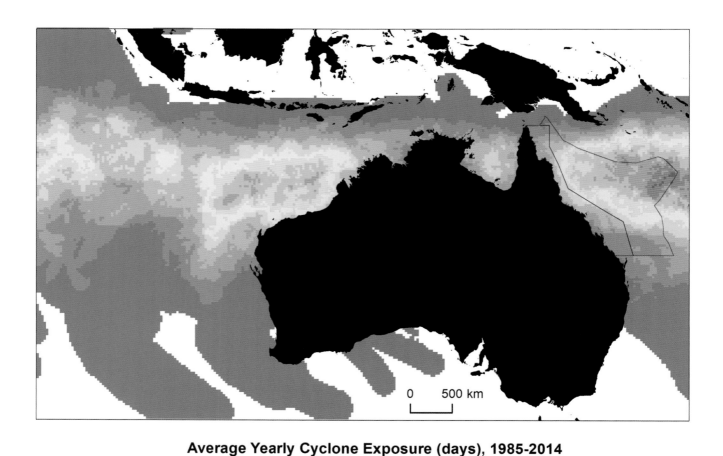

Average Yearly Cyclone Exposure (days), 1985-2014

Up to 0.5   0.5 - 1.0   1.0 - 1.5   1.5 - 2.0   2.0 - 2.5   2.5 - 3.0   3.0 - 3.5   3.5 - 4.0   4.0 - 4.5   4.5 - 5.0   5.0 - 5.5

Fig. 4.10.   Average yearly number of days of exposure to extreme wind speeds (gale force or above) generated by tropical cyclones (cyclone activity) in the Australian tropical cyclone reporting region. The Great Barrier Reef World Heritage site and the Coral Sea Australian Marine Park boundaries are outlined in black. (Adapted and updated from [18]; image credit: Marji Puotinen)

itself. This means that a stronger cyclone is not necessarily larger (see Fig. 4.11).

One of Australia's most notorious cyclones was Tracy, which devastated the Northern Territory city of Darwin on Christmas Day in 1974. Darwin suffered a direct hit from the cyclone, which was one of the smallest cyclones ever recorded around the world, with damaging winds falling within a zone that only measured 100 km in diameter. In contrast, just a few years later, a typhoon Tip, which was of similar strength, spread its damaging winds over 2200 km in the north-west Pacific. As the graphs in Fig. 4.11 show, wind speeds drop

off much more quickly with distance from the storm's eye for small rather than large cyclones. This means that winds capable of generating extreme waves are spread over a much greater distance of ocean for the latter (> 400 km versus 50 km), and thus extreme waves can occur over a much larger area. Early cyclone damage fieldwork in Australia found wave damage typically occurred within 100 km of the cyclone track. However, this was based on a typically sized, strong cyclone. More recent field data from big and strong cyclone Lua (WA, 2012) documented severely damaged coral reefs up to 850 km away

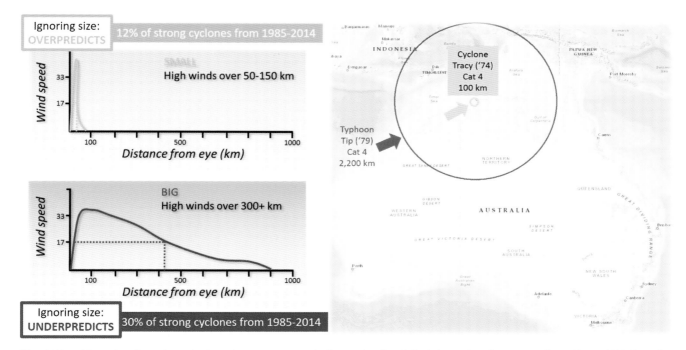

Fig. 4.11. How the size of tropical cyclones can vary independently of their intensity. Severe cyclone Tracy (1974) and typhoon Tip (1979) were similar in intensity at category 4, yet at opposite extremes in terms of size. (Image credit: Marji Puotinen)

from the cyclone track [23]. Also important to wave formation is how fast a cyclone moves along its track and its duration near reefs. Field data from the GBR show that the most damaging cyclones are typically those that are large, strong and slow-moving near reefs [24]. Luckily, cyclones with all these characteristics have been rare over the recent past (1985–2020) around Australia [23]. The chance that a particular coral colony on a given reef will be damaged by a cyclone can seem a bit like a lottery given the intermittent nature of where cyclones track, the relative position of the colony within the reef and relative to other reefs, and the physical characteristics of the colony itself.

Damaging cyclones currently revisit coral ecoregions around Australia anywhere from once every 5–10 years (Fig. 4.12, dark pink – Ningaloo and Pilbara coast; Coral Sea) to once every 35 years (dark blue – temperate southern ecoregions). Projections from global climate models suggest that intervals between cyclones

may shorten in future, causing more frequent damaging cyclones in some of Australia's ecoregions, including the southern GBR and much of the west, north-west and northern coastal ecoregions [23].

Although reefs have evolved together with cyclones over millennia, the combination of continued impacts from cyclone waves and more frequent mass coral bleaching is reducing the recovery time available between damage events. When reefs cannot fully recover, less of the reef may be covered by live coral. Moreover, such frequent damage compromises the future potential for full recovery, by changing the size and composition of corals on reefs. This has been observed for the GBR after a combination of three mass coral bleaching and mortality events in 4 years (2016, 2017, 2020) and four cyclones in 3 years (2017, 2018, 2019) [25]. Resulting changes to coral reefs may diminish the key ecosystem services they provide, such as shoreline protection as well as habitat for fish (see the box, 'What happens during a cyclone? The

Percent world reef area by return time

Reef area near the Equator, where cyclones rarely track

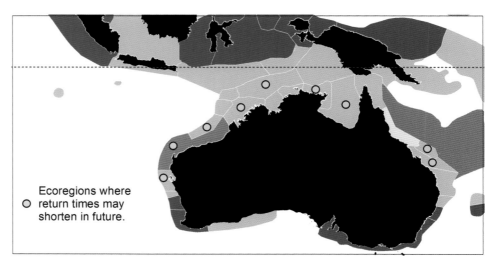

Ecoregions where ○ return times may shorten in future.

Return time (years) of big + strong tropical cyclones in coral ecoregions, 1985-2015

■ 1-5   ■ 5-10   10-15   15-20   20-25   25-30   30-35   ■ >35

Very frequent Frequent   *Occasional*   *Very rare*

Fig. 4.12. Expected number of years between large and strong (damaging) tropical cyclones (return times) in coral ecoregions around Australia, 1985–2015. Coral reef communities may not fully recover between damaging cyclones where return times are short (< 5 to 10 years). Yellow circles indicate ecoregions where return times may shorten in the future, based on projections from global climate models. The Equator is shown by the dashed line. (Adapted from [23]; image credit: Marji Puotinen)

perspective of a fish'). Indeed, such changes can shift the ecosystem balance of a coral reef such that it is no longer dominated by live hard coral. While we can neither prevent nor control cyclones or coral reef responses to them, understanding where and how often cyclone waves are likely to damage reefs is vital for planning management and conservation actions [23].

What happens during a cyclone? The perspective of a reef fish

Catheline Y. M. Froehlich

From the perspective of a reef fish, a cyclone event begins just like any other day. Then everything changes. The reef becomes dark as cyclone waves cause turbulence in the water. Waves and currents become much stronger than ever before (e.g. severe gales begin at 29 kn, while cyclone winds only start at 69 kn [26]). You are being jostled around and corals are uprooted from the reef, or their branches are damaged. You try darting from hole to hole for shelter, but the wave surge is too strong and it seems that nothing stays put. This goes on for many hours or even days [27, 28]. After the cyclone finally passes, the reef looks and feels abnormal, and the water is colder [29]. Your home is likely to be gone due to extensive damage to attached organisms, such as anemones and corals (Fig. 4.13) [27, 28], which makes the reef structurally 'flatter' [30]. You are injured from being tossed around, don't see many fish and recognise even fewer of them [27, 28]. This is what happens during and after a cyclone, and it may take years to centuries for reefs to recover to their pre-cyclonic state, depending on the extent of damage and whether other disturbances affect the reef in the meantime [27, 28].

Damage from cyclones is often patchy [28]. Does that mean that fish from undamaged areas can replenish those lost during a cyclone? Perhaps, but the impacts of cyclones on fish can extend across many scales [28, 31]. At an individual level, fish may become displaced, injured, or buried under sediment, and some may die [28]. After seas calm post-cyclone, fish that are trying to recover still must avoid predation, find mates or schools, and secure food, but such challenges are heightened post-cyclone [28]. Finding shelter is often the top priority [28] and many fish need live corals, but widespread coral loss results from cyclones (68 per cent decline from some cyclones [27]). Finding food gets harder because prey organisms themselves may be damaged or killed by cyclones [28].

Fig. 4.13. What a reef looks like after a cyclone from a fish's perspective: few live corals, rubble and debris everywhere, flatter structure and few fish. (Image credit: Matt Curnock)

Therefore, fish may grow more slowly after cyclones [28]. Looking beyond an individual level, a change in the environment or habitat can also lead to behavioural changes in fish [28, 32]. For example, for fish that live in groups, cyclones may force fish to live in smaller groups due to habitat damage [32]. On a fish community level, cyclones reduce planktivorous fish populations due to habitat loss, while larger predators, herbivorous and rubble-associated fishes flourish [27, 28]. Interestingly, the short-term cooling effects of cyclones may assist in increasing plankton production [18]. However, full recovery of fish populations is likely to require decades to centuries [29]. For example, fish abundance dropped by 15 per cent in 2011 following cyclones Hamish (2009) and Yasi (2011), but there were lag effects after 4 years (2015) at which point fish loss reached 23 per cent [27]. A local disappearance of 23 species occurred on up to 10 reefs [28]. Accordingly, we need to look many years ahead for potential recovery in fish populations from cyclones.

Marginal reefs: distinct ecosystems of extraordinarily high conservation value

Brigitte Sommer

Marginal reefs are found in suboptimal settings, such as high-latitude and deep reefs, where environmental conditions (e.g. temperature, light, salinity, turbidity, sedimentation, carbonate chemistry) are considered limiting or stressful for corals compared to reefs in more 'traditional' settings – that is, in shallow, clear, warm tropical seas [33]. As tropical coral reefs are increasingly threatened by climate change (see Chapter 8), Australia's marginal reefs have gained interest for their potential to offer refugia for corals and other organisms from climate change, and for their unique ecological values [34, 35]. Less is known about them than about 'traditional' coral reefs and this section gives an overview of marginal reefs in Australia, describes some of their key ecological features and the threats they face, and the implications for their conservation and management.

High-latitude reefs

Along Australia's western and eastern coastlines, the Leeuwin and East Australian currents transport warm tropical waters southwards towards the pole, enabling corals to grow at high latitudes. The Houtman Abrolhos Islands (28°S) in the west and Lord Howe Island (31.5°S) in the east are Australia's southernmost framework-building coral reefs, with Lord Howe Island the most southerly coral reef in the world (see 'Lord Howe Island' in Chapter 1). Although low ocean temperature, light levels and carbonate availability limit coral reef development further south, non-reef-forming coral communities flourish in pockets of suitable habitat along both coastlines – all the way to Bremer Bay (34.3°S) on the west coast and Sydney (33.8°S) in the east, with few cold-tolerant species also occurring on Australia's south coast (Fig. 4.14). In these regions, corals grow directly on the rocky seafloor, side by side with cold-water (temperate) organisms such as kelp seaweeds. This unique mix of species, living in highly seasonal conditions, makes high-latitude reefs incredibly dynamic and interesting. It also makes them vulnerable to climate change, which is already driving the expansion of corals on high-latitude reefs [34], where tropical fishes are increasingly surviving warm winters and cold-water seaweeds struggle to cope with warming oceans and the influx of

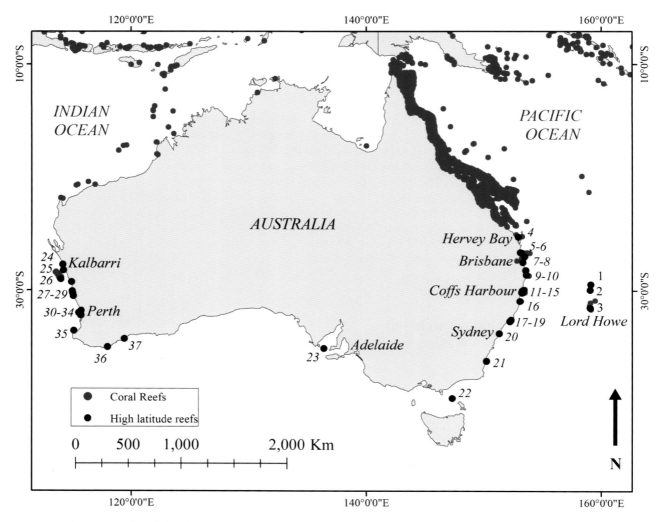

Fig. 4.14. The geographical distribution of Australia's high-latitude reefs (modified from Veron [36]). **East coast:** 1* Middleton; 2* Elizabeth; 3* Lord Howe Island; 4 Hervey Bay; 5 Mudjimba Island; 6 Gneering Shoals; 7 Flinders; 8 Moreton Bay; 9 Byron Bay; 10 Cook Island; 11–15 Solitary Islands Marine Park; 16 South West Rocks; 17–19 Port Stephens Great Lakes Marine Park; 20 Sydney; 21 southern New South Wales. **South coast**: 22 Bass Strait; 23 South Australia. **West coast**: 24 Kalbarri; 25 Port Gregory; 26* Houtman Abrolhos; 27 Port Denison; 28 Green Head; 29 Jurien Bay; 30 Cervantes; 31 Marmion; 32 Rottnest Island; 33 Garden Island; 34 Hall Bank; 35 Geographe Bay; 36 King George Sound; 37 Bremer Bay. * denotes framework building coral reefs. (Image credit: Brigitte Sommer and Sarah Hamylton)

tropical fishes eating them (see 'Coral reefs on the move?' in Chapter 8).

Although coral species richness generally declines with increasing latitude, some subtropical reefs are highly diverse (> 100 species) and have well-developed coral communities. These include the Houtman Abrolhos (28°S) in WA, Flinders Reef (27°S) near Brisbane, Queensland, and the Solitary Islands (30°S) near Coffs Harbour, New South Wales (Fig. 4.14). Other high-latitude reefs tend to have fewer coral species, but can reach high coral cover nonetheless, such as at Hall Bank (32°S) in south-west Australia (Fig. 4.15d).

Mesophotic coral ecosystems

Mesophotic coral ecosystems are typically found at water depths between 30 and 150 m and are considered marginal due to low light levels that limit the ability of resident corals to produce energy from sunlight at depth. In Australia, mesophotic reefs are estimated to occupy the same area as shallow reefs above 30 m water depth, covering large areas along the GBR, the Coral Sea and on reefs and submerged oceanic shoals in north-west Australia. Outside the tropics, diverse mesophotic coral assemblages have also been found around the subtropical Lord Howe Island (31.5°S) and are considered marginal both in terms of their high-latitude position and mesophotic depth. Given their remote locations far from land and the logistics and costs involved in studying these deep-water ecosystems, many of Australia's mesophotic reefs remain poorly documented. Nevertheless, important surveys of mesophotic habitats are currently underway to fill some of these gaps in the Coral and Tasman seas, the Ningaloo canyons and remote north-west Australia. Seafloor communities at upper mesophotic depths tend to be dominated by organisms that rely on sunlight for energy, such as hard corals and octocorals, while non-light-dependent corals and filter-feeding invertebrates are generally more abundant at lower mesophotic depths. To date 196 and 65 coral species have been recorded at mesophotic depths on the GBR and in WA, respectively, with coral diversity decreasing markedly at depths below 60 m [37].

Fig. 4.15. Examples of marginal reefs in Australia. (A) Houtman Abrolhos. (B) Solitary Islands. (C) Flinders Reef. (D) Hall Bank. (E) Solitary Islands. (F) Moreton Bay. (Image credits: Brigitte Sommer (B, C, E and F), Damian Thomson (A, D)).

Turbid reefs

Australia has extensive turbid reefs in nearshore environments along the inner GBR, the Pilbara coastline in WA, and at high-latitude locations in Hervey Bay and Moreton Bay (Figs 4.14, 4.15F). Turbid zone reefs are usually close to land and are considered marginal due to high sedimentation and low light levels. These challenging conditions for corals are further exacerbated by high nutrient loads, altered salinity and temperature during extreme disturbance from rainfall and flood events of nearby rivers (see 'Water quality' Chapter 5). Approximately one-third of the reefs on the GBR are located in shallow lagoon waters less than 20 km from the coast, where they tend to be more frequently exposed to disturbance events than clear-water reefs offshore [38]. Sedimentary and hydrodynamic regimes are considered particularly important to the distribution and survival of turbid zone reefs [38]. Some turbid reefs have many species of coral (> 50) that can withstand high levels of sedimentation and turbidity. Indeed, many corals in turbid settings have developed adaptations that enable them to cope with high sediment loads, such as the movement of tissues and tentacles to actively reject sediment. Many turbid inner-shelf reefs on the GBR are actively growing and have shown resilience to recent anthropogenic stress [35, 38].

The ecology of marginal reefs

Marginal reefs are shaped by environmental stress and, although they occur in a range of different settings, they share several key ecological characteristics. They are inhabited by stress-tolerant species that are well adapted to the environments they live in, and species that are unable to tolerate these extreme conditions tend to be rare or absent [37, 39–40]. For example, many corals on high-latitude, turbid and mesophotic reefs have flattened growth forms that allow them to capture more sunlight in these relatively darker conditions (Fig. 4.15) [37, 39–40]. Corals also tend to be slower growing and longer lived and are often not the dominant organisms on

marginal reefs, with high abundance of seaweeds, soft corals, and filter-feeding invertebrates such as sponges and ascidians. Although generally less diverse than shallow tropical coral reefs, marginal reefs have a high proportion of endemic and undescribed species and often host unusual species combinations, such as warm- and cold-water species growing alongside each other on temperate reefs [34]. Marginal reefs are ecologically distinct from reefs in more traditional settings and their unique ecological character imparts them with both intrinsic ecological and extraordinarily high conservation value. Moreover, as shallow tropical coral reefs are increasingly threatened by climate change, reefs that already exist in environmentally stressful conditions provide valuable insights about the resilience and recovery potential of corals in the face of future change [34, 35].

Threats and conservation

Although organisms on marginal reefs are well adapted to the conditions they live in (e.g. turbidity, light limitation, cold stress), many are already close to their physiological limits and might be more susceptible to other or new types of disturbance in the future. On Australia's high-latitude reefs for example, endemic species (e.g. *Pocillopora aliciae*), subtropical specialists and species close to their northern distributional limits (e.g. *Coscinaraea marshae*) have been more vulnerable to bleaching than those that are normally among the first to bleach in the tropics (e.g. *Acropora*) [35, 41]. On the GBR, some studies found dampened effects of recent heat waves on turbid inshore [42] and deeper [43] reefs, although impacts of heat stress (i.e. coral bleaching) varied geographically and among species. Marginal reefs tend to respond to disturbance differently to 'traditional' coral reefs; a better understanding of how these reefs function will therefore help us to properly assess how they might be impacted by climate change. High rates of endemism, geographical isolation, small and fragmented populations, and sporadic recruitment could make marginal reefs particularly vulnerable to altered disturbance regimes and limit their capacity for recovery [34, 35]. The prospect for marginal reefs to serve as climate change refugia is hence unclear and increasingly called into question. Ecological research on Australia's marginal reefs has to date mainly focused on documenting patterns, and deeper knowledge of the fundamental biological, ecological and physical characteristics of marginal reefs is largely lacking, including ecophysiology, life-history dynamics, species interrelationships and geographical connectivity with other coral ecosystems. This knowledge about the fundamental ecology of these important ecosystems is critical to predict how they might respond to climate change and, ultimately, to determine how to protect and manage them into the future [34, 35, 37, 38].

The basics of coral biology

Gergely Torda and Brett Lewis

Reef-building corals in the order *Scleractinia* (stony corals) are the engineers of tropical coral reefs. They are closely related to jellyfish, sea anemones and hydroids, and – with a few exceptions – form colonies made up of thousands of building blocks called polyps. These polyps are clones of each other and help build a calcium carbonate skeleton through a process called calcification to form colonies that are firmly attached to the foundational surface or substrate. There are over 400 species of reef-building corals on the GBR, with a range of colony shapes and structures that are suited to the requirements of the local environment (Fig. 4.16). It is this diversity of growth forms that creates the structurally diverse reefscape to provide shelter and habitat for a multitude of other reef organisms.

Coral reefs develop over time as the spaces between existing coral colonies are infilled by sediments, or colonised and grown over by encrusting reef organisms. This trapping and binding process grows and strengthens the existing reef framework, building larger and larger reef platforms in a process called accretion. The extent and timing of this growth, both globally and around Australia, has

Fig. 4.16. A morphologically diverse coral assemblage. (Image credit: Matt Curnock)

been dictated by environmental conditions such as sea level, temperature and water chemistry.

Coral–dinoflagellate endosymbiosis

Coral reefs are one of the biologically diverse ecosystems of our planet but exist within nutrient-poor 'blue deserts'. Corals have evolved to thrive in these nutrient-poor tropical and subtropical oceans by entering an intricate partnership with symbiotic algae and microbes that have allowed corals to form some of the largest biological structures in geological history.

The high energy needs of calcification are met by a symbiotic partnership with unicellular algae from the family Symbiodiniaceae. In this partnership, the algae live within the gastrodermal cells of their coral host (hence 'endosymbiosis'), benefiting from a stable home and using the coral's waste products as nutrition (Fig. 4.17b). In exchange, algae provide the coral with energy in the form of carbohydrates, amino acids, lipids and other compounds produced through photosynthesis. While this symbiosis is highly efficient, it is also fragile in the face of relatively small physical and chemical changes.

The symbiotic partnership through which the exchange between the coral host and its algae occurs is fine-tuned to the environment in which the coral colony lives. Even small environmental changes in water temperature, salinity or light levels can trigger changes in metabolism (biochemical processes

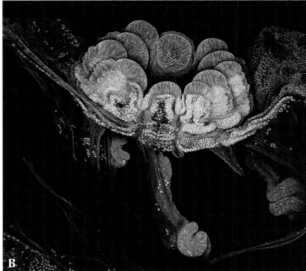

Fig. 4.17. (A) A greyscale image of clonal polyps on a coral colony. The white tissues are filled with symbiotic algae (black dots) that help modern colonies develop larger, more complex and integrated skeletons, (B) A fluorescence microscopy image of a clonal coral polyp and the interlocking surface body wall (coenenchyme). Host symbiotic algae (red) within the cells provide energy in the form of carbohydrates to the coral. The fluorescence observed in this image is produced by the coral's own fluorescent proteins and can help researchers identify individual cells and their function. (Image credits: Brett Lewis)

to break down food into energy and building blocks for growth or repair). In the case of coral bleaching, for example, the over-production of harmful by-products leads to a stress reaction in which algae are either digested or expelled from the coral tissue (Fig. 4.18). As the density of algae inside the coral tissue decreases, the coral's white calcium carbonate skeleton becomes visible through the translucent tissue, which is why this phenomenon is termed 'coral bleaching'. Bleached corals starve and become susceptible to pathogens, eventually leading to death of the coral host. However, if conditions improve, algae can repopulate the coral tissue and the colony can recover.

The coral holobiont: an evolutionary partnership

The coral holobiont is the collective term for the assemblage of the coral host and the many other microbial species living in it, which together form a discrete, functional biological unit (Fig. 4.19). This comprises not only the coral host and its algal symbionts, but also a range of other organisms, including protists, bacteria, archaea, fungi and viruses. While the function of most of these microbes is not yet known, there is increasing evidence that many of the bacterial species play critical roles in nutrient cycling and immunity [44]. Some bacteria directly interact with the coral animals, while others are associated with the symbiotic algae, leading to a complex web of metabolic interactions and interdependencies [45].

Recent technological advances, particularly the 'omics' research fields that focus on the totality of biological organisms, have helped us to better understand how some of these organisms function together in terms of coral biology. For example, transcriptomic technologies help us to understand how gene expression is regulated, thereby providing an account of active and dormant cellular processes. Metabolomic studies, which focus on the chemical fingerprints that cellular processes leave behind, have begun to unearth how coral metabolism changes to survive across environments, the stability of different symbiosis associations and the roles of different

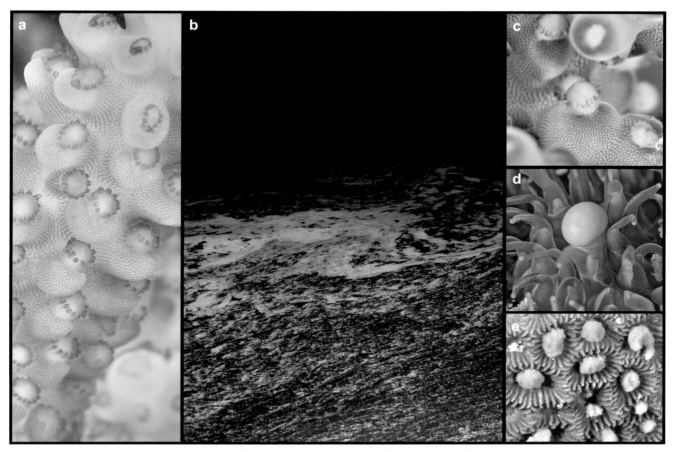

Fig. 4.22. Sexual reproduction on the reef. (a, c) Egg–sperm bundles being released by an *Acropora*. (b) Spawn slick at the surface, Saxon Reef. (d) Egg–sperm bundle being released by *Galaxea*. (e) *Favites*. (Image credits: Phil Woodhead (a, c, e), Peter Harrison (b))

Split spawning enables a realignment of spawning dates with favourable conditions for reproduction and, in this regard, is analogous with a 'leap year' in coral reproduction [56].

Once the eggs are fertilised at the surface, the resulting coral embryos undergo a pelagic larval dispersal phase that helps to ensure the flow of individuals between sites and expansion of their geographical ranges. Many factors influence the survival of planktonic larvae and their dispersal potential, including predation rates, levels of physiological stress and energy stores. Broadcast spawning larvae rely on lipid stores, which impose an energetic constraint on the length of time before they need to settle and metamorphose. Brooding corals, on the other hand, have larvae with maternally inherited zooxanthellae (Fig. 4.24). Brooded larvae are generally larger than spawned larvae and can immediately settle and metamorphose, but the additional energy sources provided by the zooxanthellae means they can also disperse long distances.

Population connectivity, defined as the exchange of individuals among marine populations, is important for the persistence of isolated populations, the re-establishment of sites following disturbances and the flow of genetic information. For most marine organisms including corals, the ocean environment provides the potential for widespread dispersal via oceanic currents, tides and the wind. However, depending on the interplay between the biology of an organism and its physical environment, the *potential* for dispersal may be radically different from the *realised* dispersal. It is the *realised*

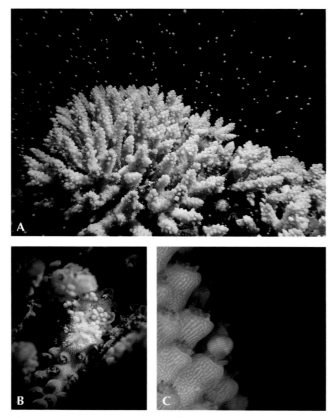

Fig. 4.23. *Acropora* spp. corals spawning on the Great Barrier Reef. (Image credits: Peter Harrison (A), Augustine J. Crosbie (B), Christopher Brunner (C))

connectivity between populations (i.e. the actual numbers of individuals that move between distant populations and survive to reproduce) that determines the distribution and abundance of marine organisms and this is especially important in the face of disturbances.

Over evolutionary timescales, larval dispersal across ocean currents has influenced the global distribution of coral species, and barriers to dispersal have influenced both speciation and extinction. On ecological timescales, connectivity is a vital demographic process underpinning the persistence and genetic diversity of populations. Two separate approaches have been used to infer larval connectivity among coral populations. The first uses Lagrangian particle tracking driven by oceanographic models to predict potential levels of dispersal among populations based on contemporary oceanographic data. The second approach uses population genetics to infer aspects of connectivity from observed genetic data.

When examining the realised connectivity on reefs, ecological, genetic and oceanographic studies alike have indicated that self-recruitment (recruitment into a population from itself) at the scale of

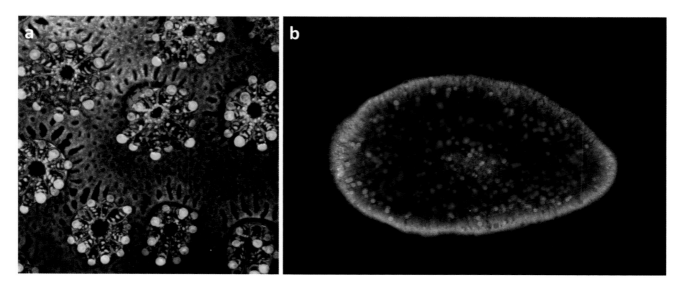

Fig. 4.24. Live corals and larvae visualised under a laser scanning confocal microscope. (a) Juvenile *Pocillopora acuta* polyps. (b) *Pocillopora acuta* larvae showing chlorophyll fluorescence of the algal symbionts (red cells) and fluorescence of the coral tissue (cyan). (Image credits: Ariana Huffmyer (a), Eva Majerová (b))

Fig. 4.25. Coral spawn at the surface of the water. (Image credit: Christina Langley)

individual reefs dominates the population replenishment of corals. Over the last two decades, routine dispersal distances amongst corals have been estimated at 10–100 km and as little as a few hundred metres. More recently, with the use of genome-wide data, dispersal distances of less than 35 km have been recorded among both spawning and brooding corals [57]. Nevertheless, occasional evidence continues to emerge that some spawning species can undergo extensive gene flow, presumably from long distance dispersal (e.g. *Acropora digitifera*). It is likely that the occasional long-distance dispersal helps to sustain the viability of coral populations in remote locations such as Lord Howe Island, Elizabeth Reef and Middleton Reef.

Overall, reproduction and connectivity are among the critical processes upon which the persistence of coral reefs depends. Dispersal determines the rate and spatial patterns of population spread, and this is a key factor in determining how a species might cope with climate change. Patterns of larval connectivity also define the size of metapopulations and guide the appropriate scale at which populations should be managed. The phenomenon of mass coral spawning demands that reefs are considered as interconnected units in their management. This new insight into the connectivity of reefs was instrumental to conservation planners when the GBR was rezoned in 2004.

Coral populations and communities are dynamic when linked linked through a complex network of local interconnections. Their survival depends fundamentally, however, upon the existence of healthy source populations. As new modelling develops our understanding of how source and sink reefs are interconnected, these types of functional links are increasingly being accounted for in reef management. The problem currently facing corals is that the increasing frequency of major disturbance events (specifically thermal stress from climate change driving mass coral bleaching and mortality) is jeopardising the diversity, health, reproductive viability and recruitment success of coral communities [58]. Hence, even large, well-connected reefs like the GBR face a growing risk that species, and at the very least populations, are quietly slipping away. This problem is further exacerbated at geographically isolated locations such as high-latitude or isolated oceanic reefs where there is growing concern that the density of coral colonies belonging to a given species may become so low that fertilisation success is compromised as too few larvae are produced to facilitate local recovery and renewal.

Successful reproduction and the survival of new recruits are key to the ability for coral communities to maintain and renew themselves. Synchronised spawning is a strategy that has served corals well for millions of years; however, unpredictable changes to environmental conditions threaten reproductive success through the breakdown of spawning synchrony [59]. Furthermore, the loss of reproductive fitness, leading to impaired stock-recruitment relationships and population fragmentation, is emerging as an additional threat to the resilience of Australian coral reefs.

Coral reef ecology
Andrew S. Hoey, Eva McClure and Carrie Sims

Coral reefs are one of the world's most taxonomically diverse marine ecosystems, including thousands of species of hard and soft corals, fishes, mobile and sessile invertebrates, algae, fungi and microbes (Fig. 4.26). This diversity is surprising given the fact that tropical and sub-tropical oceans are typically nutrient-poor (oligotrophic), with low numbers of species and rates of primary production. Coral reefs by comparison contain some 550 000–1 330 000 species [60], equating to more than 25 per cent of all marine life despite covering less than 1 per cent of the ocean floor. However, only ~100 000 coral reef species have been described, leaving most species yet to be discovered. The seemingly endless interactions among these coral reef species and with their environment, and the scale and exceptional diversity of coral reefs, have fascinated scientists for centuries (including Charles Darwin). This has culminated in a rich history of coral reef ecology: the study of the interaction of organisms with one another and their physical surroundings.

A brief history of coral reef ecology

Initially, it was widely held that coral reefs were formed by stony plants, or 'lithophytes', as outlined by the Italian naturalist Luigi Fernando Marsigli di Bologna, who received a Fellowship of the Royal Society in 1691 for his treatise, *Du Corail*. This long-held view was contested by the French physicist and naturalist Jean-André Peyssonnel,

Fig. 4.26. Coral ecosystems form a complex network of biological and environmental interactions. (Image credit: Martin Colognoli)

who believed coral reefs were not formed by plants, but by animals he termed *insectes* living in large colonies. Through exhaustive research and experimentation, Peyssonnel (*Traité du Corail*, 1744) successfully demonstrated that corals were in fact animals, and was the first to differentiate between soft coral with eight 'petals' and stony reef-building corals with six 'petals'.

By the end of the 18th century, investigations shifted towards the features and formation of atolls. During a voyage aboard the Russian ship *Rurik* to the Hawaiian, Marshall and Mariana islands of the north-west Pacific, German botanist Adelbert Von Chamisso made the important observations that corals tend to thrive on more turbulent, windward reef fronts, and that the size and distribution of atolls probably reflects that of underwater mountain tops on which they are built. The British naturalist Charles Darwin contributed significantly during the voyage of the HMS *Beagle* when he investigated the formation of the circularly formed coral islands in the Pacific. Darwin developed his theory of coral reef formation [61] where he differentiated among the principal kinds of coral reefs (i.e. atolls, barrier reefs, and fringing reefs) and their origin or formation, leading to the proposal of the subsidence hypothesis of coral reef development. Deduced from observations that reef-building corals flourish only within a limited and shallow depth range, and that atolls and barrier reefs do not rise above that of the surrounding sea, Darwin proposed ([61], p. 4): '… that in both atolls and barrier-reefs, the foundation on which the coral was primarily attached, has subsided; and that during this downward movement, the reefs have grown upwards.' Darwin's work provoked geologists and naturalists to further investigate the geomorphology, distribution and origin, and taxonomic composition of the fauna and flora of coral reefs.

The 1928–29 Great Barrier Reef Expedition to Low Isles jointly organised by the Royal Society of London and Great Barrier Reef Committee, led by marine zoologist Maurice Yonge, resulted in a step change in coral reef science, and marked the

Future directions of coral reef ecology

Advances in technology and its accessibility are transforming coral reef ecology. Emerging genetic tools are being increasingly used to resolve taxonomic uncertainty, revealing the widespread occurrence of species complexes from previously recognised cosmopolitan species and suggesting that coral reef diversity may be even greater than previously thought. The use of technologies such as satellites, environmental sensors, remotely operated underwater vehicles, artificial intelligence, electronic tags, chemical and molecular techniques, and high-resolution time-resolved imaging are enabling animal movements and habitat connections to be incorporated into a more holistic seascape approach to coral reef ecology. In addition, numeric modelling is improving the capacity to understand and predict how environmental and socio-economic change might shape coral reefs into the future.

Sex, baby fish, connectivity and recruitment to reefs

Michael J. Kingsford

As the days lengthen from spring to summer, reef fish feel a growing urge to spawn. Groupers (family Serranidae), snappers (Lutjanidae) and surgeon fishes (Acanthuridae) aggregate to spawn in large groups, often close to dusk and when the tide is right. Females loop upward and, chased by eager males, they release their eggs and the males respond with clouds of sperm (Fig. 4.30). Both sexes then quickly return to the bottom to avoid getting eaten by predators. Fertilised eggs drift away from the reef, and the danger of hungry mouths, and to the open ocean where they grow into larvae. Other small fish, such as damselfishes and gobies, deposit eggs that stick to the sea floor, to which these so-called 'demersal eggs' stay attached until each embryo is ready to hatch as a tiny larva. Males generally guard a nest of eggs spawned by females, but in the case of cardinal fish the eggs are brooded by males in their mouths. For demersal eggs and those of mouth brooders, the eggs hatch after a few days and generally at night. Given that the majority of reef fishes never leave home, the only potential for dispersal and connectivity among reefs is during the larval stage. Many tropical reefs in Australia are made up of a mosaic of individual reefs that provide homes for fishes. Accordingly, reef managers are keen to know how fish larvae and adults disperse, spreading their genes, among local reef fish populations. This is perhaps best understood by examining some of the different early life stages of reef fish.

Only a few decades ago scientists had no idea what happened to larvae when they left a reef, so much so that this part of a fish's life was referred to as the 'black box'. Partly based on a knowledge of northern hemisphere larvae such as herring, we thought that larvae drift aimlessly in the ocean and, through the vagaries of currents and with some luck, may find a reef where they would settle to commence life as a juvenile reef fish. This was not unreasonable because when larvae from pelagic or demersal eggs hatch, they are generally 2–4 mm long. Able to survive on nutritious, buoyant yolk provided by the mother, they often float upside down in a somewhat uncoordinated fashion, with limited swimming abilities. Because they are so small, finding food is an issue and swimming in water is like swimming in a highly viscous fluid. This situation for larvae is analogous to a ping-pong ball making its way through treacle! Needless to say, swimming in mid-water surrounded by predators comes with great risk and early on it was recognised that only a tiny fraction of larvae, probably less than 0.1 per cent, survive the larval phase to recruit onto a reef. Despite the pioneering efforts of Jeff Leis from the Australian Museum to identify fish larvae of the GBR [71], our knowledge of the ecology of fishes during early life was scant up until the late 1970s.

Self-contained underwater breathing apparatus (SCUBA) did not become widely available to scientists until the 1960s, bringing greater access to reefs and the capacity to document the recruitment of baby fish. It was clear that recruitment of different reef-dependent species was patchy among days and

Fig. 4.30. Twinspot snapper (*Lutjanus bohar*) in a spawning frenzy releasing clouds of eggs and sperm. (Image credit: Tony Wu)

between sites, but scientists' understanding of what determined successful survival was largely speculation. In the 1980s, a marine biologist from Cornell University, Ed Brothers, looked at the ear bones (otoliths) of fish and found tiny growth rings. Through validation in the laboratory, scientists realised that these rings were formed daily – new recruits therefore had a kind of birth certificate indicating when they had hatched. The spacing of the rings was even useful as an indication of how well they had grown. It turned out that anemone fish only spent 9–10 days in the plankton as larvae, while others such as surgeon fishes spent as long as 3 months.

Coral reef managers thought this was all very interesting, but they wanted to know where the larvae came from; in other words, how 'connected' are reefs? Given the strength of currents on the GBR and their high transportability, larvae could have travelled tens, even hundreds, of kilometres from their source reef to finally 'settle' on a so-called 'sink' reef. The pioneering work on the GBR and at other locations that James Cook University's Geoff Jones undertook in the 1990s helped greatly with this problem [72]. Geoff's team used 'intergeneration tags' that allowed the eggs sourced from females on reefs to be tagged and the tag could subsequently identify fish that had survived to recruitment. These studies showed that a substantial percentage of fish (often 30 per cent or more) recruited back to the reef from which the adults had spawned, while other larvae 'spilled over' to other reefs [71]. This was just the sort of information that the managers needed to know. It has now been demonstrated experimentally that protected areas subsidise fished areas with recruits and therefore help to make fishing sustainable. Finding a home close to where they spawned makes sense from an evolutionary point of view as, clearly, their parents survived to reproduce at the same place. While settling at another reef may be obligatory if a fish cannot find a home nearby, it is also beneficial for the geographical expansion of a species and to reduce the risk of local extinctions at reefs, perhaps through major disturbance events such as cyclones and coral bleaching.

A reasonable question would be: if larvae are so helpless, how do such a high proportion of successful recruits get home? It turns out that larvae can be excellent swimmers, especially late in their planktonic phase, and this could only be established by collecting larvae in their later stages. Different types of equipment were used to collect larvae, including light traps developed by Peter Doherty of the Australian Institute of Marine Science and nets that were deployed at night on the reef crest. Plankton mesh purse seines were also used to minimise avoidance of nets by fishes. These innovative approaches caught large pelagic juveniles that looked similar to new settlers rather than helpless larvae (Fig. 4.31). Swimming trials with these pelagic stages in the 1990s and early 2000s by

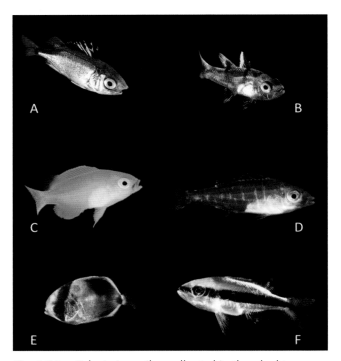

Fig. 4.31.   Pelagic juveniles collected in the plankton before they settle as recruits on a reef; these individuals are 8–15 mm long. At this stage they can swim well and have excellent sensory abilities. (A) Squirrelfish (*Holocentridae*). (B) Threespot cardinalfish (*Apogonidae*). (C) Lemon damselfish (*Pomacentridae*). (D) Emperor (*Lethrinidae*). (E) Butterflyfish (*Chaetodontidae*). (F) Snapper (*Lutjanidae*). Prior to the pelagic juvenile stage many larvae have specialised structures such as long spines that not found in juveniles. (Image credits: Kynan Hartog-Burnett)

James Cook University students IIona Stobutski and Rebecca Fisher demonstrated the larvae could swim for kilometres, even tens of kilometres, without food, often cruising at a speed of 1 km h^{-1} or more. But the question remained: how did they know where to go?

Sophisticated orientation in a fish that only measured 1–4 cm in length seemed unlikely, but scientists were shocked to find this was the case. Early experiments by the University of Boston's Jelle Atema demonstrated that fish could not only recognise water chemistry from different reefs, but that they also generally preferred the water chemistry of their home or 'natal reef'. Sound could also play a role as reef noises can be detected up to 4 km from reefs. Remarkably, these tiny fish are also capable of using sun compass orientation. Henrik Mouritsen from the University of Oldenburg in 2013 showed that fish orientated best when the sun is at low angles early and late in the day. Moreover, if you changed their biological clock through sustained periods in darkness, their ability to use the sun was lost. But the sensory marvels did not stop there. Michael Bottesch and his team from the University of Olderburg demonstrated that pelagic juveniles also have magnetics senses that can assist with orientation. There are even suggestions that celestial navigation is possible. So how would these abilities have evolved? Research on the southern GBR suggested that fish swim against predictable currents that transport them away from the natal reef [73]. In other cases, fish may be attracted to coastal vegetation and alternative cues that indicate suitable habitats for settlement. Some species of fish are less reef-dependent and treat the whole environment between the mainland and the edge of the continental shelf, including reefs, spanning tens of kilometres as home. In the case of red emperor fish (*Lutjanus sebae*), young larvae that spawned offshore are attracted to coastal habitats such as seagrass beds in their early stages of development, moving seaward to reefs and inter-reefal areas as they grow and to outer shelf reefs when mature (Fig. 4.32).

For those species that depend on reefs, does finding the natal reef have to be that hard? Finding home on some parts of the GBR is easier than in other places. Eric Wolanski from the Australian Institute of Marine Science pointed out that the nature of flows on the GBR is greatly influenced by the mosaic of reefs in an area [74]. For example, on the GBR and in the Pompeys and Swains reefs water is channelled through reef passes and a complex array of eddies are formed. The term 'sticky water' indicates that particles transported in water, such as larvae, may not be swept far from home and the chances of recruiting at or close to the natal reef are much greater than for reefs exposed to strong currents [75]. Wolanski used biophysical models to further suggest, from a knowledge of larval behaviour and oceanography, that larval returns to natal reefs in strong currents could be 20 per cent or more.

Finding a home when you get to the natal or another reef is the next part of the story. Distributional studies from the 1990s used multiple techniques to better understand larval movements towards the end of the planktonic stage. These included examining the chemistry of ear bones and the behaviour of larvae in floating chambers. Findings suggested that larvae remain in the vicinity of a reef for some time before settling, slowing down their food intake while choosing the right place and time to settle. My own larval studies have shown that fish primarily settle at night. This is when the many plankton-feeding fishes and other small predators who feed on larvae, known collectively as the hungry 'wall of mouths', are asleep.

Remarkably, fish that are about to settle can find suitable habitat and the presence of their own species in the dark! Hugh Sweatman of the Australian Institute of Marine Science showed experimentally that humbug damselfish preferentially selected certain types of coral and their own kind at night. He tricked some fish into settling by pumping water from patches of favoured coral with humbugs to coral that did not have these characteristics. Clearly, a night-time environment of smell and sounds can be navigated by settling fishes.

Fig. 4.32. Crossing the 'blue highway' – the image shows habitat use by red emperor. The fish are not restricted to one reef. Adults spawn at outer reefs, and larvae are transported and swim to inshore habitats such as estuaries and seagrass beds. Juveniles move offshore to reefs and inter-reefal habitats before they mature. (Photo supplied with kind permission by the Australian Coral Reef Society; image credit: © Russell Kelley, BYOGUIDES)

But the 'hard work' is not over once a fish has settled. The narrow width of daily rings in the otoliths of new recruits tells us that they often feed poorly, possibly because they are hiding from predators and competitors. The high density of fishes on reefs indicated that space may be saturated, such that competition would dictate the group of species in an area, as was the prevailing ecological theory of the 1950s to 1970s. At this time, Peter Sale from the University of Sydney revealed that the turnover of species was high. This suggested that fish present on a reef may perhaps reflect a 'lottery' of who arrives first and 'holds' the space. Studies by students of Sale, Peter Doherty and David Williams demonstrated that space often

wasn't limited and that the 'supply' of recruits often dictated the nature of the assemblage [76]. But studies by Jones and others on the GBR have shown that it is not always that simple. Competition among species and the density of fish can influence the outcomes of which recruits survive the juvenile stage to spawn as adults.

We now have a better idea of which reefs act as producers (i.e. source reefs) and which act as receivers (i.e. sink reefs) of fish larvae. Armed with this knowledge, scientists can assist managers who need to explain to fishers and other stakeholders the likelihood of spillover from protected areas to fished zones [77]. Researchers working on Australian coral reefs can identify individual reefs, or clusters of reefs, that may be a source of new species as the emigration of larvae from these areas may be small. Further, habitat changes on reefs will affect larval connectivity and supply. For example, tropical cyclones can wipe out habitats and even populations of adults, influencing in turn the likelihood of natal recruitment and the level of spillover to other reefs.

Science on Australian reefs has contributed greatly to the ecological detective work required to probe the early life of fishes to recruitment. Some of the questions are old, such as where the larva come from, but technological advances including intergenerational tagging, modern genetics, otolith studies, biophysical modelling and habitat mapping are providing important tools for developing an even better understanding of the recruitment and movements of baby reef fish.

The sharks, rays, whales and dugongs of Australia's coral reefs

Helene Marsh and Colin Simpfendorfer

The terms 'marine wildlife' and 'marine megafauna' typically include sharks and rays, marine mammals, marine reptiles (sea snakes and turtles) and seabirds. This section illustrates how sharks, rays, whales and dugong intersect with Australia's coral reefs.

Sharks and rays

Sharks and rays are ubiquitous inhabitants of coral reefs. The species that are commonly encountered on Australian coral reefs are the same as those seen on most reefs in the Indian and Pacific oceans and are highly conserved within ocean basins [78]. The most well-known of the species are the grey reef (*Carcharhinus amblyrhynchos*), blacktip reef (*Carcharhinus melanopterus*) and whitetip reef (*Triaenodon obesus*) sharks. Divers frequently observe these sharks, which regularly take baits from fishers.

There are a wide variety of other species that occur on coral reefs (Fig. 4.33). These include cryptic shark species such as the epaulette shark

(*Hemiscyllium ocellatum*) and several species of wobbegong (*Orectolobus* spp.) that hide among the corals, as well as larger, transient species such as tiger (*Galeocerdo cuvier*), bull (*Carcharhinus leucas*) and great hammerhead (*Sphyrna mokarran*) sharks that regularly visit reefs on their journeys around the ocean. The largest shark in the ocean, the whale shark (*Rhincodon typus*), is also a regular sight on Australian coral reefs, especially Ningaloo where their aggregation in autumn supports a significant ecotourism industry.

Rays are also common on coral reefs, especially in lagoon habitats with their soft sediments that are home to important food resources. Rays that occur on Australian coral reefs include small colourful species such as the blue-spotted lagoon rays (*Taeniura lessoni*) and blue-spotted mask rays (*Neotrygon* spp.), as well as the larger cowtail ray (*Pastinachus ater*) and pink whipray (*Pateobatis fai*) that are often seen swimming in large groups. Reefs are also used by pelagic rays, including the majestic reef manta ray (*Mobula alfredi*) that often visits cleaning stations, and white spotted eagle rays (*Aetobatus ocellatus*) that are seen leaping from the water. The species mentioned above are but a representative sample of the diversity of sharks and rays seen on Australian coral reefs. In the GBR World Heritage site alone more than 130 species of sharks and rays have been reported, representing ~10 per cent of the global diversity of this taxa.

Sharks and rays fulfil important ecological roles on coral reefs. The larger resident shark species are often mistaken for apex predators on reefs, but in reality fill roles as mid-level predators. The apex predator role is taken by the larger, transient species (e.g. tiger and bull sharks). In addition to their direct predatory functions, sharks create 'landscapes of fear' on coral reefs, with some of the consequences being the haloes of algae-free benthos around lagoon bommies that can be observed from space [79]. These haloes form because herbivores will only venture a short distance from the cover of their coral bommie homes to graze on seafloor algal turf when predator levels are high on a reef.

Fig. 4.33. A wide variety of sharks and rays inhabit Australia's coral reefs. (A) Blacktip reef shark, *Carcharhinus melanopterus*. (B) Bull shark, *Carcharhinus leucas*. (C) Epaulette shark, *Hemiscyllium ocellatum*. (D) Reef manta ray, *Mobula alfredi*. (E) Cowtail ray, *Pastinachus ater*. (Image credits: Colin Simpfendorfer)

Rays fill different roles on reefs, mostly feeding on invertebrates buried in the sand, and so are also important in oxygenating substrates. The large, transient apex predators also connect coral reefs to other distant habitats. For example, bull sharks have been documented to move between the central GBR and New South Wales, thus connecting tropical and temperate regions. Similarly, tiger sharks connect reefs to open ocean habitats and even reefs in other nations as they wander over thousands of kilometres of ocean.

Globally sharks and rays are facing a growing crisis, with more than 32 per cent of species now having an elevated risk of extinction, including those that live on coral reefs. Increasing human pressure on coral reefs, especially as the result of fishing, but also climate change and habitat degradation, have led to these declines [78]. Australian reef shark and ray populations are much healthier than in most other nations [80], a result of comparatively low fishing pressure across most of the northern coast where coral reefs are common, well-enforced science-based fisheries management, protection for a small number of threatened species and extensive marine protected areas encompassing the two major reef systems (the GBR and Ningaloo). In fact, a recent global survey identified the northern GBR as the location globally with the highest diversity of coral reef-dwelling sharks and rays, and as having some of the healthiest populations of reef sharks [81]. This is not to say that fishing has not affected reef shark populations in some Australian locations, but that management interventions such as introducing no-take marine reserves or fishing regulations have led to population recoveries.

Whales

Many species of marine mammals spend all or part of their lives in coral reef environments. Most of these species are members of the order Cetacea (whales and dolphins). More than 30 species of whales and dolphins spend at least part of their lives in the GBR World Heritage site [82], with similar numbers likely visiting the other major coral reef regions of northern Australia.

In the northern GBR, the Ribbon Reefs #3–10, north-east of Cooktown support an aggregation of hundreds of dwarf minke whales (*Balaenoptera acutorostrata* ssp.; Fig. 4.34) between late May and late August each year [83]. This is the only known aggregation of this whale, which migrates each year between Antarctic waters to the south and tropical waters. Alistair Birtles and his co-workers have been studying this aggregation for many years and have documented individual adults returning to this area over several years. Some females are accompanied by young calves, suggesting that calving may occur in the deeper waters adjacent to the outer barrier reefs. These whales initiate interactions with boats and divers, behaviour that has led to the development of an active

Fig. 4.34. Part of the aggregation of dwarf minke whales that occurs close to Ribbon Reefs #3–10, north-east of Cooktown in the northern Great Barrier Reef between late May and late August each year. (Image credit: Matt Curnock)

'swim with' whale-watching program [83]. The importance of this region to marine mammals has been recognised by its declaration as the 'Great Barrier Reef Ribbon Reefs and Outer Shelf Important Marine Mammal Area' [83]. The area is also important for the Omura's whale (*Balaenoptera omurai*), humpback whales (*Megaptera novaeangliae*) and false killer whales (*Pseudorca crassidens*). Satellite tagging indicates that dwarf minke whales are likely to migrate along the edge of the GBR, but then move inshore off south-east Queensland and closely follow the coastline southwards [81].

Humpback whales are the most reported by visitors to Australia's coral reefs. Visitors marvel at their large size and spectacular acrobatic displays. Humpback whales migrate each year up the west and east coasts of Australia from their Southern Ocean feeding grounds to winter in warm tropical waters. These populations were drastically reduced by whaling, including coastal whaling in Australia in the 1950s and early 1960s, but have recovered strongly [83].

Although humpback whales are sighted in many places in the GBR World Heritage site in winter, two core areas have been identified: (1) part of the inner reef lagoon region (19.5–21.5° S) off Proserpine and Mackay is an important wintering area, particularly the area ~100 km off Mackay (21° S), and (2) the Capricorn and Bunker groups of islands and reefs ~100 km east of Gladstone, which is the northern end of their east coast migration route [84]. The two most important environmental predictors of the distribution of humpback whales in the GBR are (1) sea surface temperatures between 21 and 23°C, and (2) water depths between 30 and 58 m. These values are within the ranges reported for humpback whales wintering grounds in other parts of the world. Thus, the presence of coral reefs *per se* is not a predictor of humpback whale wintering habitat, but rather the whales seek a relatively large expanse of warm, shallow, protected water, as is offered by the central GBR lagoon.

Dugong

Australian waters support most of the world's dugong, or sea cows (members of the order

Sirenia). The global importance of coral reefs to dugong, which are listed by IUCN as Vulnerable at a global scale, has been recognised in the World Heritage Listing of the GBR and Ningaloo Coast World Heritage sites, and more recently by their inclusion in Australia's Important Marine Mammal Areas [83]. Dugong intersect with coral reefs that are associated with seagrass meadows, eating most of the seagrass species that occur in their range [85]. As Indigenous people have recognised for thousands of years, large senile planar reefs, such as those that occur in Princess Charlotte Bay and in Torres Strait (especially Orman Reef and the Warrior Reefs), are significant dugong habitats.

Dugongs also frequent seagrass beds associated with fringing reefs (Fig. 4.35). A skilled Aboriginal or Torres Strait Islander hunter can tell from feeding trails in the seagrass where a dugong will return to feed over the next few nights. Armed with this knowledge, hunters used to build a platform over such a site from which to spear the dugong guided by its phosphorescent glow in the water [85]. Inter-reefal seagrass beds are also globally significant dugong habitats, especially the area of shallow reefs of the GBR between 14 and 15 °S and the outer barrier reef. Dugongs are occasionally sighted in outer barrier reef locations that support seagrass, including Raine Island (see the section 'Australia's reef islands' earlier in this chapter).

Sea snakes: a unique group of marine reptiles

Vinay Udyawer and Harold Heatwole

Sea snakes are found in clear offshore reefs, shallow sandy bays, inter-reefal depths with sandy or muddy bottoms, mudflats and estuaries. They belong to the family Elapidae and are directly related to Australasian terrestrial snakes, such as taipans and tiger snakes. In evolutionary terms, they have invaded the marine environment twice, first ~13 million years ago as amphibious sea kraits (subfamily Laticaudinae, found only as waifs in Australia) and then again ~6–8 million years ago as fully marine 'true' sea snakes (subfamily Hydrophiinae) (Fig. 4.36).

More than 30 species of true sea snakes have been recorded in Australian waters, with at least nine breeding on Australian coral reefs. Populations of high densities and diversity occur in the central and southern GBR, with species like the olive sea snake (*Aipysurus laevis*) and spine-bellied sea snake (*Hydrophis curtus*) commonly found in inshore bays and fringing reefs along the Queensland coast. Divers on the GBR often encounter olive sea snakes at popular dive sites including the *SS Yongala* east of Cape Bowling Green, the offshore Ribbon Reefs east of Lizard Island, and at the Keppel Islands east of Rockhampton.

True sea snakes spend their entire life at sea and have a range of unique biological capacities

Fig. 4.35. (A) Dugong feeding on seagrass. (B) Dugong feeding trails in a seagrass bed associated with a fringing reef. (C) Dugong swimming above shallow reef, Ashmore Reef. (Image credits: Doug Perrine (A), Len McKenzie (B), Scott Whiting (C); (A) and (C) © Great Barrier Reef Marine Park Authority, supplied with kind permission)

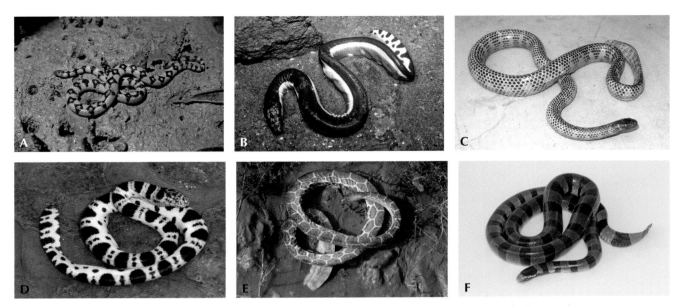

Fig. 4.36.   Diversity of species of sea snakes found in the tropical Australian coastline. (A) Black-ringed Mangrove sea snake, *Hydrelaps darwiniensis*. (B) Pelagic yellow-bellied sea snake, *Hydrophis platurus*. (C) Brown-lined sea snake, *Hydrophis tenuis*. (D) Juvenile olive-headed sea snake, *Hydrophis major*. (E) Geometrical sea snake, *Hydrophis czeblukovi*. (F) Yellow-lipped sea krait, *Laticauda colubrina*, which are sometimes seen as waifs in Australian waters. (Image credits: Ruchira Somaweera)

that enable a fully marine lifestyle, including giving birth to live young. In addition to their paddle-shaped tails, small ventral scales, and valves in their nostrils that keep water out, they have many physiological peculiarities. They can withhold breathing for up to several hours, dive to extreme depths and have been observed foraging for prey at water depths of more than 250 m. Unlike terrestrial snakes, some get rid of excess CO_2 and take up to a fifth of their oxygen needs through their skin.

Some species of sea snake have light-sensitive receptors in the tips of their tails that have been mapped and their physiology studied in the laboratory. These assist the eyes in letting the snake know when both ends of its body are safely hidden among the corals out of sight of predatory fish during the day. By day, one never sees tails hanging out of the coral. By night, tails are frequently exposed but, when illuminated by a diver's light, are immediately pulled out of sight.

Curiously, although sea snakes possess good sight, at least some species do not use vision to identify prey. Rather, they ignore bite-size fish swimming within centimetres of their snout unless they happen to touch them with their tongue, which transfers odours to an odour-sensitive vomeronasal organ in the roof of their mouth. Many species hunt by investigating crevices among corals, or burrows in the sand, for fish, or sometimes other prey (Fig. 4.37). When they detect the odour of prey, they lunge and bite, then drag the captured, envenomated prey onto the sand where they flick their tongue against its body as though measuring its size and 'swallowability'. If a surge of water separates snakes from their prey, they must find it again, not by sight, but by tongue-flicking and moving at random, until they contact it by chance.

Sea snakes do not drink seawater, even when severely dehydrated, and can go for months without water. At sea they drink from the thin freshwater layer briefly upon the surface following heavy rainfall. Unlike the true sea snakes, the related sea kraits are amphibious, and spend half of their time on land. They have somewhat expanded ventral

Fig. 4.37. Reef associated sea snakes can often be observed hunting for prey on reefs by investigating crevices. (A) Stokes' sea snake, *Hydrophis stokesii*. (B) Olive sea snake, *Aipysurus laevis*. (C) Turtle-headed sea snake, *Emydocephalus annulatus*. (Image credits: Blanche Danastasi (A, B), Vinay Udyawer (C))

scales and locomote well on land, even climbing trees. They forage for fish, mainly eels, in the sea but, like sea turtles, come ashore to lay eggs as well as rest, digest meals and mate.

Research on the population connectivity and movements of true sea snakes on the GBR has shown that some species are highly attached to specific reefs. Ultrasonic tracking studies have shown that olive sea snakes do not venture far from their home reef. Males increase movements during the mating season in search of females within the same reef. Seasonal aggregations of sea snakes are often observed during mating seasons in locations with large populations. Despite the increased seasonal activity, the limited ability of sea snakes to disperse between reefs has resulted in highly fragmented populations in the Swains and Pompey reef complex in the southern GBR.

Movements of other species highlight a strong seasonal pattern in their use of specific habitats. For example, shallow coastal bays serve as important nursery habitats for spine-bellied sea snakes (*Hydrophis curtus*) and elegant sea snakes (*Hydrophis elegans*). Gravid females of these species enter shallow bays to give birth with juveniles remaining for at least 2 years in these refuges where prey is abundant. These shallow habitats also isolate them from incidental by-catch of trawlers that exacts a high mortality in some sites.

Most sea snakes are highly venomous, but only two species account for almost all human fatalities globally (beaked sea snake, *Hydrophis schistosa*, and annulated sea snake, *Hydrophis cyanocinctus*). Envenomation by sea snakes is rare in Australia compared to other tropical regions of the world, such as Sri Lanka and the Gulf of Thailand. Unless provoked or disturbed while mating, most sea snakes do not attack. Of the more than 60 species, three eat only fish eggs, which do not need to be subdued and do not counterattack; these snakes have tiny fangs and minute venom glands that produce a small amount of very weak venom. Clearly, sea-snake venom functions to quickly subdue prey, rather than as a defensive mechanism.

The diversity and abundance of sea snakes on Australia's reefs highlight their ecological importance. Information about the long-term trends in their populations is currently severely limited. Surveys from some sites along the GBR, such as Marion Reef east of Airlie Beach, show a drastic reduction in sightings of sea snakes over the past decade, whereas populations in other areas (e.g. Cleveland Bay) have stable populations. As predators, sea snakes play an important role in maintaining a healthy ecosystem and often are used as important bio-indicators for cryptic species of fish. Sea snakes are under increasing threat throughout their global range and Australia's reefs play an important role in their conservation.

For more comprehensive and detailed accounts of sea snakes, readers are referred to three general texts [86, 87, 88].

Marine Science **6**, 12–26. doi:10.3389/fmars.2019. 00012

79. Madin EMP, Harborne AR, Harmer AMT, Luiz OJ, Atwood TB, Sullivan BJ, Madin JS (2019) Marine reserves shape seascapes on scales visible from space. *Proceedings, Biological Sciences* **286**, 20190053. doi:10.1098/rspb.2019.0053

80. Simpfendorfer CA, Chin A, Kyne PM, Rigby CL, Sherman CS, White WT (2019) *A Report Card for Australia's Sharks*. James Cook University, Townsville.

81. MacNeil MA, Chapman DD, Heupel M, Simpfendorfer CA, Heithaus M, Meekan M, Harvey E, Goetze J, Kiszka J, Bond ME, *et al.* (2020) Global status and conservation potential of reef sharks. *Nature* **583**, 801–806. doi:10.1038/s41586-020-2519-y

82. Marsh H, Heatwole H, Lukoschek V (2018) Marine mammals and reptiles. In *The Great Barrier Reef Biology, Environment and Management*. 2nd edn. (Eds P Hutchings, M Kingsford and O Hoegh-Guldberg) pp. 407–418. CSIRO Publishing, Melbourne.

83. Marine Mammal Protected Areas Task Force (2013) 'Important marine mammal areas', <https://www. marinemammalhabitat.org/>.

84. Smith JN, Grantham HS, Gales N, Double MC, Noad MJ, Paton D (2012) Identification of humpback whale breeding and calving habitat in the Great Barrier Reef. *Marine Ecology Progress Series* **447**, 259–272. doi:10.3354/meps09462.

85. Marsh H, O'Shea TJ, Reynolds JE, III (2011) *The Ecology and Conservation of Sirenia: Dugongs and Manatees*. Cambridge University Press, Cambridge, UK.

86. Cogger HG (2014) *Reptiles & Amphibians of Australia*. CSIRO Publishing, Melbourne.

87. Dunson WA (1975) *The Biology of Sea Snakes*. University Park Press, Baltimore MD.

88. Heatwole H (1999) *Sea Snakes*. University of New South Wales Press, Sydney.

5
Managing Australia's coral reefs

The management of Australia's coral reefs brings together long-held Traditional Owner connections with the latest technology for mapping and monitoring corals in real time. Indigenous ranger programs follow on from a history of land dispossession for Australia's Aboriginal and Torres Strait Islander peoples to provide a cultural interface for Traditional Owner groups through which traditional knowledge is applied across reef management initiatives.

Some major threats to Australia's coral reefs have provided a focal point for management activity, including outbreaks of the invasive crown-of-thorns starfish (COTS), poor water quality on coastal reefs and alterations to floodplain wetlands. Attempts to manage COTS rely on a detailed understanding of their life cycles, including specialised traits such as high fecundity, larval cloning, juvenile resilience, resilience to starvation and extraordinary plasticity.

The Queensland coastline that runs along the Great Barrier Reef (GBR) comprises rainforests, saltmarsh and mangrove areas that have been partially cleared, drained and fertilised for sugarcane farming. Agricultural activity has brought sediments, high turbidity, nitrogen, phosphorus, disease susceptibility and algal blooms to reefs; these issues are being addressed through catchment-level land management and wetland restoration.

Zoning the Great Barrier Reef Marine Park (GBRMP) into a network of multi-use zones was a major, internationally acclaimed management exercise that brought together scientific experts and the public to account for biophysical and social aspects of park use. The use of increasingly advanced technology including satellite remote sensing, artificial intelligence and recent three-dimensional photogrammetric survey has enhanced our ability to track the health of reefs, often by observing individual corals, in real time. This has driven a detailed knowledge of large

areas of Australia's coral reefs. In addition, much information on GBR marine communities has been collected via the Australian Institute of Marine Science's Long-Term Monitoring Program (LTMP). Established in 1985, this regular survey of coral communities has enabled critical observations to be made regarding long-term trends in coral cover and COTS outbreaks, and provided a basis for reef status reports such as the Great Barrier Reef Outlook Reports. The growth of citizen science programs has complemented this initiative by building public participation into information collection, enhancing the data available while supporting educational, social and capacity outcomes.

The Great Barrier Reef as a cultural landscape: continuing our connection through the eyes of a Traditional Owner ranger

Gavin Singleton

Country is our identity, it brings us knowledge. [1]

Since time immemorial, Traditional Owners have held connections with the coral reefs, islands, beaches and coastline that make up the GBR. Our knowledge systems and way of life have been shaped by the ongoing interactions and relationships with the changing environment and landscapes. It has sustained us for thousands of years and continues to be a vital part of our lives.

Traditional Owners have stories and beliefs about the 'postglacial rise of sea level and the formation of the Great Barrier Reef' [2]. Long ago in the creation time, we were told, it was all dry land out to the continental shelf, where the original coastline once stood. It was covered in forests, wetlands, mountains, and waterways abundant in food and resources. However, when the rising ocean covered the land, our people were forced to

Fig. 5.1. A boomerang on the beach frames *Wangal Djungay* (Double Island) within Yirrganydji sea country. The island has been used by the Djabugay people (particularly the Yirrgay group) for thousands of years. (Image credit: Bernard Singleton)

relocate to higher ground and adapt to the new environment.

The GBR is rich in natural and cultural values (Fig. 5.1). It encompasses a unique array of habitats, plants and animals. The waters provide important habitat for marine and terrestrial species, including turtles and dugongs, which are of worldwide significance. This region has many significant cultural heritage sites and places of significance including shell middens, fish traps and camp sites to name a few. Our people still hold a wealth of knowledge and recall stories about the changes in the local environment.

The GBR is a cultural landscape. There are creation storylines that are interlinked and overlapping each other across the GBR coast [3]. These stories are part of lore and are embedded in our core philosophy guiding our daily livelihood activities, resource use and cultural practices. They not only provide us with messages about connectivity between people and places, but also serve as a pathway for sustainability, conservation, balance and wellbeing for people and the environment.

Remnants of the original landscape can still be found today in what is now the GBR. Some of our forests have become coral reefs, while mountains

have become islands. Cultural sites have been submerged by the rising of the sea, some 10 000 to 15 000 years ago.

Our people have continued to live on the GBR and still continue our culture and connection today. 'Our cultural values are inseparable with the environment and bind to all land, sea, water, plants, animals, reef, sky and people' [3]. Our identity is closely tied to caring for places on country, continuing our song, dance, arts and crafts, traditional use of marine resources and being actively involved in looking after animals and plants on country. It is important to us that we preserve and maintain our knowledge, innovations, rights and practices of our people and promote an understanding of our heritage and culture to the wider community.

However, in the past two centuries, the way of life for Traditional Owners had dramatically changed. The relationships and custodial roles of Traditional Owners with the GBR had been deeply distressed by early conflicts of settlement, assimilation and protection policies, and other events throughout the history of Aboriginal and Torres Strait Islander peoples in Queensland. Nevertheless, Traditional Owner communities have begun their own journeys of self-determination, thereby forming a pathway to 'cultural resurgence' across the GBR [4].

Traditional Owner land and sea management

Working on country uplifts the spirit and wellbeing, particularly when working on own country. [1]

The Indigenous ranger program has been a popular model across Australia for social, cultural, economic and environmental change. An independent study by Social Ventures Australia for the Department of the Prime Minister and Cabinet found that the ranger program activities have 'enhanced skills and capabilities, improved confidence and wellbeing, career pathways, employment and financial security, and role models for their community'. [5].

The ranger program is an important contact point and resource for the community, creating a 'cultural interface' where Traditional Owners and external parties can interact, engage with and exchange knowledge on projects [6]. It is also important to acknowledge that Traditional Owner communities tailor the ranger program by embedding their *ways of knowing, being and doing* [7] and taking ownership when administering in their respective regions.

In Australia, there are now over 200 Indigenous ranger programs operating and contributing to the local, state and national priorities across all ecosystems; 20 of these programs are specific to the GBR Region. In addition, Traditional Owner communities utilise other management approaches such as land and sea country plans, Indigenous Protected Areas or voluntary agreements to help manage land and sea country. They each share a collective aim to create long-term social, cultural and economic prosperity for their community.

The GBR landscape is very diverse from the Torres Strait to south-east Queensland. There are ~70 Traditional Owner communities who engage with the GBR at different levels of capacity, capability and governance structures. They are faced with different issues and varying accessibility to the reef, and have different community values, goals and aspirations. In recent times, most community programs have been focused within the catchments but are now becoming increasingly involved in activities on the GBR.

We know that the environment, particularly the GBR, is under threat by a range of issues as outlined in the Reef 2050 Long-Term Sustainability Plan and Great Barrier Reef Outlook Reports (see Chapter 1). Given the range of management issues that our environment faces, we need to apply two sets of knowledge to land and sea management – Indigenous ecological knowledge and contemporary, science-based knowledge. We are concerned that our Indigenous knowledge is not recognised or acknowledged. For thousands of years, we have had a unique cultural and spiritual connection with the environment and an intimate knowledge

of the habitats, plants and animals. We still hold a wealth of knowledge and perspective of biota, landscapes, weather and linkages with culture and land uses. We want to ensure that Traditional Owners benefit from research and monitoring activities including through employment, training and capacity building opportunities.

As stated by the Dawul Wuru Aboriginal Corporation, 'there are gaps in scientific knowledge about our marine and coastal environment' [3] across the GBR coast as well as effects of threats and other drivers. Therefore, to prevent long-term irreversible damage, 'a precautionary approach to managing our environment should be taken' [3].

Traditional Owner management on the GBR is imperative if we are to meet priorities and plans at the state and national level. The positive work being achieved at the community level often goes unnoticed and is not given the recognition it deserves, including efforts to address management targets such as the Reef 2050 Long-Term Sustainability Plan, Great Barrier Reef Blueprint for Resilience, climate change actions and sustainable development goals.

The Indigenous ranger program on the Great Barrier Reef

To Traditional Owners, the GBR is part of our identity. Our knowledge, lores, customs and practices are interconnected with the GBR landscape and we continue to uphold our cultural responsibilities to manage and sustainably use marine resources for customary and communal purposes.

We had strict protocols in place to govern everyday activities. We had our own systems of agriculture and farming, aquaculture, health and education, which all had elements of environmental management embedded within them. For example, the fish traps on the coast not only provide food, but also act as shelter and habitat for fisheries, benefiting both people and the environment.

The Yirrganydji (Irukandji) Land and Sea Ranger Program between Cairns and Port Douglas has collaborated with a range of partners to apply traditional knowledge to conservation by delivering on-ground activities on the GBR. Improvements to water quality have been made through the removal of marine debris from urban waterways, beaches and sand cays, and restoration of riparian forest in the Barron River catchment to reduce sediment runoff and restore habitat for flora and fauna. Natural areas have been restored through weed and pest management and cultural burning. Programs have been established to monitor threatened species and ecosystems such as coastal birds, sharks, marine turtles, sea grass, mangroves, oyster beds, and coral reef health and spawning, using technology such as aerial and underwater drones, virtual reality videos and GIS software. Underwater activities have focused on enhancing coral reef resilience, reef restoration, and COTS control (Fig. 5.2). An active communications program has also operated through school engagement, media, technology and events. All of this activity has positioned Yirrganydji rangers to be involved in policy, planning and decision-making for the GBR.

A model for social, cultural, economic and environmental change

It brings a sense of achievement of doing something where my ancestors roamed, and continuing their connection and history, and preserving their cultural sites … Our sites connect us to that time and to the present where we are today. [1]

Fig. 5.2.   Yirrganydjii rangers carry out an underwater survey of reef health. (Image credit: Gus Burrows)

The GBR is renowned for its outstanding universal values. As a World Heritage site, it is of global significance and a popular destination for all across the world. The GBR is a landscape rich in cultural values. It is also home to Traditional Owner communities who continue to live and maintain their cultural responsibilities and connections to the GBR.

The GBR is part of our identity. It has sustained our people for generations and continues to be a vital part of our lives. It is important that we preserve and maintain our knowledge, innovations, rights and practices of our people and to promote an understanding of our heritage and culture to the wider community.

The Indigenous ranger program is a great model for social, cultural, economic and environmental change. It combines traditional knowledge with science and conservation practices and engages external partners to achieve broader outcomes for both people and the environment.

It is critical that Traditional Owners are given leadership roles, with recognised responsibilities for management of country. It is vital that government authorities, research organisations and industry actively support Traditional Owners in creating real, meaningful management roles to meet our customary responsibilities, particularly for our special places like the GBR.

The connectedness of reefs, whales and people: a Yuin teaching
Anthony McKnight

When asked to write something about coral reefs, I went to Gurawill, the Whale, a Dreaming story held by Uncle Max Harrison a senior Yuin Elder who has been teaching this Dreaming to keep a promise to his Grandfather of 'giving it away to keep it'. The reader may ask the question, why a Whale Dreaming, when the book is about reefs? What a great question! Why and how do whales take care of reefs? I am sharing what I have learnt, in a similar manner to how I have been taught.

An Aboriginal story or teaching derives from Country and can only be truly initiated in a person's mind, body and spirit when the stories of spiritual embodiment are observed on Country as a living, breathing story. Uncle Max Harrison calls this reading the 'text of the land' [8, p. 39]. Reefs are sacred. They provide protection, sanctuary, communication, food and ceremony, and contribute to the patterns required by the sea to maintain life. The reefs hold their own teachings and stories. These stories tell of how reefs connect to other animals and places. It is not only humans who manage, or look after, reefs.

A crucial aspect of Aboriginal culture is living in connectedness, identifying and taking care of the many diverse relationships between everything in the natural world, including coral reefs [8, 9, 10]. The contemporary teaching shared here is bound to an ancient traditional story that connects Gurawill, the whale, with people who research and manage coral reefs, to reduce separation that can damage connectedness. With permission from Uncle Max Harrison, the teaching is shared to invite a two-way knowledge relationship and a conciliation between people and Country, reducing the focus on the human entity. Aboriginal people have walked where these reefs lay. We still observe the reefs from the shore and boats, as do non-Aboriginal people. Every year 'we' observe the whales going up Australia's eastern coast to spend winter months in the warmer shallows of the reefs. Observing reefs, whales and people connects with our responsibility to take care of them.

However, we must also be aware of the differences in Aboriginal and non-Aboriginal knowledge systems as the appropriation of Aboriginal knowledge is very well documented and deeply felt as pain by Aboriginal peoples, which damages spirit. One of these differences is in how these observations are directed, shared or taught and how the knowledge is deepened by people and other entities that have been observing and taking care of the reefs for thousands of years. Aboriginal people still utilise our heart and spirit through song and dance to take care of Country:

The notion of Caring challenges Western management assumptions of being able to control Country and alerts resource managers to alternate ways of relating to Country, including nontangible relationships of nourishment and care, for example through the songs, dances and cryings which are critical to the relationships of care between Yolŋu at Bawaka and miyapunu. [10, p. 191]

Songs are still sung by Aboriginal people today about reefs and entities that are connected to the reef. To guide the reader in understanding the teaching and its message: humans are not the only living entity to have responsibility for the reefs. Whales too Sing up and Dance the reefs (Fig. 5.3).

This teaching is demonstrating and putting into practice 'another way' that is just as valid as Western approaches, for the reader to learn about reefs and their interconnectedness to all living things. The reader has an opportunity to engage in an Aboriginal way of thinking, knowing and learning. As previously mentioned, it is more meaningful to sit with this teaching while observing the whale going past a reef.

Think of it as a stride towards reciprocating to Country and Aboriginal people, who take the time to learn through Western knowledge. This same stride is a contribution to conciliation and reconciliation. If you do take this stride, an open mind is required, as well as time for it to unfold in its own time within you.

Enjoy the brain tease, explore what is there, so you can be respectfully challenged and invited on board to conciliate with entities that provide us humans with so much. Please take the time to engage with a knowledge system that may be unfamiliar to you, thereby showing respect to Aboriginal knowledge: Country. Yarn and learn. Be patient and tolerant. I too am still learning from this ditty as the spiritual meaning unfolds in its time, not mine.

The Gurawill Teaching

Anthony McKnight

The whale in which we invite, including the Humpback and the Southern Right, to speak through a story that might bite. To bight a memory of permission as we place out a mat that welcomes a being that is often seen as a visitor who completely belongs because it was chased by a 'cat'. This cool cat still exists but it looks a bit different in its current posit.

In this moment we invite you to think, not of a poem or a thought, or a skit, but a feeling that you truly know more than an instinct. So let's leave the rhythm to the whale as this is its tale/tail, that gives it lift and provides a force of a movement of a course. As rough as it may be this tail/tale of the whale, oh I have not left the cleft of the rhythm as it always exists, but we truly invite you to just simple sit to see as the whale does more than move its tale, it sings, teaches and speaks with its tail. As the whale moves along the coast she/he refreshes a memory that is steeped in a spiritual deepness so when it returns they clearly tell a speech in silence that cannot be impeached. The memory is as long as the path from the South to North and back. This song is sung as they walk the path that has many a reef to go past and glide along. To complete this song, mmm yes I'm back with a beet, as you might think whales have no feet to beet. But to complete their long journey of each breath they hold this responsibility of great depth.

To be charged with such clout these magnificent whales have no doubt as they hold this responsibility of more than tale, oh yes have I repeated on purpose so I am not defeated. They bring up a truth that does more than feed us, they suckle so many with their bubbles for those that live along the whales' roof which is inclusive of the reef. Proof is right there through a responsibility of holding a pattern of movement that saved many who mattered if they held each movement in a way of respect and atonement.

Conceive being a whale before being a whale in which you stayed so still, but you held it so something could tell that you know how to hold something that is not nothing and often not believed. Again I repeat this is not just a ditty to be seen as silly, there is a reason of course, that you may have to seek. As we know the whale companies took so long for pity. Can I now change my style to suit an academic article, oh well I might just be a smart arsiticle: or was that Eristical. Let's go back to the article's brief of examining the reef and as I hope you have seen this cannot be done with the range being separate or emancipated to a separated leaf. The leaf is nearly always connected to a little branch, as we see the oneness of when it falls to the ground as it is always bound.

Do we ask what is the cultural or spiritual significance of a reef, is there one true answer to this, only if seen as a whole, which cannot be told in full in this human space. That is what I mean, have we today ever stopped and actually ask the reef who they are and what they hold. There may well be similarities in what the 'scientist' and an 'old fulla' observed, but did they ask the same question with a certain intent to which the reef deserved. Who and how are you (my Country woman/man) reef on this day? Have I only come to see you speak as a script of a thinking belief, or a text full of essence within a reef? Each reef deserves the same question and respect just as we do to every single person we meet? When observing the reef after asking politely to voice an answer without a sound, well you better stay close to the ground as there is no need to include the pound. Man I am still in a rhythm which is where I am meant to be if I want to see the spirit of this living entity. I just do not think as we must go beyond the mental and physical links. Spirit will communicate if we give it time to settle from within and without. Being open to being bitten means a lot more than being smitten, oh okay I might stop for a bite. Not really as this is not right and will stop me from seeing the answer I have sought, not bought.

Sure, if you are sitting on the ground and as luck may have it you get bitten by an ant, you are most likely feel this physical affect/effect, maybe just the word oh … that really hurt. As you can see this is the emotional effect of a reaction in fear, anger or regret. Science is seeing this too the mental – emotional aspect. Then you may ask on what does this mean on my spiritual connection of me and the ant. From there it is between you and the ant and whenever you have the guts to bare what has been shared between two living entities that exist and entwine on so many levels that we as humans often forget. Thereby who are you reef, in so many layers that will take some time to collect? Who were you reef when you were above sea level and not knowing the answer may well be the beginning of or initiating the gift of giving and receiving an answer which may take time to decipher. This time is important for spirit to move from the soul of the body to the mind of yourself and maybe it might never be clear to speak but is felt with an intensity that cannot be told by voice. The question 'why' is not needed because if we take our time our mind will be superseded and filled with spirit, and shown by vision. Each level it may take but there is no mistake time is required not as with two stakes driven through its self.

Can a reef give you an answer who it is, well yes who else could or should? Other living entities will help contribute to the answer who are connected to the reef, through many connections a whale can share to show bare who we are. Stories of quintessence can help, stories told by the old people who used to, and still do, live and walk along the shelf that holds these things now known as reefs, that provide life for so many.

Fig. 5.3. Humpback whale breaching. (Image credit: M. Simmons)

Management based on a sound understanding of the Great Barrier Reef
Russell Reichelt

Since the late 1970s, the capabilities and scope of Australian coral reef research to support

management has expanded rapidly as the number of researchers increased and research vessels arrived at the newly commissioned Australian Institute of Marine Science (AIMS) and James Cook University (JCU). These were accompanied by automated oceanographic equipment, field survey and experimental capacity, and new technological capabilities in geochemistry, microbiology, physical and biological oceanography, and newly invented computational capability.

During this time, coral reef science transitioned from discovery to explaining how ecosystems function in ecological and physiological terms. Descriptions have also focused on individual species, population biology, biodiversity, rare species, and then threatened and endangered species, accompanied by a growing awareness of the geological origins and dynamics of many reefs, as well as the role of ocean currents in connecting reefs within enormous provinces, such as the GBR.

The early tracking of flood plumes using aerial photography and satellite data initially documented drivers of oceanographic currents and the fate of runoff sediments originating on the land. The concerns about endangered species such as turtles and dugong, and about fragile coastal systems such as wetlands, mangroves and seagrasses, grew with their documentation and was usually driven by one or a few researchers, who were either working in a university but closely linked to management authorities, or directly employed by those organisations charged with a management responsibility for the Reef, namely the Great Barrier Reef Marine Park Authority (GBRMPA), AIMS and various museums. Notable examples include, but are not limited to, studies of marine plants by Tony Larkum;

turtles by Colin Limpus; dugong by Helene Marsh; polychaetes by Pat Hutchings; and corals by Charlie Veron, Carden Wallace and Terry Done.

How to do the science necessary to understand the Reef?

The taxonomy of Reef flora and fauna was accepted by all as a valuable early point of focus. Opinions varied, however, on how to do the science needed to understand the function and dynamics reef ecosystems.

Some experts preferred to test hypotheses about reef functions through controlled field experiments, whereas others preferred to analyse the numerical patterns in large ecological datasets for clues about how the GBR worked. Debates among these experts could be quite energetic during coral reef meetings.

It turned out that both approaches could be very useful. For example, Terry Hughes and colleagues built experimental cages to exclude herbivorous fish at Orpheus Island, as did Tony Larkum and Bruce Hatcher on One Tree Island in 1983 – see 'Coral reef ecology' in Chapter 4). These yielded the first quantitative assessments of herbivory rates, revealing the keystone role those fish play in sustaining the GBR's diverse benthic communities. The field experiment approach was championed across a range of ecosystems by Tony Underwood at the University of Sydney in Australia, and Paul Dayton and others in the United States of America, and derived from the pre-World War II work by botanists and statistician Ronald Fisher. Field experiments were later taken up by coral reef scientists and added to our understanding of Reef ecosystem function and system integrity. In turn, the integrity of the Reef's ecosystem was used as one of the natural values for which the GBR was listed as the first marine World Heritage site in 1981.

The value of the 'pattern analysis' approach became evident after vessel-based research delivered data about the spatial distribution of plants and animals over extensive areas of the Reef. Bill Williams from CSIRO, Roger Bradbury and John Bunt from the AIMS and Mike Dale (CSIRO) were early leaders of pattern analysis among coral reefs and mangroves.

Analysing the patterns of distribution and abundance of species across areas subject to distinct environmental conditions enabled important questions to be answered about how form and function of coral reef communities change in different parts of the reef. Inshore, mid-shelf and outer Reef communities were described in the early 1980s. By the early 2000s, this type of analysis enabled 70 bioregions across the 345 000 km^2 of the entire Reef to be described – a key step in designing the Great Barrier Reef Marine Park Zoning Plan in 2003.

'What is there?' A fundamental management question

During the 1980s, management-generated questions about the Reef were posed by the GBRMPA and the Queensland Government. The first documented decision of the newly formed GBRMPA board was to 'make a map of the Reef'. JCU's David Hopley, with Peter Isdale and others, worked initially with aerial photos and then with early Landsat satellite imagery – each pixel 90 square metres – to create the Reef Gazetteer.

The first field survey of the Reef over large areas was undertaken in 1985. It was managed by Roger Bradbury, who had been tasked by the then Director of AIMS, John Bunt, to do a 1-year survey of as many reefs as possible, with financial support from a one-off grant from a Commonwealth Government employment program. The design included 'manta tow' surveys, employed a few years earlier by the GBRMPA's Richard Kenchington working with the Royal Australian Navy, and line transect methods established by collaboration between Roger and Yossi Loya from Tel Aviv University in Israel. Those 1985 surveys produced the first data in the AIMS LTMP. After some initial methodological refinement, this was formally established in 1992 and is still running today.

Declines in coral abundance and reef biodiversity

With the accumulation of data about the state of the Reef in the 1980s and 1990s, it became clear that the

Reef was, and had been, changing. Terry Done at AIMS published data showing that large corals such as *Porites* (which were often > 500 years old) and fringing reefs along the mainland were declining in abundance.

Hypothesised causes of coral decline included the impact of coral predators such as the COTS (*Acanthaster* sp.) and poor water quality. On a contrarian note, the geomorphologists observed that the fringing reefs along the inshore coast originated more than 5000 years ago, when the sea level was higher than today. It therefore followed that all mainland fringing reefs were at risk of dying from over-exposure by being stranded above the current low tide level.

The water quality and, to a lesser extent, COTS outbreaks became dominant concerns relating to the impacts on coral communities. Flood plumes extended offshore from catchments where the coastal rainforest had been cleared for farming. Drain-like channels formed in creeks that had previously been full of natural obstacles like logs. Bunded pastures along coastal floodplains were preventing the wetlands from trapping sediment and providing habitat for wildlife. Land use was affecting water quality. Layers of fine red sediment and bacterial flocs known as marine snow were appearing on reefs adjacent to intensive agriculture. Flood plumes were reducing coral abundance and possibly also amplifying the reproductive success of *Acanthaster* sp.

During this period, Helene Marsh was reporting a significant decline in dugong, the iconic reef mammal, along the Queensland coast where inshore gill-net fisheries had expanded. Human impacts on the Reef were becoming increasingly apparent.

Building science and management partnerships to promote reef health

COTS increased as a risk in the 1960s, as identified by Bob Endean and Ann Cameron at the University of Queensland (see the box, 'Robert Endean: crown-of-thorns hero' later in this chapter). Bob Endean attracted public attention and raised concerns for

the Reef, but was also soundly attacked and undermined by leaders in the science community at the same time. Such attacks are a common feature in the long history of science on the GBR. Support for Endean's views, however, grew among the scientific community. Endean initially favoured the hypothesis that overfishing of the predators of *Acanthaster* sp. was causing population increases and outbreaks of the starfish. He also supported the American reef ecologist Chuck Birkeland's hypothesis that nutrient runoff in floodwaters could cause the outbreaks to increase in intensity and frequency.

Although the science was increasingly pointing to a human origin of outbreaks of COTS, GBRMPA's position on COTS outbreaks in the 1980s was that it is not an introduced species and that outbreaks were natural phenomena, so there was largely no need to interfere, although localised control around significant sites for tourism was desirable. In the fullness of time, a U-turn in this position would eventuate as the GBRMPA began to lead government intervention to reduce starfish numbers, while simultaneously working on improving water quality.

Awareness of risks to the Reef from destructive fishing grew as scientists and managers, including John Tanzer, Ian McPhail, Helene Marsh, Virginia Chadwick at the GBRMPA and the Queensland Department of Environment and Primary Industries, began to focus on other risks to the reef. Research shed light on the impact that nets and traps have as they entangle dugong, turtles, and rare species of sharks, as well as damage the extensive seafloor communities of plants and animals living on the seabed between the shallow coral reefs. These periods of management action on risks to the Reef were respectively known as 'the trawl wars' and the 'dugong wars', adopting and expanding the poet Judith Wright's analogy of the Reef being a conservation battleground.

Concerns about the impact of trawl nets were addressed when the GBRMPA and the Queensland Department of Primary Industries and Fisheries created areas in the GBRMP and Coastal Marine Park where bottom trawling was prohibited so that the seabed flora and fauna were left undisturbed.

Fishing licences were consolidated, resulting in fewer trawlers (Fig. 5.4). Rules were introduced to mandate the fitting of 'turtle excluder devices' to trawl nets, thereby preventing turtles being caught and drowned. Such practical actions established clear links between the science that revealed risks to the reef and corresponding management responses.

Helene Marsh had monitored dugong abundance for a long time and found them to be in sharp decline. The decline was caused partly by flood damage to seagrass beds, the dugong's food, but mainly to their entanglement and drowning in monofilament fishing nets that are almost invisible underwater. The GBRMPA worked with JCU and the Queensland Government to design dugong protected areas where they were most abundant along the entire Reef. In these areas, net fishing was prohibited. Rules were established by Queensland Fisheries to limit the size and number of nets and require fishers to 'attend' their nets and release entangled turtles and dugong.

Although the issue of water quality on the Reef was raised in the 1960s, it wasn't until the 1980s that science began to reveal the implications of this. Management intervention would not follow until the late 1990s. Large-scale experiments such as the ENCORE (Effect of Nutrient Enrichment on Coral Reefs) demonstrated the physiological to ecological

Fig. 5.4. Trawler boat in the Port Douglas region, Far North Queensland. (Image credit: Dominic Chaplin © Great Barrier Reef Marine Park Authority, supplied with kind permission)

changes that arise from eutrophication on reefs (see Chapter 3).

Scientific publications outlined the extent of coastal rainforest and wetlands loss, and the degree to which floodwaters had become loaded with excessive amounts of eroded sediments, fertilisers, and agrochemicals. The work led to the 2003 Reef Water Quality Protection Plan, driven by Jon Brodie and Sheriden Morris in the GBRMPA, Scott Spencer in the Queensland Government and Conall O'Connell in the Commonwealth Government, with strong support from World Wildlife Fund's Imogen Zethoven.

The long-term goal of the Reef Water Quality Protection Plan was to ensure that the quality of water entering the reef had no detrimental impact on the health and resilience of the GBR by 2020. The plan set out measures to monitor water quality flowing into the Reef from the major river catchments and coastal waters, targeting changes in fine sediment loads, excess nutrients from fertilisers used by farmers and harmful pesticides.

Since 2003, governments and industry have successfully reduced these pollutants by improving farming and grazing land use practices. However, the goal of no detrimental impact is yet to be met and coastal water quality remains 'poor'. The Reef Water Quality Report Card was launched by the Australian and Queensland governments in 2009, to assess the results of all Reef 2050 Water Quality Improvement Plan actions. So far, this has revealed encouraging progress at regional and catchment levels.

Since the 1970s, scientists have played a central role in helping managers to address issues such as COTS, destructive fishing, and poor water quality on the Great Barrier GBR (Fig. 5.5). Through partnerships supported by research facilities and by employing scientists within management authorities and a shared goal of providing effective Reef stewardship, decisions regarding some of the key challenges to the Reef have been based on the best available information and expertise. These partnerships have delivered a sound understanding of the Reef and the processes occurring on it, a critical foundation for reef management.

Fig. 5.5. The Reef Ranger Marine Parks compliance vessel at Raine Island, northern Great Barrier Reef. (Image credit: Andrew Simmonds © Great Barrier Reef Marine Park Authority, supplied with kind permission)

Filling in the pieces of the crown-of-thorns starfish puzzle

Maria Byrne

Seastars in the genus *Acanthaster* sp. have a strikingly large (> 40 cm diameter), multi-armed profile (> 20 arms) and, due to their dense cover of venomous spines, are called the crown-of-thorns starfish (COTS) (Fig. 5.6). These seastars are common throughout the Indo-Pacific and are well known as coral predators and for the injuries their spines inflict on people. It is now known that the species previously called *Acanthaster planci* is a complex of at least four species. The identity of the species widely occurring on the GBR is uncertain and herein is referred to as *Acanthaster* sp. or COTS.

As with ecologically important predatory seastars in temperate waters, *Acanthaster* sp. exhibits boom–bust population cycles [11]. These cycles can drive coral reefs to contrasting ecological conditions. In low densities, COTS can enhance coral diversity because their preference for feeding on fast-growing, dominant coral species provides opportunities for slow growing, less numerous species to become established [12]. During population outbreaks, COTS reach high densities, causing mass mortality in corals. *Acanthaster* sp. are one of the few seastars that can eat coral because their specialised stomach enzymes allow them to digest waxy coral tissue by extending their exceptionally large stomach over the coral.

The crown-of-thorns starfish on the Great Barrier Reef

When the outbreak phenomenon on the GBR first raised the alarm in the 1960s at valuable tourist sites such as Green Island near Cairns, there was great concern from the public, tourism operators, scientists and managers as to what could be done to 'save the reef' from the devastation of COTS. The debate that followed on what causes outbreaks and how they should be managed was intense, fractious and highly political. One of the first scientists to highlight the problem was Robert Endean (see the box on Robert Endean later in this chapter), whose predator removal hypothesis suggested that COTS outbreaks were due to human interference with the balance of nature on the reef, specifically overfishing of their predators. Others suggested that, as COTS is a native species, outbreaks were a natural part of the reef's ecological cycle and there was no reason to panic, as corals would recover.

Predation has been a significant evolutionary selective force on COTS, as indicated by their venomous spines, cryptic nature and nocturnal foraging behaviour. Fishes and invertebrates prey on COTS at various life stages and the presence of

Fig. 5.6. Crown-of-thorns starfish on a patch of coral. (Image credit: Joanna Hurford © Great Barrier Reef Marine Park Authority, supplied with kind permission)

COTS DNA in the stomachs of reef fish indicates that they are an integral part of the food chain [13]. Marine protected areas that are closed to fishing are less prone to outbreaks, lending credibility to the predator-removal hypothesis [10].

As scientists started to monitor the GBR from the 1970s, it became evident that repeated cycles of COTS outbreaks were occurring, but their causes continued to be debated. In 1981, Charles Birkeland analysed records of COTS outbreaks alongside weather records, revealing that outbreaks occur several years after tropical cyclones and heavy rainfall events [14]. Cyclones and heavy rainfall are a natural occurrence in tropical regions and often coincide with the time when COTS larvae are in the plankton. Birkeland suggested that the nutrients washed into the sea by storms stimulated algal blooms that, in turn, provided food for COTS larvae, thereby promoting their success and leading to outbreaks. The proposition that outbreaks are stimulated by terrestrial runoff and enhanced nutrients implies that COTS larvae starve in non-runoff years.

Despite the popularity of the enhanced nutrients hypothesis and the potential to manage COTS by improving agricultural practices and water quality, the link between runoff events and outbreaks on the GBR is tenuous. The timing of flood events and the expected (~3 years) detection of outbreaks appears unconnected [15]. Moreover, outbreaks occur in locations such as the Elizabeth and Middleton reefs in the Coral Sea, which are distant from terrestrial nutrient input [16].

Despite its reputation as a pest, *Acanthaster* sp. is a native species on Indo-Pacific reefs and outbreaks are likely to be a natural occurrence. After decades of research, the causes of COTS outbreaks and their increasing frequency remain unresolved [15]. To address outstanding critical questions for COTS management, there has been a renewed focus on the biology of the starfish. As Birkeland pointed out, the extraordinary success of COTS in comparison with other starfish that do not undergo population fluctuations is likely to be due to a suite of 'Faustian traits' [17]. How can the biological and ecological traits of COTS that underlie its disproportionate success be better understood?

Acanthaster's unique and specialised traits: a recipe for success

Acanthaster's unique traits support an opportunistic boom–bust lifestyle. These include high fecundity, large chemically protected eggs, larval cloning, juvenile resilience, a highly defended adult body, resistance to starvation and extraordinary plasticity across all life stages.

The extremely high fecundity of a single female (more than 100 million eggs spawned in its summer period) is clearly a critical factor underlying COTS outbreaks [15]. *Acanthaster* spp. is unusual among starfish in that male gonads express female genes and vice versa, while some COTS have hermaphroditic tendencies with eggs and sperm in the same gonad [18]. This creates the possibility of reproduction without a mate. COTS eggs are unusually large for seastars and so their offspring start life with an extra boost of maternal provisions. These eggs also contain chemicals that deter predators. Considering the high regenerative capacity of echinoderms and use of asexual propagation, it is not surprising that COTS larvae exhibit cloning and that the juveniles and adults are well able to recover from misadventure such as losing an arm to a predator.

The impressive abilities of COTS larvae to multiply via cloning may help maintain larval population levels and enhance settlement rates (Fig. 5.7A). The larvae clone by either budding off small portions of their body that grow into a new larva, or by splitting the body in half with the two portions forming two larvae [19]. Thus, the larvae have the potential to be eternal. The propensity for larval cloning challenges our ability to model sources and sinks of COTS propagules in connectivity models.

COTS larvae are well adapted to the low-nutrient conditions typical of coral reefs. In times of low food availability, they can grow to the juvenile stage under low phytoplankton levels [20]. They may also have nutrient-providing symbiotic bacteria in their gut [21]. The larvae are also able to adjust

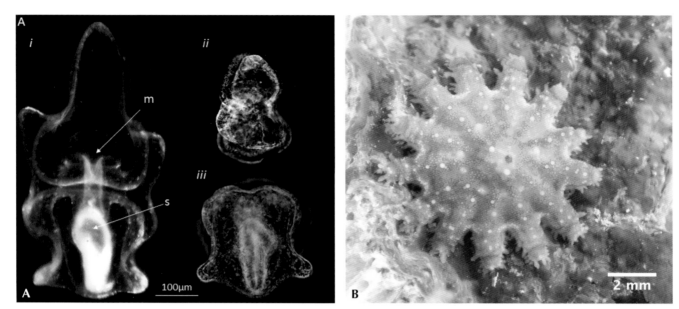

Fig. 5.7. (A) Larval cloning in the crown-of-thorns starfish. *i.* Fully formed bipinnaria larva showing the complete digestive tract with the anterior mouth (m) and posterior stomach (s). Cloning occurs by a split across the body to make two portions that regenerate to make two complete larvae. *ii.* Anterior portion clone. *iii.* Posterior portion clone. (B) Juvenile crown-of-thorns starfish, or *Acanthaster* sp. which can persist as herbivores for years. (Image credits: Matthew Clements (A), Dione Deaker (B))

the length of their feeding device (ciliary band) with respect to phytoplankton levels to enhance feeding efficiency [20]. These traits would enhance COTS larval survival and are undoubtedly important in facilitating population outbreaks. In contrast to the notion of larval starvation in non-runoff years, COTS larvae are very robust to low food conditions. However, Birkeland's 'enhanced nutrients driven by runoff' hypothesis does not contradict the idea that COTS larvae are very resilient to low nutrients. The robust and adaptive nature of the larvae lies at the core of both of them.

Acanthaster sp. larvae settle onto surfaces in the reef where they are well hidden. They start their benthic life as five-armed herbivorous juveniles, utilise a range of algae with a preference for coralline algae and can survive on very little food (Fig. 5.7B). The juveniles can persist for years in the herbivorous phase and so have the potential to build up in number over multiple years before becoming adults [22]. These findings prompted Dione Deaker to propose the 'juveniles-in-waiting' hypothesis that juvenile accumulation over many

years may help to drive outbreaks [22]. That the juveniles can persist in the herbivorous phase for years may explain the disconnect between the timing of heavy rain–nutrient runoff events and outbreaks. This presents a caveat for COTS management. The adult killing program may release the juveniles from competition with the adults, thereby stimulating their transition to the coral-eating stage. The presence of a juvenile 'hidden army' many explain why some reefs require multiple culling visits to clear them of COTS.

Management of the crown-of-thorns starfish problem

For many years, COTS have been managed through mass-killing programs that control populations via lethal injection of substances such as vinegar (Fig. 5.8). While this may be effective in protecting closely managed reefs such as those important for tourism, the ability to protect the 2000+ reefs on the GBR by killing COTS is questionable.

We are beginning to fill in pieces in the COTS puzzle. A better understanding of the traits that

Fig. 5.8. A diver administers a fatal injection to a crown-of-thorns starfish. Mass-killing programs are a population control strategy for managing localised outbreaks on reefs that are of importance for tourism. (Image credit: Matt Curnock)

underpin their remarkable success will better inform management of the COTS problem. It seems that eutrophic plankton blooms driven by nutrient-rich water runoff during extreme weather may not be required to generate outbreaks, but, together with other factors such as larval cloning, juvenile accumulation and overfishing of predators, would contribute to population increase. Deliberations over 60 years, and the hypotheses proposed, have shaped investigations and contributed greatly to our understanding of COTS. Going forward, to effectively managing the causes of COTS outbreaks, it will be important to incorporate the plasticity and resilience of the larval and juvenile stages and better understand the multifaceted Faustian traits of this opportunistic coral predator.

Water quality
Barbara Robson

Since European settlement, human activities have affected Australian coastal water quality in a variety of ways, with obvious and more subtle impacts on reef health. In many coastal catchments, there have been dramatic increases in the delivery of sediments, as well as inorganic nitrogen and phosphorus to coastal waters. In the GBR for example, the mean annual sediment loads delivered by rivers are estimated to have increased by a factor of at least 5, nitrogen loads by a factor of around 6 and

Robert Endean: Crown-of-thorns hero
Sarah Hamylton

Professor Robert Endean (1925–97) was a marine scientist with an interest in echinoderms who worked at University of Queensland's Department of Zoology (Fig. 5.9). In the early 1960s, Endean was one of the first marine scientists to observe large aggregations of *Acanthaster* sp. feeding on the soft bodies of scleractinian corals and leaving stark white skeletons on many of the platform reefs on the GBR. From this point on, Endean played a leading part in publicising this issue [23].

In the decade 1965–75, Endean was almost superhumanly active in surveying *Acanthaster* sp. populations, visiting many of the 308 inshore reefs along the central and southern GBR [20]. Deeply dismayed by what he saw, Endean anticipated catastrophic damage to the GBR, and worked hard to arouse public concern and government action to address the COTS problem.

In a study carried out in 1965, Endean surveyed 92 reefs between Lizard Island and Palm Island, finding that 32 of these reefs had far larger than normal *Acanthaster* sp. populations [24]. From field observations and laboratory work, Endean theorised that *Acanthaster* sp. outbreaks were caused because of the removal of the natural predators of the starfish, notably two species of mollusc, including the giant triton (*Charonia tritonis*), that were subsequently declared protected species [25].

Endean was critical in establishing the COTS scientific advisory committee, which funded the first systematic surveys in 1973–74 of whole reef perimeters and lagoons to understand the within-reef distribution of active starfish aggregations and the extent of coral death.

Fig. 5.9. Robert Endean on the Great Barrier Reef, 1954. (Image supplied with kind permission from the State Library of Queensland, photographer unknown)

Endean thought that infestations would ultimately embrace the whole ecosystem and that recovery would take 20–40 years, with residual starfish populations potentially leaving the reef indefinitely impoverished. Calling for methods of control, Endean formed the 'Save the Reef' Committee, bringing together a hundred divers in 1972. The program of hand-harvesting that he initiated at this time remains the principal action for managing COTS outbreaks today.

Robert Endean's personal endeavours to survey and control the *Acanthaster* sp. infestations did much to draw attention to, and encourage public funds to be expended on, this problem. Endean was also a tireless council member and, for some time, Chairman of the Great Barrier Reef Committee, who also played an active role in establishing Heron Island Research Station, where he died while attending the 75th Australian Coral Reef Society conference in 1997.

phosphorus loads by a factor of nearly 9 [26]. This is primarily because of agricultural development and extensive land clearing. Pesticides from agricultural activities are also regularly detected in nearshore Reef waters at potentially harmful concentrations.

Water quality is also affected by coastal engineering and shipping (Fig. 5.10). In the waters of tropical Western Australia, river loads have changed to a lesser extent than in the GBR, due to lower river flow and runoff, and also due to less intensive agricultural development of the catchments of Western Australia. Dredging to support oil and gas development and shipping in Western Australia, however, have produced local areas of acute or chronic high turbidity that have affected both the physiology and health of adult corals, and the suitability of substrate for settlement and establishment of juveniles.

Over the past 100 years, and especially the past 20 years, our understanding of the impacts of sediments, nutrients and pesticides on coral reefs has evolved considerably. Coral reefs in nearshore waters are exposed to river flood plumes during the summer wet season, and it is here that the impacts of reduced water quality are most clear. This includes exposure to pesticides, as well as elevated nutrient concentrations and very high sediment loads, the latter of which are associated with low light conditions. Several lines of evidence indicate that exposure to pesticides, nutrients and additional sediments reduces the growth, survival and recruitment of corals and increases their susceptibility to disease. Such evidence includes experiments in aquaculture facilities, observations of nearshore-to-offshore gradients in reef health and diversity, and long-term monitoring and correlational analysis of both water quality and reef condition in nearshore waters [27].

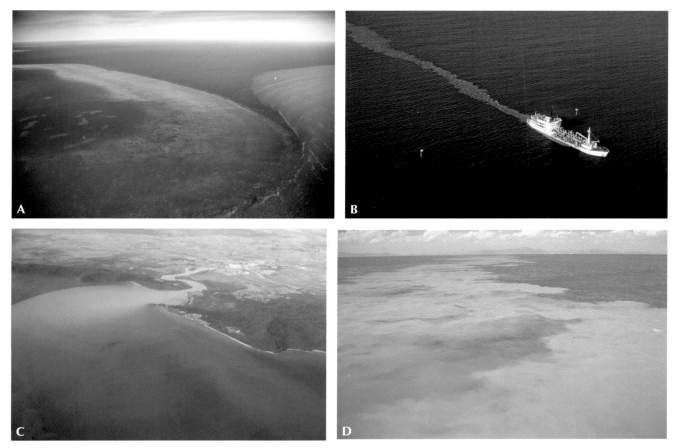

Fig. 5.10. (A) Old Reef following a flood plume from Burdekin River in 2019. (B) Plume trailing behind a boat dredging Townsville Port channel. (C) Johnstone River flood plume. (D) Muddy plume caused by dredging, Townsville. (Image credits: Matt Curnock (A), Elliot (B), Honchin (C), Harrigan (D))

Beyond the immediate impacts of flood-plume exposure, reduced water quality continues to have a longer-term effect over time. In nearshore waters, materials delivered during the wet season are continually resuspended and organic nutrients remineralised to release nutrients that can drive increased algal and phytoplankton production for months after a flood event. Turf algae on reef surfaces compete with coral recruits for habitat space, while both phytoplankton and fine, suspended sediments reduce the penetration of light through water to corals. Long-term satellite observations indicate that water clarity, and hence the potential for photosynthesis by coral symbionts, is reduced even in some mid-shelf GBR waters for several months after flood events.

Transects along turbidity gradients have confirmed that coral biodiversity as well as coral cover is reduced in areas with elevated turbidity; however, other work has also shown that most coral species can survive chronic exposure to reduced light, within limits.

It has been proposed that more frequent COTS outbreaks may also be attributable to nutrients delivered from GBR catchments to the so-called 'COTS initiation zone' near Cairns [14]. Both field observations and laboratory studies have supported a role of increased phytoplankton concentrations, boosted by an elevated nitrogen supply, in enhanced survival of COTS larvae. COTS population modelling also supports this as a possible driver of outbreaks. If this mechanism is confirmed, it may well be the most serious impact of water quality on coral cover in the GBR.

In response to the increasingly clear picture of the impacts of reduced water quality, the GBRMPA, Queensland Government and regional natural resource management groups have implemented a series of policy initiatives and regulations designed to reduce loads of sediments, nutrients and pesticides delivered to the GBR from catchment activities. However, there is limited evidence of the efficacy of these policies in improving nearshore water quality to date. The most recent, and in regulatory terms the strongest of these strategies, is the Reef 2050 Water Quality Improvement Plan, which aims to reduce sediment, nutrient and pesticide loads throughout GBR catchments, primarily through catchment-level improvements in land management practices.

In addition to the impacts of suspended sediments, nutrients and pesticides, a recent review identified six contaminants of emerging concern for GBR and Torres Strait marine environments [28]. These are antifouling paints, coal dust and particles, heavy/trace metals and metalloids, marine debris (Fig. 5.11) and microplastics, pharmaceuticals and personal care products, and petroleum hydrocarbons. All of these contaminants can affect the health of corals at high concentrations, but they are not yet routinely monitored and their impacts in known environmental concentrations in

Australian reef waters are generally not well understood. Of these emerging contaminants, marine microplastics have received increasing attention in recent years and have been found throughout marine food webs globally. This includes corals and other organisms on the GBR, with current research focusing on the potential ecological impacts of such exposures.

Impacts and opportunities for floodplain wetlands on the Great Barrier Reef

Catherine E. Lovelock, Maria Fernanda Adame, Valerie Hagger, Damien T. Maher and Nathan Waltham

Floodplains support a range of ecosystems that contribute to the health and productivity of adjacent coral reefs. An intact and functioning network of floodplain wetlands, including freshwater rivers and saline tidal estuaries, is particularly important to the water quality that enters the GBR lagoon.

The water that enters reef lagoons transits through the floodplain. Thus, the ecosystems and activities on the floodplain contribute to the quality of the water that eventually bathes the reef, and therefore influences reef health. Floodplain ecosystems, including wetlands, support high terrestrial and marine biodiversity, many species of which are currently vulnerable to extinction (e.g. cassowaries *Casuarius casuarius*, and water mice *Xeromys myoides*). A wide range of marine fauna utilise connected habitats for foraging, or over parts of their life cycles. For example, mud crabs (*Scylla serrata*) spend their adult phases in estuaries, returning to the ocean to spawn. Further, barramundi (*Lates calcarifer*) return from coastal waters to the rivers and creeks of floodplains to reproduce, and giant shovel-nose rays (*Glaucostegus typus*, on the IUCN Red List as critically endangered) are found foraging on coral reefs in the high intertidal zones of estuarine wetlands as juveniles.

The ecosystems of floodplains are also important for global carbon sequestration (capturing and storing atmospheric carbon dioxide) and locally for

Fig. 5.11. Noddy tern among marine debris that has washed up on to beach. (Image credit: Scott Whiting)

nutrient cycling. Carbon stocks in soils are particularly high for coastal wetlands, where waterlogged soils, arising from tidal inundation and flooding, support the high productivity of specially adapted plant communities, leading to the accumulation of large stocks of soil organic matter, which is resistant to decomposition. In these ecosystems, soil organic matter has accumulated over thousands of years. It is these fertile soils, rich in both carbon and nitrogen, which have underpinned agricultural production of the floodplains after European colonisation and clearing of wetlands.

Floodplains past and the present

The floodplains adjacent to the GBR, before European colonisation and development of agriculture and urban settlements, were composed of rainforests, as well as herbaceous and forested (palustrine) wetlands and estuarine wetlands. After European colonisation many of these ecosystems were cleared, and the land was drained and isolated from the tide by bunds and tidal gates to enable increased agricultural production. For grazing beef production in lower rainfall areas of the GBR catchments, land was cleared, and estuarine wetlands were separated from the tide by bund walls to form 'ponded pastures' that filled with fresh floodwaters, increasing the extent of freshwater pasture vegetation on the floodplain. For sugarcane production, land was cleared, drained and fertilised. Drainage works were initiated in the 1920s and continued to the 1990s [29]. Clearing of the floodplain has resulted in loss of rainforests, which occurred on higher elevation land in the floodplain and forested freshwater wetlands at lower elevations, which declined to 79 per cent of their original distribution [30]. The widespread conversion of forested wetlands has led many plant communities and species to be declared as vulnerable, and the Queensland Government to institute laws to protect the remaining fragments of wetland vegetation and related ecological and hydrological processes [31].

Clearing of vegetation and the modification of waters in and around floodplains, including groundwater extraction, has led to functional changes that have negatively affected the resilience and condition of the GBR [32]. The CO_2 emissions associated with the loss of biomass and soil carbon on floodplains could be conservatively estimated to be in the order of 260 Tg, which is equivalent to 30 per cent of emissions associated with the 2019–20 megafires on the east coast of Australia. Most of these emissions were due to losses of floodplain eucalypt forests (50 per cent) and melaleuca forested wetlands (34 per cent).

In addition to loss of habitat and CO_2 emissions, the reduction in wetland area and connectivity has decreased the capacity of wetland ecosystem functions that support the GBR, including carbon sequestration, nutrient cycling and export of alkalinity that buffers ocean acidification and may contribute to carbon sequestration. The limited remaining floodplain vegetation has a reduced capacity to trap sediments and nutrients. The vegetation takes up nutrients, while the structure of vegetation traps sediment and microbial processes associated with soils and vegetation facilitate nutrient cycling, including the process of denitrification, which reduces levels of nitrogen in the water and soils. Reduced natural vegetation cover and canalisation of watercourses through drainage decreases water residence times, which further reduces the capacity for capture and processing of materials on the floodplain [31]. Finally, reduced connectivity among wetlands and estuaries reduces access to critical habitat for related marine fauna [33].

While fish abundance and the capacity to retain and process nutrients and sediments declined with clearing and conversion of wetland vegetation, the addition of nutrients and pesticides to the floodplain increased exponentially with agriculture, giving rise to poor water quality and depauperate faunal communities in the coastal waters adjacent to catchments and on the GBR [32]. Globally, simultaneous loss of wetland vegetation and expansion of agriculture have led to negative outcomes for adjacent marine ecosystems including coral reefs.

The future

Climate change will influence the floodplain, its ecosystems and agriculture. Sea-level rise, intense storms and drought all contribute to the intrusion of saltwater into floodplain ecosystems and agricultural land, resulting in ecosystem change and reduced agricultural production. Drought contributes to the intrusion of saltwater because a reduction in the mass of fresh groundwater allows saltwater to push further inland. Examples of ecosystem change include the large recent dieback of melaleuca on Yarrabah and expansion of mangrove fern, dieback of melaleuca due to saltwater intrusion onto the floodplain in the Northern Territory, and the expansion of mangroves into saltmarsh and salt flat ecosystems in south-east Queensland, Northern Territory and GBR catchments. Intense storms and flooding can overtop or damage infrastructure that protects agricultural land from tidal inundation, leading to inundation and salt damage on agricultural lands and loss of production (Fig. 5.12) [34]. The changes in floodplain ecosystems that have already been observed are likely to increase as sea-level rise accelerates in conjunction with other extreme climatic events.

Floodplain wetland restoration has the potential to limit the damage to agriculture and human settlements caused by rising sea levels and intense storms, as well as increase the sequestration of carbon, reduce greenhouse gas emissions and retain nutrients (Fig. 5.13). Coastal wetlands form natural barriers that protect coastlines and their economies. They reduce flood damage as they slow water movement, enhancing infiltration of floodwater into the soil as well as limiting tidal reach by attenuating tidal waters (Fig. 5.14). Restoring wetlands on floodplains will improve water quality through increasing nutrient retention and cycling, and enhanced carbon sequestration, as well as supporting biodiversity. Restoring lost wetlands could remove ~30 per cent of the nitrogen delivered to the reef between 2016 and 2018. The restored East Trinity Inlet wetlands, which were cleared and isolated from the tide for agriculture, now support a high diversity of birds [35]. The restoration of coastal wetlands is also likely to increase populations of fish, particularly where duration and frequency of access are improved, as well as providing more habitat for freshwater species, including water birds [33].

While restoration of floodplain wetlands holds potential for increasing ecological functions that support the GBR, achieving this goal is challenging. This may be assisted by a range of economic incentives for landholders to restore wetlands on relatively unproductive lands [34]. An analysis of economic feasibility of restoration of coastal wetlands on grazing and sugarcane land in the Wet Tropics World Heritage site found that restoration could provide returns in the form of payments for

Fig. 5.12.   (A) Drained agricultural land affected by saltwater intrusion. (B) Mangroves encroaching onto saltmarsh at Insulator Creek, Herbert River. (Image credits: Nathan Waltham (B), Maria Fernanda Adame (B))

BEFORE RESTORATION

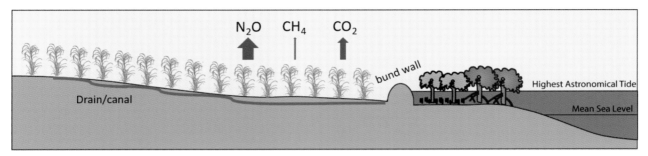

AFTER RESTORATION

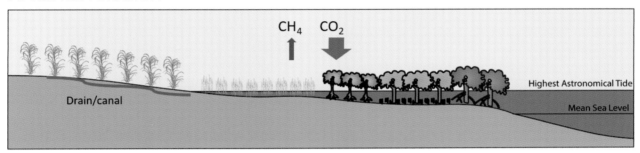

Fig. 5.13. Restoration of coastal wetlands on the floodplain by removal of bund walls. The illustrations show greenhouse gas emissions from sugarcane cropping before restoration (CO_2 from soil carbon loss, nitrous oxide (N_2O) from fertilisation and methane (CH_4) from flooded drains and field burning), and removals and emissions after restoration from reinstatement of tidal flows (CO_2 removals in above-ground biomass and soil carbon, and CH_4 emissions from brackish water flooding). (Image credit: V. Hagger)

the carbon sequestered in restored wetlands, but that returns would be higher with the addition of nutrient as well as biodiversity payments [36]. Additionally, 'Reef Credits' for nitrogen removal and biodiversity payments through the Queensland Government's Land Restoration Fund may increase incentives for farmers to transition land back to restored wetlands.

The wetlands of the floodplain of the GBR have been removed and degraded, reducing their capacity to enhance water quality, mitigate climate change, and assist with climate change adaptation and fisheries production in the manner of intact, connected and diverse faunal communities. Restoration of floodplain wetlands offers a wide range of benefits to the reef, as well as to coastal communities. Carbon and nutrient trading from restoring wetlands can incentivise restoration on private lands, providing new income sources for landholders that diversify their

activities. Emerging pilot restoration projects, Indigenous-led projects, the development of methods and guidance, the availability of appropriate finance and markets, and capacity-building all have a role to play in increasing restoration of floodplain wetlands.

Fig. 5.14. A *Rhizophora* mangrove sapling on the edge of a reef at Houghton Island, northern Great Barrier Reef. (Image credit: Paul Jones)

Zoning the Great Barrier Reef

Jon C. Day and Leanne Fernandes

Early zoning of the Great Barrier Reef

At the time it was proclaimed in 1975, the *Great Barrier Reef Marine Park Act* (the Act) was pioneering federal legislation. Zoning was foreseen as the key management tool for the GBR, and the Act specified the process for zoning to be undertaken, recognising the dual objectives of both conservation and reasonable use and entry. The spectrum of zones ranges from strict biodiversity protection, through no-take 'Marine National Park' zones that allow public access, to zones allowing recreational and commercial activities including fishing and collecting.

Zoning plans were subsequently developed for various sections of the GBRMP. For example, a small prototype section in the Capricorn-Bunker Group of the southern GBRMP came into effect in 1979, and subsequent sections were sequentially declared building on that experience. By 1988, plans had been completed for six sections that collectively comprised virtually the entire GBR Region [37]. Approximately 4.6 per cent of the GBR was zoned within no-take zones ('green zones'). Coral reefs were the primary focus of zoning as they were considered the most important part of the GBR ecosystem; far less consideration was given to protecting non-coral reef habitats. However, one large cross-shelf area east of Cape York stretched from the coast to the outer reefs, included non-reef and reefal areas, and comprised 72 per cent of all no-take zones in the GBRMP. Between 1992 and 2002, two existing zoning plans were revised, and a new small coastal section was zoned [37].

The GBRMP had no precedents and introduced the concept of a multiple-use marine park. This allowed for conservation and the reasonable use of natural resources, including many types of fishing, provided that reasonable use did not compromise the long-term conservation of the GBR.

The need to protect a representative range of biodiversity

In 1994, the 25 Year Strategic Plan for the Great Barrier Reef World Heritage site recognised the need to increase protection of a greater range of biodiversity across the GBR [38]. The 25 Year Plan recommended protecting representative biological communities for protection, including non-reef areas, across the GBR. Consequently, between 1998 and 2004, the GBRMPA undertook a comprehensive planning program to develop a single zoning plan for the entire GBRMP [39, 40].

The primary aim of the Representative Areas Program (RAP) was to ensure that the full range of biodiversity was protected by including representative examples of all major habitat types in a network of no-take zones [39, 40]. Concurrently, a review of all existing zones across the GBRMP was undertaken in conjunction with the RAP, leading to a comprehensive rezoning of the entire GBRMP [38].

Today, the Great Barrier Reef Marine Park Zoning Plan 2003 [41] provides for seven marine zones allowing different types and levels of park use. The zones range from the least restrictive General Use Zone (light blue), allowing most reasonable uses including various commercial and non-commercial fishing activities, through 'no-take' Marine National Park Zones (green), allowing public access but prohibiting extractive activities (e.g. fishing or collecting), to Preservation Zones, very small 'no-go' areas (pink; e.g. One Tree Island), set aside as undisturbed 'scientific baselines' in which only approved scientific research is permitted.

In addition, a Commonwealth Islands zone covers all Commonwealth islands in the GBRMP. Figure 5.15 identifies what activities are allowed in different zones, what activities are prohibited, and specifies activities that can occur only if an appropriate permit has been issued.

The Representative Areas Program

The RAP/rezoning process evolved from lessons learned from previous zoning experiences,

GBRMP Zoning
(see relevant *Zoning Plans* and *Regulations* for details)

	General Use Zone	Habitat Protection Zone	Conservation Park Zone	Buffer Zone	Scientific Research Zone *	Marine National Park Zone	Preservation Zone
Aquaculture	Permit	Permit	Permit*	✗	✗	✗	✗
Bait netting	✓	✓	✓ *	✗	✗	✗	✗
Boating, diving, photography	✓	✓	✓	✓	✓ *	✓	✗
Crabbing (trapping)	✓	✓	✓ *	✗	✗	✗	✗
Harvest fishing for aquarium fish, coral and beachworm	Permit	Permit	Permit*	✗	✗	✗	✗
Harvest fishing for sea cucumber, trochus, tropical rock lobster	Permit	Permit	✗	✗	✗	✗	✗
Limited collecting	✓ *	✓ *	✓ *	✗	✗	✗	✗
Limited spearfishing (snorkel only)	✓	✓	✓ *	✗	✗	✗	✗
Line fishing	✓ *	✓ *	✓ *	✗	✗	✗	✗
Netting (other than bait netting)	✓	✓	✗	✗	✗	✗	✗
Research (other than limited impact research)	Permit	Permit	Permit	Permit	Permit	Permit	Permit
Shipping (other than in a designated shipping area)	✓	Permit	Permit	Permit	Permit	Permit	✗
Tourism programme	Permit	Permit	Permit	Permit	Permit	Permit	✗
Traditional use of marine resources	✓ *	✓ *	✓ *	✓ *	✓ *	✓ *	✗
Trawling	✓	✗	✗	✗	✗	✗	✗
Trolling	✓ *	✓ *	✓ *	✓ *	✗	✗	✗

PLEASE NOTE: This guide provides an introduction to Zoning in the Great Barrier Reef Marine Park. Relevant Great Barrier Reef Marine Park Zoning Plans should be consulted for confirmation of use or entry requirements.

* Additional restrictions / conditions apply.

ACCESS TO ALL ZONES IS PERMITTED IN AN EMERGENCY.

Fig. 5.15. Activities matrix indicating which activities can occur in which zone, which are prohibited, and which activities need a permit. (Image © Great Barrier Reef Marine Park Authority, supplied with kind permission)

although some aspects were more complicated and far-reaching compared with previous GBR zoning programs. The process was innovative and systematic, as well as underpinned by the best available biophysical and sociological science.

The GBRMPA, in conjunction with scientific experts, used 40 datasets of existing biophysical data to map 70 bioregions (30 reef bioregions, 40 non-reef bioregions) covering the entire GBR Region. This bioregionalisation was a fundamental building block for the design of the no-take area network [39].

The GBRMPA's scientific committees developed 11 Biophysical Operational Principles and four Social, Economic, Cultural and Management Operational Principles to design the zoning network [42]. A combination of expert opinion, stakeholder involvement and analytical tools was used to identify possible network options. Options were then fine-tuned based on additional knowledge, expertise and public input to achieve the final, politically acceptable zoning network that maintained the priority scientific advice of at least 20 per cent of each bioregion being protected in no-take zones [39, 43].

The RAP/rezoning process followed, and exceeded, the requirements of the Act. The systematic, mandated process of public engagement was comprehensive and unprecedented compared to previous GBR and Australian environmental planning programs [44]. As required by the Act, two formal phases of community participation occurred during the RAP/rezoning, and over 31 650 written public submissions were received [44]. The first phase provided input to develop the draft zoning plan while the second phase commented on the draft plan. This participatory planning was among the key influencing factors contributing to the final rezoning outcome. Figure 5.16 shows an example of changes between the draft and final zoning plans due to new information provided, including that in public submissions. In some locations, particularly inshore coastal areas where there were higher levels of extractive use, fewer options were available to modify the proposed no-take zones while retaining the recommended minimum levels of protection per bioregion.

The politics of marine park zones

Politics, at many levels, played a fundamental role during planning, with different politics associated with the Federal Government (at that time, a coalition of several political parties) and the Queensland

(State) Government (then Labour). There were several instances when political concerns nearly derailed the planning process [38]. Fortunately, strong leadership at multiple levels meant that the priority biophysical principles were achieved in the final zoning plan.

The maps in Fig. 5.17 compare the GBR zoning before the implementation of the RAP/rezoning (Fig. 5.17A), with the zoning plan after it was rezoned. Rezoning came into effect on 1 July 2004 and remains current today (Fig. 5.17B). Comparing these two maps shows the major expansion of the no-take zones in the 2003 Zoning Plan. When the new zoning plan became law in July 2004, the no-take zones increased from 4.6 per cent of the GBRMP to 33.3 per cent (or 114 530 km^2). More importantly, the new network protected at least 20 per cent of all the 70 bioregions in representative no-take zones. There were numerous benefits beyond the new no-take zones for additional biodiversity protection; for example, doubling the Habitat Protection Zone significantly increased the protection of benthic habitats.

The final plan (Fig. 5.17B) was tabled in Federal Parliament in December 2003. The Minister determined the plan would formally come into effect on 1 July 2004; the intervening period enabled new interpretive materials to be prepared and disseminated (e.g. by TV advertising, free zoning maps and coastal boat-ramp signs). At this time, the Federal Government also introduced a structural adjustment package to assist those adversely affected by the rezoning, such as commercial fishers and fishery-related businesses.

The complementary approach between jurisdictions enabled the Queensland Government to largely 'mirror' the adjoining federal zoning in adjacent Queensland waters, ensuring similar zoning provisions for all waters within the GBR Region irrespective of the jurisdiction.

Bringing together science, public participation, effective leadership and political support

While there was an initial reticence to undertake such an untested 'representative areas' approach across the whole of the GBR, the outcome today is

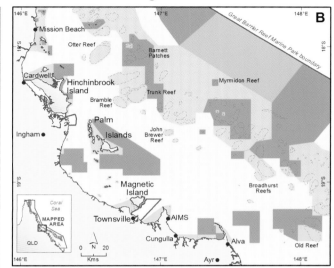

Fig. 5.16.   Zoning changes between the draft (A) and final (B) zoning plans for a small part of the Great Barrier Reef Marine Park (changes addressed the concerns raised in public submissions, particularly from fishers). Figure 5.15 provides an explanation of the zone colours. (Image © Great Barrier Reef Marine Park Authority, supplied with kind permission)

widely acknowledged as 'best practice' [38]. The RAP/rezoning demonstrates that effective marine biodiversity outcomes can be achieved by using the best available science, comprehensive public participation, and effective leadership if underpinned with a sound legislative framework and political support. The rezoning process is still recognised by many marine experts as one of the most comprehensive and innovative global advances in the protection of marine biodiversity. The program and the outcome subsequently received numerous national and international awards.

As was intended, the revised zoning plan maximised the protection of the full range of biodiversity while minimising the impacts on other users, particularly fishers. The overall planning costs may appear high, and certainly the structural adjustment program that the Federal Government introduced was expensive. These costs, however, need to be compared to the $6 billion per year normally generated by the GBR and the value of $56 billion that the Australian public places upon the GBR [38].

Research and monitoring in the GBR continue to show the benefits of the amended zoning network. These include more and larger fish both inside and outside no-take zones, less disease, fewer outbreaks of COTS, increased resilience and quicker recovery after damage [38]. Benefits beyond the network of no-take zones include spillover to adjacent fished areas and greater protection of benthic habitats [38].

Many factors contributed to the success of the RAP/rezoning, but the four most significant factors were using the best available scientific knowledge, effective leadership (both within the agency and politically), a high level of public participation, and political support and will [38]. The 2003 Zoning Plan remains in place today and in conjunction with many other management 'layers' has considerably enhanced the resilience of the GBR. Successive federal environment ministers on both sides of politics have retained the current zoning plan. However, given the various pressures facing the GBR, especially from climate change, water quality and unsustainable fishing (Chapters 5 and 8), a review is probable in the future and many lessons learned in the rezoning are likely to apply.

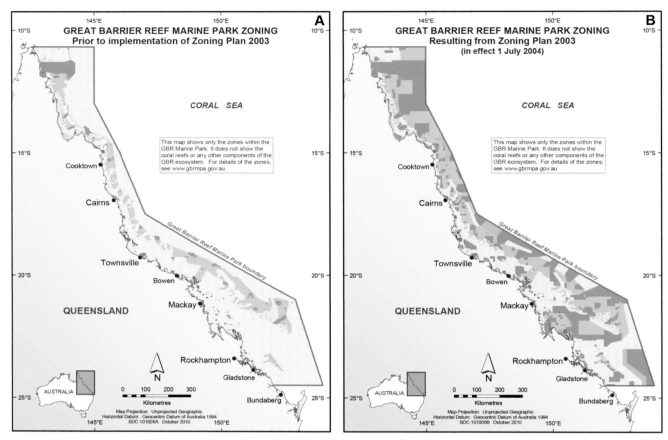

Fig. 5.17. Comparison of GBRMP zoning before implementation of the 2003 Zoning Plan (A) and with the zoning in effect after 1 July 2004 (B). Figure 5.15 provides an explanation of the zone colours. (Image © Great Barrier Reef Marine Park Authority, supplied with kind permission)

Mapping Australia's coral reefs

Sarah Hamylton, Chris Roelfsema, Robin Beaman and Emma Kennedy

Mapping is a fundamental activity that underpins how we understand, and manage, coral reef environments. The synthesis and expression of information through maps has a long history of not only delineating the geographical distribution of coral reefs, but also guiding coral reef management decisions by helping us to see how environmental processes shape and maintain coral reefs. At a very basic level, mapping a coral reef is the first step to managing it: effective management relies on spatial planning and community agreement, and maps are a critical tool for communicating within this process. Detailed maps of coral reefs, including the habitat they provide for other marine life, have clear management applications, for example in evaluating reef connectivity for the dispersal of coral and COTS larvae, modelling water quality and evaluating reef restoration sites [45].

The approximate distribution of Australia's reefs was recorded on the hydrographic charts of early European explorers. In 1606, the Spanish explorer Luis Váez de Torres was among the first known Europeans to chart a navigable corridor between the Pacific and Indian oceans through what would later become known as the Torres Strait. As he journeyed along Australia's northwestern coastline, Dutch seafarer Abel Tasman combined observations from two East India

Company voyages to chart reefs for the Tasman map (1644). In 1770, James Cook mapped portions of the GBR while travelling up the east coast of New Holland (modern-day Queensland). Matthew Flinders (1802–03) and Joseph Jukes (1842–46) would chart many reefs as they circumnavigated the entire Australian continent, including naming the GBR in the case of Flinders.

By the mid-20th century, the complexity of reefs was increasingly acknowledged and the scale of mapping campaigns transitioned from ocean basin, shipborne approaches to more detailed mapping of individual reefs via field surveys. Field techniques played an increasingly important role in mapping coral reef environments, while the use of aerial photography became popularised after World War II.

Field approaches to mapping at Low Isles

Low Isles on the northern Great Barrier Reef has a long history of mapping, to which a range of field cartographic approaches have been applied (Fig. 5.18). Some of the earliest cartographic maps of Low Isles were made during the 1928–29 Great Barrier Reef Expedition, when the major island features were captured using a series of field-intensive and time-consuming methods (see Chapter 3). These included field sketches supplemented by theodolite triangulations; the use of aerial photography to fill in details of shingle, boulders, sand, and mud across reef flats; and compass-traverse surveys to measure direction and distance to notable features, such as the lighthouse, sand cay and mangrove forests.

All of this provided an early understanding of how coral reef organisms are zonally distributed across reef platforms, which informed later management initiatives. With the advent of remote sensing technology, particularly the arrival of Earth Observation satellite images in the 1960s, the scope of coral reef mapping initiatives expanded to cover larger geographical areas and the focus was increasingly applied to management needs.

Mapping the underwater shoals of the Coral Sea

Swathe mapping is an underwater approach to seabed mapping that involves mounting a multibeam sonar system onto the hull of a survey ship to transmit a wide fan (swathe) of acoustic energy down to the seafloor to measure water depths [49]. Sonar instruments transmit a pulse of sound to measure water depths, usually to within several centimetres of accuracy, based on the travel time for the reflected seafloor echoes to return to the instrument. In the case of single beam echosounders, depths are measured from directly under the hull, one sounding at a time. Underwater reefs can be mapped in their entirety as the survey ship moves back and forth along survey lines to over the reef. Such mapping approaches reveal the existence and shape of reefs that are largely unknown because of the deep waters they lie in (up to several thousand metres depth).

A similar approach called light detection and ranging (LiDAR) can map the three-dimensional detail of shallow reefs from a low-flying plane above the reef using pulses of laser beams to scan the seafloor. In clear water, light can penetrate water down to depths of 50–70 m; thus, many shallow reefs can be mapped using LiDAR instruments (typically to within several decimetres of accuracy). Moreover, a combination of LiDAR and sonar technology can map the topographic character of both the shallow and deep areas of reef platforms in detail.

Figure 5.19 shows a three-dimensional map of Flinders Reefs from swathe data collected by the Schmidt Ocean Institute's research vessel *Falkor* and LiDAR data collected by the Australian Hydrographic Office Laser Airborne Depth Sounder (LADS). The Flinders Reefs sit on the Queensland Plateau within the Coral Sea Marine Park, offshore of north-eastern Australia. The deeper flanks of Flinders Reefs extend down to over 1000 m beneath the sea surface. Such approaches allow the full complexity of deeper seafloor environments to be captured, including submerged reefs, drowned reef pinnacles and submarine canyons. Increased accessibility of both LiDAR and multibeam data enables Australia's massive and ecologically important marine estate to be better characterised.

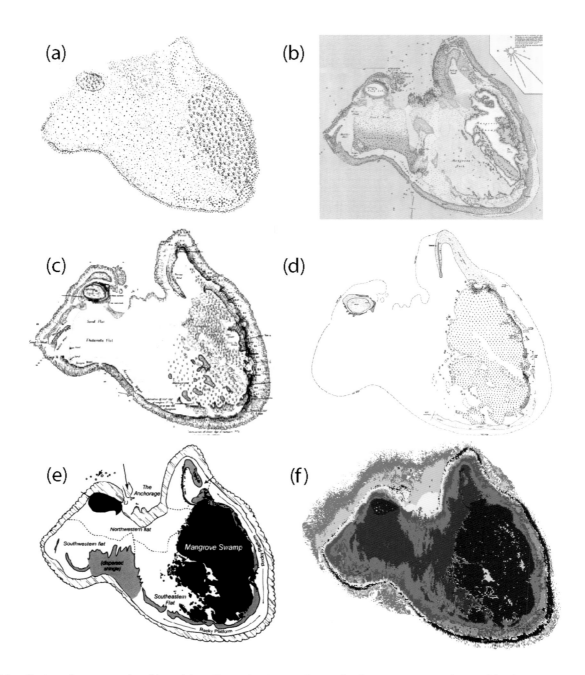

Fig. 5.18.   Series of maps made of Low Isles, Great Barrier Reef over the last century. (a) Plane-table sketch by E. Marchant, 1928. (b) Theodolite triangulation plane-table sketch (Michael Spender, 1929). (c) Aerial photograph trace (Rhodes Fairbridge, 1948, [46]). (d) Compass-traverse survey (David Stoddart, 1974). (e) Aerial photograph sketch (Tracy Frank, 2006 [47]). (f) Satellite image classification (Sarah Hamylton, 2017 [48]). (Image credits: Sarah Hamylton)

Modern mapping of Australia's coral reefs: the Allen Coral Atlas

The *World Atlas of Coral Reefs* provided the first global overview of the location, extent and distribution of reef systems based on remotely sensed information published by the United Nations Environment Programme-World Conservation Monitoring Centre (UNEP-WCMC) [50].

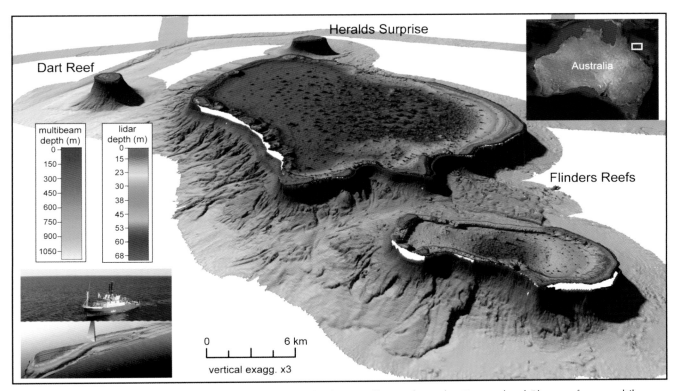

Fig. 5.19.   Three-dimensional map of Flinders Reefs and other smaller reefs on the Queensland Plateau, from multibeam data acquired by the Schmidt Ocean Institute's research vessel *Falkor* and with LiDAR data collected by the Australian Hydrographic Office Laser Airborne Depth Sounder (LADS). Inset shows a composite figure of a hull-mounted swathe system collecting dense bathymetry data. (Image credit: Robin Beaman)

The level of detail was largely determined by the availability and quality of marine charts across different reef regions. The later Millennium Coral Reef Mapping Project (MCRMP), established in 2004, took advantage of the consistency offered by satellite remote sensing for mapping coral reef environments. This examined more than 1600 Landsat 7 ETM+ satellite images, employing visual interpretation techniques to manually digitise the reef systems of the major reef provinces of the world using a globally applicable and consistent typology of 800 classes [51]. The resulting digital maps were freely downloadable in the UNEP-WCMC 2018 dataset.

Artificial intelligence is an exciting technological frontier for analysing images of coral reef environments. This has been especially exciting when combined with remote-sensing technologies, such as the emergence of machine-learning algorithms for mapping and detecting and analysing features from aerial images of coral reef environments. The use of machine-learning algorithms in coral reef remote-sensing applications is founded on several technological advances. These include the spatial resolution of remote-sensing images, which has increased incrementally since Earth Observation images were first collected in the late 1960s so that greater detail and smaller features are now visible in coral reef environments. In addition, the increased capture of Earth Observation images has delivered the information on which artificial intelligence relies to recognise environmental patterns and trends. Global repositories are now continuously updated to provide real-time satellite images for observing coral reefs. Computational advances have made low-cost machines widely available, notably through virtual high-end processing facilities so that these images can be easily processed into maps. These advances are fundamentally shifting the way that the remote sensing community are interpreting images, with important implications for coral reef managers. Here, we illustrate the example of the Allen Coral Atlas, which has embedded machine learning algorithms into image classification routines for mapping coral reef habitat at large geographical scales.

The Allen Coral Atlas is an international collaboration that provides detailed publicly accessible geomorphic and benthic habitat maps and monitoring products for all shallow coral reefs globally. The reef extent map was derived using convolutional neural networks applied to a global

mosaic of Planet Dove satellite imagery that had been trained using the MCRMP maps [52]. Geomorphic and benthic habitat maps were created for all reefs within 30° of latitude from the equator that were visible from space and located in shallow waters (up to 15 m water depth for geomorphic maps and 10 m water depth for benthic habitat maps). A major part of this mapping initiative was the derivation of a basic global reef classification scheme (Fig. 5.20, [53]).

The habitat-mapping process involved dividing reef areas into 28 major mapping regions, within which a globally consistent mapping approach was applied within Google Earth Engine using a combination of a random forest machine learning

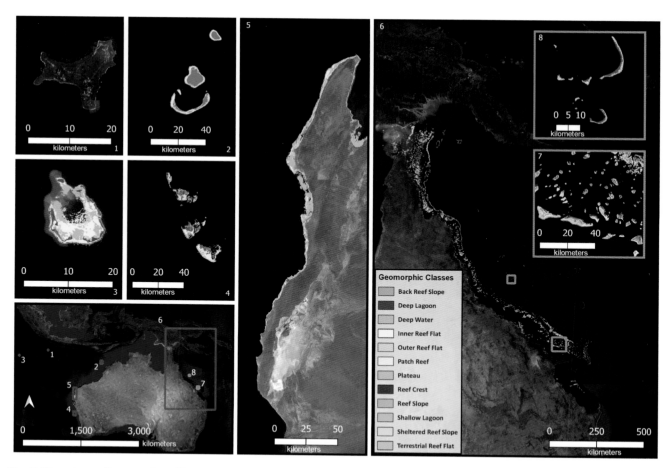

Fig. 5.20. Australian coral reef habitat maps generated from the application of machine learning algorithms to high resolution satellite imagery. Areas include 1. Christmas Island, 2. Scott Reef, 3. Cocos (Keeling) Atoll, 4. Houtman Abrolhos Islands, 5. Ningaloo, 6. The Great Barrier Reef, 7. The Coral Sea, 8. The Swains Reefs. (Image credits: Chris Roelfsema, Allen Coral Atlas)

classifier and an object-based clean-up of the resulting classified map [54]. Google Earth Engine provided both a repository of high-resolution Earth Observation imagery and a remote super-computer for processing. The input data included satellite image reflectance of the reef at low tide (the Planet Dove mosaic), satellite-derived water depth (Sentinel 2, Landsat 8 or Planet Dove) [52], slope, seabed texture and wave data. To train the classifier and validate the map, reference data were sourced from government and non-government agencies, research institutes and local teams, and included field data or existing maps for each mapping region (Fig. 5.21).

The application of artificial intelligence technology through machine learning to satellites is part of a trend of increased automation of coral reef mapping that is enabling some of Australia's remote coral reefs to be characterised accurately in detail. This is an exciting technological advance that opens up a wealth of potential applications for coral reef managers.

Citizen science for managing Queensland's coral reef habitats

Chris Roelfsema and Jen Loder

The Australian Citizen Science Association defines 'citizen science' as public participation and collaboration in scientific research, with the aim of increasing scientific knowledge (Fig. 5.22).

Citizen science programs engage members of the community in collecting, sharing and applying scientific information across various marine habitats. These programs, in turn, offer a range of benefits: providing data that can complement government and academic research; supporting educational, social, and capacity outcomes; and enabling collaborative policy and Reef management goals

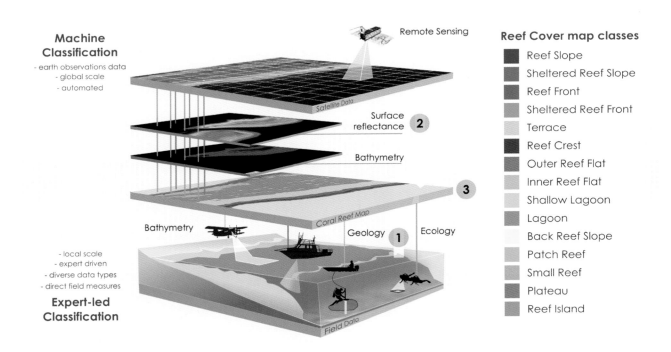

Fig. 5.21. Coral reef habitat maps, like the Allen Coral Atlas, integrate decades of local scientific knowledge on coral reef systems, with the global-scale information that is becoming more accessible from satellite sensors and information derived from these products, to generate map classes that can inform understanding or management of coral reefs. (Image credit: Emma Kennedy, University of Queensland)

Fig. 5.22.   Enthusiastic citizen scientists participating in the Reef Check and Coral Watch experience as part of ReefBlitz. (Image credit: Gus Burrows)

[55, 56]. Here, we provide some examples of citizen science focused on Queensland's coral reef habitats and how they deliver some of these benefits, and discuss some of the challenges and opportunities for further growth.

Filling data gaps

Citizen science efforts have yielded coral reef datasets at spatial and temporal scales not possible within the often more limited scope of coral reef monitoring typically employed by local marine authorities. Moreover, Earth Observation-based monitoring cannot always provide a suitable level of detail; hence, citizen science fills critical spatial, thematic and temporal data gaps.

A range of active citizen science groups monitors Queensland's coral reefs, often mobilising volunteers to swim or dive above coral reefs and collect information on the reef and the marine communities that live around it. Some of these programs require very minimal training (e.g. CoralWatch, Great Reef Census, Eye on the Reef sightings) and provide basic early warning information of coral bleaching and benthic cover changes that prompt further investigation by reef managers. Other globally recognised programs require specific training (e.g. Reef Life Survey, Reef Check Australia) and coordinate the collection of more comprehensive information on substrate composition, invertebrates, fish and impacts and are well placed to enable long-term monitoring.

Citizen science programs collect data from the Torres Strait in the north to the southern end of Queensland. Some programs have established long-term sites that are visited annually (e.g. Reef Check

Australia, Reef Life Surveys), where others carry out opportunistic surveys (e.g. Coral Watch, Great Reef Census), but often offer greater spatial and temporal coverage through ease of participation.

The GBRMPA's Eye on the Reef monitoring and assessment program enables anyone who visits the Great Barrier Reef to contribute to its long-term protection by collecting valuable information about reef health (e.g. coral cover, the presence of bleaching), marine animals (e.g. turtles) and incidents (e.g. oil spills). This is subsequently used to establish a bigger picture that informs Reef management. The program has several levels, ranging from options for untrained citizens (e.g. animal sightings), to comprehensive training to be a Reef health impact surveyor.

Trained Reef Check Australia volunteers conduct annual ecological monitoring at long-term sites throughout Queensland using a standard global monitoring protocol whereby trained divers assess reef community characteristics along 4 × 20 m transects focusing on invertebrates, substrate type, fish and impacts. The resultant coral cover data form an integral part of two regional report cards (Dry Tropics and Mackay Whitsunday Isaac).

Citizens of the Great Barrier Reef have initiated the Great Reef Census, for which citizen scientists can upload and analyse underwater images to calculate the percentage cover of different benthic community components. The project was designed to address a gap in wide-scale reconnaissance monitoring across the GBR and validation of habitat maps. In 2016 Reef Check volunteers collected information about the distribution of reef habitats that informed an update of the detailed 2004 reef habitat map for Moreton Bay.

Enabling collaborative science and management

Citizen science groups support capacity-building in science literacy [55] and community conservation leadership [57]. Genuine community engagement is a critical enabling factor for reef management [55], and citizen science activities provide an effective tool for communities to engage meaningfully in science and conservation. Activities are carried out through a range of ambassador programs, workshops, specialised initiatives, curriculum materials, datasets, papers, and educational material. In 2018, CoralWatch organised a specialised initiative called 'Corals in the Outback', an event at which local outback communities were educated about the importance of coral reefs through school visits, displays, presentations and virtual reality.

Multiple citizen science programs coordinated annual ReefBlitz events in 2015–19, through which 1170 participants collected information (> 76 000 data points) on catchment and reef health through 47 community events. Participant surveys indicated that the programs increased the number of people communicating about reef conservation and related science; 91 per cent of participants reported a willingness to share information learned about reef conservation. Eighty-seven per cent of those surveyed also reported increased support for marine science and conservation. Participating in citizen science may catalyse behaviour changes that are useful for reef management; ~51 per cent of participants indicated that they would change their behaviour to encourage conservation [56].

Citizen science promotes collaboration across a range of partners, and can foster greater trust, enable social licence and facilitate shared management goals [55, 58]. Local programs that build understanding about ecological values and management may also help communities provide informed comment on local management issues and influence legislative change. For example, habitat maps created by citizen science groups of Wolf Rock on Double Island Point helped to define marine park boundaries during the review of zoning plans for south-east Queensland marine parks.

Citizen science offers three key benefits: data provision; delivering educational, social, and capacity outcomes; and enabling collaborative science and management. These are applicable in any coral reef environment, including Australia's remote coral reefs.

Monitoring coral reefs at the Australian Institute of Marine Science

Michael J. Emslie

Coral reef monitoring at the AIMS was originally established to document the extent and impacts of the second wave of COTS (*Acanthaster* spp.) outbreaks that began in the 1970s, using broad-scale manta tow surveys (Fig. 5.23). However, coordinated large-scale monitoring on the GBR did not commence until 1985. This was the first time a standard technique was used to survey COTS abundance and impacts on hard corals across the entire geographic range of the GBR in a single year. By the early 1990s, it became clear that many reef systems world-wide were degraded and only the provision of ongoing high quality monitoring data could inform management about the status of the GBR. To fill this knowledge gap, AIMS established the LTMP in 1992 to provide detailed information on the spatio-temporal variation in the status of coral reef communities beyond what was supplied by manta tow, namely quantitative data on benthic and fish assemblages, and agents of coral mortality (e.g. coral bleaching, coral disease, storms and COTS; Fig. 5.23b). The program was supported by the government of the day, and AIMS has committed ongoing annual funding for an in-depth monitoring program to assist policy-makers and managers in effective decision-making with regards to the status and health of the GBR. Government funding has been supplemented by in-kind contributions from AIMS for ship time and personnel.

While constrained by resources and logistics, the LTMP sampling design allows the quantification of spatio-temporal change across several key indicators of coral reef status. It is a trade-off between intense sampling over small scales and less intensive sampling over broad scales, and although a key objective was to sample over most of the GBR it was also vital to be able to detect changes over time using appropriate experimental designs. Hence, the LTMP focused on intensive annual surveys in a standard reef slope habitat with three sites of five permanently marked transects ($n = 15$ reef^{-1} year^{-1}). The LTMP combines broad-scale manta tow surveys, which record cover of hard and soft corals and algae and counts of COTS, coral trout and sharks around whole reef perimeters with quantitative surveys of fish, benthic communities and agents of coral

Fig. 5.23. Survey methods used by the LTMP include (A) manta tow surveys, which are a broad-scale technique for quickly assessing the status of coral reefs where an observer is towed behind a small boat around entire reef perimeters and records the categorical percentage of hard and soft coral cover, as well as numbers of crown-of-thorns starfish and coral trout, and (B) fixed sites surveyed along permanently marked transects where the benthos is quantified using digital imagery, and reef fishes and agents of coral mortality are recorded using visual census. (Image supplied with kind permission from the Australian Institute of Marine Science (AIMS LTMP), photographer unknown)

mortality. These surveys are carried out on fixed sites at reefs spread across the length and breadth of the GBR.

To date, information from the LTMP has provided a key source of knowledge for the high level strategic management of Australian reefs (e.g. the 5-yearly Outlook Report for the GBR, State of the Environment Report). Additionally, LTMP data are regularly reported via the AIMS website and AIMS monitoring methodologies have been widely adopted across the world through the Global Coral Reef Monitoring Network (GCRMN) capacity-building program. The 35-year manta tow record has enabled the status of corals on the GBR to be tracked over time, revealing that the GBR had lost half its coral cover since the 1980s (Fig. 5.24) [59].

Manta tow data captured detailed information about three of the four recorded waves of outbreaks of COTS and brought to light the southward propagation of outbreaks from the initiation box between Cairns and Lizard Island, and that reefs inside no-take marine reserves had fewer outbreaks (Fig. 5.25A) [60].

Intensive fixed-site surveys have demonstrated the influence of environmental gradients across the continental shelf in shaping coral reef communities. These surveys also allow inferences into changes in community structure that go beyond coarse measures of reef health such as changes in the percentage of coral cover or total fish abundance. Fixed-site surveys on reefs in no-take marine reserves and those open to fishing demonstrated that the rezoning of the GBR in 2004 has led to rapid and large increases in coral trout populations, a key GBR fishery species (Fig. 5.25B) [61, 62], and confer resilience to coral reefs inside no-take marine reserves in the face of disturbances [63].

The long-term data have also allowed an appraisal of the recovery and reassembly of GBR reefs after disturbances, and have demonstrated that the recovery of hard coral cover does not necessarily equate to the reassembly of coral communities, that corals recover faster following COTS outbreaks than other disturbances, and that coral bleaching (i.e. thermal stress) can impair the rate of coral reef recovery from disturbance [64, 65, 66].

While core LTMP surveys continue, AIMS is actively seeking technological solutions to increase the efficiency of sampling, analysis and reporting, while maintaining the highest standards of data integrity.

Fig. 5.24. The 35 year record has captured the impacts and recovery from disturbances to Great Barrier Reef coral reefs, and unequivocally demonstrated the loss of half of the coral cover on the Great Barrier Reef from 1985 to 2012. (Image supplied with kind permission from the Australian Institute of Marine Science (AIMS LTMP), photographer unknown.)

Fig. 5.25.   The AIMS Long-term Monitoring Program has (A) captured the southward progression and impacts of multiple waves of crown-of-thorns starfish outbreaks on the GBR and (B) demonstrated the rapid and substantial benefits of the 2004 rezoning of the Great Barrier Reef Marine Park, especially for increases in populations of important fishery species like the coral trout (*Plectropomus* spp.). (Image supplied with kind permission from the Australian Institute of Marine Science (AIMS LTMP), photographer unknown)

The use of advanced technology for monitoring coral reefs

Gal Eyal, Matan Yuval and Tali Treibitz

Technology plays a key role in advancing global research. Since the 1960s, environmental scientists have progressed from human observations and paper-based calculations to remote sensing and computer-based data analysis. For example, remote sensing from satellites has become an efficient tool for monitoring large-scale terrestrial environments and shallow aquatic-habitats. Nevertheless, the data from satellites and other airborne methods are limited for the majority of coral reef areas, since observing through the water column is not straightforward. Satellites, drones and other airborne methods are capable of capturing images of large areas of shallow reefs down to a depth of 20 m in clear waters, representing a relatively small fraction of entire reef ecosystems [67]. Furthermore, sea surface conditions (e.g. cloud cover, waves and light conditions) can impact the timing in which these observations can be made.

Traditionally, coral reef ecological monitoring surveys were limited to a narrow spatial region that is in reach of snorkellers or SCUBA divers in shallow reefs (mostly 0–30 m depths) and analysed manually by experts. Even today, most assessments of baseline ecological states of coral reef benthic communities are conducted by *in situ* observational surveys using various transect-based or plot-based approaches that were adapted from terrestrial ecology [68].

Recently, three-dimensional (3D) model reconstructions and two-dimensional (2D) orthomosaics (Fig. 5.26) from high-resolution photos have been developed as a more efficient methodology that substantially increases survey area and data volume per unit time and effort invested, as well as being effectively repeatable [69]. By conducting 3D-photogrammetric surveys at multiple sites, researchers can uncover spatial patterns and quantify structural complexity, abundances, size-frequency distributions, diversity, and live cover of coral reef benthic organisms. Furthermore, repeated surveys enable detailed assessments of coral communities to be made, including demography, growth-rates, survival/mortality, ecosystem dynamics and the correlation of these observations with environmental parameters and ecological changes. Three-dimensional reconstructions and 2D orthomosaics can be generated by acquiring

Fig. 5.26.   A three-dimensional model of a 12 m² area of shallow coral reef. The model is composed of ~1000 images taken by a diver underwater. (Image credit: Matan Yuval)

images from various research platforms, such as diver-based surveys, robotic surveys (where cameras are attached to underwater vehicles – remotely operated vehicles (ROVs) and autonomous underwater vehicles (AUVs)), and submersibles, or by capturing photos with towed cameras. These can then be processed using structure from motion (SfM) photogrammetry [70].

Photogrammetry by SfM gives great advantages to surveying large areas in fine detail for seascape ecology and terrain mapping. However, annotating such large and complex data to individual species level requires interpretation with specialised expertise and is time consuming. Hence, new artificial intelligence (AI)-based machine learning algorithms for point annotation [71] and semantic segmentation of images for reef surveys [72] have gained significant traction. Using AI for ecological monitoring can reliably speed up the process of meaningful annotation and achieve near real-time interpretation of images for decision-makers [70] (Fig. 5.27). This stands in stark contrast to the situation we see today, in which we act retrospectively to manage and conserve reefs in crisis.

The ultimate goal of incorporating computer-vision AI in coral reef monitoring is to achieve complete segmentation of 3D reef models – that is, to make detailed and accurate statements about reef character in terms that are useful for decision-makers and reef managers (e.g. alive versus dead corals). Although deep learning has shown impressive results with a plethora of tasks in many scientific, social and industrial domains, coral reefs are some of the most intricate environments on the planet, and full 3D semantic segmentation of coral reefs is still in the initial stages of research. Much work remains to be done on meaningful classification and segmentation of benthic images and underwater 3D models [73].

There are several recent technological developments in remote sensing that hold promise for monitoring coral reefs. Micro-scale sensing has seen equipment adaptation and modification for working in aquatic environments, for example underwater microscopy [74], micro-sensors (e.g. Unisense®) and underwater spectroscopy (e.g. Ocean Optics®). Using micro-scale methods enables processes of relevance to coral reefs to be tracked from the molecule to organism level. Furthermore, over the last two decades, an increased awareness of deeper marine habitats, such as the mesophotic coral ecosystems, has led to the development of specialised research

Fig. 5.27.   Photo of a coral reef automatically analysed by deep learning. Artificial intelligence programs are able to identify as many as 40 categories of corals, sponges, algae and other organisms. (Image provided with kind permission from Underwater Earth/XL Catlin Seaview Survey; image credit: Lorna Parry)

platforms and methods to increase both the time we can spend in deeper depths and the depth of operation. Technical diving methods such as closed-circuit rebreathers and relatively low-cost ROVs and AUVs have improved access to and research capabilities in deeper environments [75, 76].

References

1.   Dawul Wuru Aboriginal Corporation (2018) *Yirrganydji Traditional Use of Marine Resources Agreement*. Great Barrier Reef Marine Park Authority, Cairns.

2.   Hiscock P, Kershaw P (1992) Palaeoenvironments and prehistory of Australia's tropical Top End. In *The Naïve Lands: Prehistory and Environmental Change in Australia and the Southwest Pacific*. (Ed. J Dodson) pp. 43–75. Longman Cheshire, Melbourne.

3.   Dawul Wuru Aboriginal Corporation (2014) *Yirrganydji Kulpul-wu Mamingal Looking After Yirrganydji Sea Country Yirrganydji Sea Country Plan*. Dawul Wuru Aboriginal Corporation and Queensland Government, Cairns.

4.   Alfred T (2015) Cultural strength: restoring the place of Indigenous knowledge in practice and policy. *Australian Aboriginal Studies* **1**, 3–11.

5.   Department of the Prime Minister and Cabinet (2016) 'SVA Consulting Consolidated report on Indigenous Protected Area following social return on investment analyses'. SVA Consulting, Sydney.

6.   Martin K, Mirraboopa B (2003) Ways of knowing, being and doing: a theoretical framework and

methods for indigenous and indigenist re-search. *Journal of Australian Studies* **27**, 203–214. doi:10.1080/14443050309387838

7. Nakata MN (1997) The cultural interface: an exploration of the intersection of Western knowledge systems and Torres Strait Islanders positions and experiences. PhD thesis. James Cook University, Australia.

8. Harrison MD, McConchie P (2009) *My People's Dreaming: An Aboriginal Elder Speaks on Life, Land, Spirit and Forgiveness*. Finch Publishing, Warriewood.

9. McKnight AD (2015) Mingadhuga Mingayung: respecting Country through Mother Mountain's stories to share her cultural voice in Western academic structures. *Educational Philosophy and Theory* **47**, 276–290. doi:10.1080/00131857.2013.860581

10. Bawaka Country including Suchet-Pearson S, Wright S, Lloyd K, Burarrwanga L (2013) Caring as country: towards an ontology of co-becoming in natural resource management. *Asia Pacific Viewpoint* **54**, 185–197. doi:10.1111/apv.12018

11. Uthicke S, Schaffelke B, Byrne M (2009) A boom-bust phylum? Ecological and evolutionary consequences of density variations in echinoderms. *Ecological Monographs* **79**, 3–24. doi:10.1890/07-2136.1

12. Done TJ, Potts DC (1992) Influences of habitat and natural disturbances on contributions of massive *Porites* corals to reef communities. *Marine Biology* **114**, 479–493. doi:10.1007/BF00350040

13. Kroon FJ, Lefèvre CD, Doyle JR, Patel F, Milton G, Severati A, Kenway M, Johansson CL, Schnebert S, Thomas-Hall P, *et al.* (2020) DNA-based identification of predators of the corallivorous crown-of-thorns starfish (*Acanthaster* cf. *solaris*) from fish faeces and gut contents. *Scientific Reports* **10**, 8184. doi:10.1038/s41598-020-65136-4

14. Birkeland C (1982) Terrestrial runoff as a cause of outbreaks of *Acanthaster planci* (Echinodermata, Asteroidea). *Marine Biology* **69**, 175–185. doi:10.1007/BF00396897

15. Pratchett MS, Caballes CF, Wilmes JC, Matthews S, Mellin C, Sweatman HPA, Nadler LE, Brodie J, Thompson CA, Hoey J, *et al.* (2017) Thirty years of research on crown-of-thorns starfish (1986–2016):

scientific advances and emerging opportunities. *Diversity (Basel)* **9**, 41. doi:10.3390/d9040041

16. Hutchings PA (Ed.) (1992) *Biology – A Survey of Elizabeth and Middleton Reefs, South Pacific*. Australian National Parks and Wildlife, Canberra.

17. Birkeland C (1989) The Faustian traits of the crown-of-thorns starfish. *American Scientist* **77**, 154–163.

18. Guerra V, Haynes G, Byrne M, Yasuda N, Adachi S, Nakamura M, Nakachi S, Hart MW (2020) Nonspecific expression of fertilization genes in the crown-of-thorns *Acanthaster* cf. *solaris*: unexpected evidence of hermaphroditism in a coral reef predator. *Molecular Ecology* **29**, 363–379. doi:10.1111/mec.15332

19. Allen JD, Richardson EL, Deaker D, Agüera A, Byrne M (2019) Larval cloning in the crown-of-thorns sea star, a keystone coral predator. *Marine Ecology Progress Series* **609**, 271–276. doi:10.3354/meps12843

20. Wolfe K, Graba-Landry A, Dworjanyn SA, Byrne M (2017) Superstars: assessing nutrient thresholds for enhanced larval success of *Acanthaster planci*, a review of the evidence. *Marine Pollution Bulletin* **116**, 307–314. doi:10.1016/j.marpolbul.2016.12.079

21. Carrier TJ, Wolfe K, Lopez K, Gall M, Janies DA, Byrne M, Reitzel AM (2018) Diet-induced shifts in the crown-of-thorns (*Acanthaster* sp.) larval microbiome. *Marine Biology* **165**, 157. doi:10.1007/s00227-018-3416-x

22. Deaker DJ, Agüera A, Lin H-A, Lawson C, Budden C, Dworjanyn SA, Mos B, Byrne M (2020) The hidden army: corallivorous crown-of-thorns sea-stars can spend years as herbivorous juveniles. *Biology Letters* **16**, 20190849. doi:10.1098/rsbl.2019.0849

23. Hill D (1984) The Great Barrier Reef Committee, 1922–82 Part II: The last three decades. *Historical Records of Australian Science* **6**, 195–221. doi:10.1071/HR9850620195

24. Endean R (1969) 'Report on investigations made into aspects of the current *Acanthaster planci* (Crown-of-Thorns) infestations of certain reefs of the Great Barrier Reef'. Queensland Department of Primary Industry, Fisheries Branch, Brisbane.

25. Endean R (1974) *Acanthaster planci* on the Great Barrier Reef. *Proceedings of the Second International Coral Reef Symposium* **1**, 563–576.

26. Kroon FJ, Kuhnert PM, Henderson BL, Wilkinson SN, Kinsey-Henderson A, Abbott B, Brodi JE, Turner RDR (2012) River loads of suspended solids, nitrogen, phosphorus and herbicides delivered to the Great Barrier Reef lagoon. *Marine Pollution Bulletin* **65**, 167–181. doi:10.1016/j.marpolbul.2011.10.018

27. Waterhouse J, Schaffelke B, Bartley R, Eberhard R, Brodie J, Star M, Thorburn P, Rolfe J, Ronan M, Taylor B, Kroon F (2017) *Scientific Consensus Statement: Land-use Impacts on Great Barrier Reef Water Quality and Ecosystem Condition.* The State of Queensland, Brisbane.

28. Kroon FJ, Berry KLE, Brinkman DL, Kookana R, Leusch FDL, Melvin SD, Neale PA, Negri AP, Puotinen M, Tsang JJ, *et al.* (2020) Sources, presence and potential effects of contaminants of emerging concern in the marine environments of the Great Barrier Reef and Torres Strait, Australia. *The Science of the Total Environment* **719**, 135140. doi:10.1016/j.scitotenv.2019.135140

29. Griggs PD (2018) Too much water: drainage schemes and landscape change in the sugar-producing areas of Queensland, 1920–90. *The Australian Geographer* **49**, 81–105. doi:10.1080/00049182.2017.1336965

30. Department of Environment and Science, Queensland (2019) Great Barrier Reef contributing catchments – facts and maps. WetlandInfo, <https://wetlandinfo.des.qld.gov.au/wetlands/facts-maps/study-area-great-barrier-reef/>.

31. Adame MF, Arthington AH, Waltham N, Hasan S, Selles A, Ronan M (2019) Managing threats and restoring wetlands within catchments of the Great Barrier Reef, Australia. *Aquatic Conservation* **29**, 829–839. doi:10.1002/aqc.3096

32. Lewis SE, Bartley R, Wilkinson SN, Bainbridge ZT, Henderson AE, James CS, Irvine SA, Brodie JE (2021) Land use change in the river basins of the Great Barrier Reef, 1860 to 2019: a foundation for understanding environmental history across the catchment to reef. *Marine Pollution Bulletin* **166**, 112193. doi:10.1016/j.marpolbul.2021.112193

33. Abbott BN, Wallace J, Nicholas DM, Karim F, Waltham NJ (2020) Bund removal to re-establish tidal flow, remove aquatic weeds and restore coastal wetland services – North Queensland, Australia. *PLoS One* **15**, 0217531. doi:10.1371/journal.pone.0217531

34. Waltham NJC, Wegscheidl A, Volders JCR, Smart S, Hasan E, Ledee E, Waterhouse J (2021) Land use conversion to improve water quality in high DIN risk, low-lying sugarcane areas of the Great Barrier Reef catchments. *Marine Pollution Bulletin* **167**, 112373. doi:10.1016/j.marpolbul.2021.112373

35. Luke H, Martens MA, Moon EM, Smith D, Ward NJ, Bush RT (2017) Ecological restoration of a severely degraded coastal acid sulfate soil: a case study of the East Trinity wetland, Queensland. *Ecological Management & Restoration* **18**, 103–114. doi:10.1111/emr.12264

36. Hagger V, Waltham NJ, Lovelock CE (2022) Opportunities for coastal wetland restoration for blue carbon with co-benefits for biodiversity, coastal fisheries and water quality. *Ecosystem Services* **55**, 101423. doi:10.1016/j.ecoser.2022.101423

37. Day JC (2016) The Great Barrier Reef Marine Park – the grandfather of modern MPAs. In *Big, Bold and Blue: Lessons from Australia's Marine Protected Areas.* (Eds J Fitzsimmons and G. Wescott), pp. 65–97. CSIRO, Melbourne.

38. Day JC (2020) Ensuring effective and transformative policy reform: lessons from rezoning Australia's Great Barrier Reef, 1999–2004. PhD thesis. James Cook University, Australia.

39. Fernandes L, Day JO, Lewis A, Slegers S, Kerrigan B, Breen DA, Cameron D, Jago B, Hall J, Lowe D, Innes J (2005) Establishing representative no-take areas in the Great Barrier Reef: large-scale implementation of theory on marine protected areas. *Conservation Biology* **19**, 1733–1744. doi:10.1111/j.1523-1739.2005.00302.x

40. Day JC, Fernandes L, Lewis A, Innes J (2004) RAP – an ecosystem level approach to biodiversity protection planning. In *Proceedings of the 2nd International Tropical Marine Ecosystem Management Symposium.* 24–27 March 2003. Manila, Philippines. pp. 251–265. International Coral Reef Action Network.

41. Great Barrier Reef Marine Park Authority (2004) *Great Barrier Reef Marine Park Zoning Plan 2003.* Great Barrier Reef Marine Park Authority, Townsville.

42. Fernandes L, Day J, Kerrigan B, Breen D, De'ath G, Mapstone B, Coles R, Done T, Marsh H, Poiner I,

Ward T (2009) A process to design a network of marine no-take areas: lessons from the Great Barrier Reef. *Ocean and Coastal Management* **52**, 439–447. doi:10.1016/j.ocecoaman.2009.06.004

43. Day JC, Kenchington RA, Tanzer JM, Cameron DS (2019) Marine zoning revisited: How decades of zoning the Great Barrier Reef has evolved as an effective spatial planning approach for marine ecosystem-based management. *Aquatic Conservation* **29**, 9–32. doi:10.1002/aqc.3115

44. Great Barrier Reef Marine Park Authority (2005) 'Report on the Great Barrier Reef Marine Park Zoning Plan 2003'. GBRMPA, Townsville.

45. Roelfsema CM, Kovacs EM, Ortiz JC, Callaghan DP, Hock K, Mongin M, Johansen K, Mumby PJ, Wettle M, Ronan M, Lundgren P (2020) Habitat maps to enhance monitoring and management of the Great Barrier Reef. *Coral Reefs* **39**, 1039–1054. doi:10.1007/s00338-020-01929-3

46. Fairbridge RW, Teichert C (1948) The Low Isles of the Great Barrier Reef: a new analysis. *The Geographical Journal* **111**, 67–88. doi:10.2307/1789287

47. Frank TD, Jell JS (2006) Recent developments on a nearshore, terrigenous-influenced reef: Low Isles Reef, Australia. *Journal of Coastal Research* **22**, 474–486. doi:10.2112/03-0127.1

48. Hamylton SM (2017) Mapping coral reef environments: a review of historical methods, recent advances and future opportunities. *Progress in Physical Geography* **41**, 803–833. doi:10.1177/0309133317744998

49. Beaman RJ (2011) Swathe mapping. In *Encyclopedia of Modern Coral Reefs: Structure, Form and Process.* (Ed. D Hopley) pp. 1067–1070. Encyclopedia of Earth Sciences. Springer, Dordrecht, The Netherlands.

50. Spalding M, Spalding MD, Ravilious C, Green EP (2001) *World Atlas of Coral Reefs.* University of California Press, Oakland CA.

51. Andréfouët S, Muller-Karger FE, Robinson JA, Kranenburg CJ, Torres-Pulliza D, Spraggins SA, Murch B (2006) Global assessment of modern coral reef extent and diversity for regional science and management applications: a view from space. In *Proceedings of the 10th International Coral Reef Symposium* **2**, 1732–1745. 28 June–2 July, Okinawa, Japan.

52. Li J, Knapp DE, Fabina NS, Kennedy EV, Larsen K, Lyons MB, Murray NJ, Phinn SR, Roelfsema CM, Asner GP (2020) A global coral reef probability map generated using convolutional neural networks. *Coral Reefs* **39**, 1805–1815. doi:10.1007/s00338-020-02005-6

53. Kennedy EV, Roelfsema C, Lyons M, Kovacs E, Borrego-Acevedo R, Roe M, Phinn S, Larsen K, Murray N, Yuwono D, *et al.* (2021) Reef Cover, a coral reef classification for global habitat mapping from remote sensing. *Scientific Data* **8**, 196. doi:10.1038/s41597-021-00958-z

54. Lyons M, Roelfsema C, Kennedy E, Kovacs E, Borrego-Acevedo R, Markey K, Roe M, Yuwono D, Harris D, Phinn S, Asner GP (2020) Mapping the world's coral reefs using a global multiscale earth observation framework. *Remote Sensing in Ecology and Conservation* **6**, 557–568. doi:10.1002/rse2.157

55. Kelly R, Fleming A, Pecl GT, Richter A, Bonn A (2019) Social license through citizen science: a tool for marine conservation. *Ecology and Society* **24**, 16. doi:10.5751/ES-10704-240116

56. Dean AJ, Church EK, Loder J, Fielding KS, Wilson KA (2018) How do marine and coastal citizen science experiences foster environmental engagement? *Journal of Environmental Management* **213**, 409–416. doi:10.1016/j.jenvman.2018.02.080

57. Cigliano JA, Meyer R, Ballard HL, Freitag A, Phillips TB, Wasser A (2015) Making marine and coastal citizen science matter. *Ocean and Coastal Management* **115**, 77–87. doi:10.1016/j.ocecoaman.2015.06.012

58. McKinley DC, Miller-Rushing AJ, Ballard HL, Bonney R, Brown H, Cook-Patton SC, Evans DM, French RA, Parrish JK, Phillips TB, *et al.* (2017) Citizen science can improve conservation science, natural resource management, and environmental protection. *Biological Conservation* **208**, 15–28. doi:10.1016/j.biocon.2016.05.015

59. De'ath G, Fabricius KE, Sweatman H, Puotinen M (2012) The 27-year decline of coral cover on the Great Barrier Reef and its causes. *Proceedings of the National Academy of Sciences of the United States of America* **109**, 17995–17999. doi:10.1073/pnas.1208909109

60. Sweatman H (2008) No-take reserves protect coral reefs from predatory starfish. *Current Biology* **18**, R598. doi:10.1016/j.cub.2008.05.033

61. Russ GR, Cheal AJ, Dolman AM, Emslie MJ, Evans RD, Miller I, Sweatman H, Williamson DH (2008) Rapid increase in fish numbers follows creation of world's largest marine reserve network. *Current Biology* **18**, 514–515. doi:10.1016/j.cub.2008.04.016

62. Emslie MJ, Logan M, Williamson DH, Ayling AM, MacNeil MA, Ceccarelli D, Cheal AJ, Evans RD, Johns KA, Jonker MJ, *et al.* (2015) Expectations and outcomes of reserve network performance following re-zoning of the Great Barrier Reef Marine Park. *Current Biology* **25**, 983–992. doi:10.1016/j.cub.2015.01.073

63. Mellin C, Aaron MacNeil M, Cheal AJ, Emslie MJ, Caley JM (2016) Marine protected areas increase resilience among coral reef communities. *Ecology Letters* **19**, 629–637. doi:10.1111/ele.12598

64. Johns KA, Osborne KO, Logan M (2014) Contrasting rates of coral recovery and reassembly in coral communities on the Great Barrier Reef. *Coral Reefs* **33**, 553–563. doi:10.1007/s00338-014-1148-z

65. Osborne K, Dolman AM, Burgess SC, Johns KA (2011) Disturbance and the dynamics of coral cover on the Great Barrier Reef (1995–2009). *PLoS One* **6**, e17516. doi:10.1371/journal.pone.0017516

66. Osborne K, Thompson AA, Cheal AJ, Emslie MJ, Johns KA, Jonker MJ, Logan M, Miller IR, Sweatman HPA (2017) Delayed coral recovery in a warming ocean. *Global Change Biology* **23**, 3869–3881. doi:10.1111/gcb.13707

67. Foo SA, Asner GP (2019) Scaling up coral reef restoration using remote sensing technology. *Frontiers in Marine Science* **6**, 79-92. doi:10.3389/fmars.2019.00079

68. Loya Y (1978) Plotless and transect methods. In *Coral Reefs: Research Methods*. (Eds DR Stoddart and RE Johannes) pp. 197–217. UNESCO, Paris.

69. Yuval M, Alonso I, Eyal G, Tchernov D, Loya Y, Murillo AC, Treibitz T (2021) Repeatable semantic reef-mapping through photogrammetry and label-augmentation. *Remote Sensing* **13**, 659–682. doi:10.3390/rs13040659

70. González-Rivero M, Beijbom O, Rodriguez-Ramirez A, Bryant DEP, Ganase A, Gonzalez-Marrero Y, Herrera-Reveles A, Kennedy EV, Kim CJS, Lopez-Marcano S, *et al.* (2020) Monitoring of coral reefs using artificial intelligence: a feasible and cost-effective approach. *Remote Sensing* **12**, 489. doi.org/10.3390/rs12030489

71. Beijbom O, Edmunds PJ, Kline DI, Mitchell BG, Kriegman D (2012) Automated annotation of coral reef survey images. In *2012 IEEE Conference on Computer Vision and Pattern Recognition*. 16–21 June. Providence, Rhode Island. pp. 1170–1177.

72. Alonso I, Yuval M, Eyal G, Treibitz T, Murillo AC (2019) CoralSeg: learning coral segmentation from sparse annotations. *Journal of Field Robotics* **36**, 1456–1477. doi:10.1002/rob.21915

73. Rossi P, Ponti M, Righi S, Castagnetti C, Simonini R, Mancini F, Agrafiotis P, Bassani L, Bruno F, Cerrano C, *et al.* (2021) Needs and gaps in optical underwater technologies and methods for the investigation of marine animal forest 3D-structural complexity. *Frontiers in Marine Science* **8**, 591292. doi:10.3389/fmars.2021.591292

74. Mullen AD, Treibitz T, Roberts PL, Kelly EL, Horwitz R, Smith JE, Jaffe JS (2016) Underwater microscopy for in situ studies of benthic ecosystems. *Nature Communications* **7**, 1–9. doi:10.1038/ncomms12093

75. Armstrong RA, Pizarro O, Roman C (2019) *Underwater Robotic Technology for Imaging Mesophotic Coral Ecosystems Mesophotic Coral Ecosystems*. pp. 973–988. Springer, Cham, Switzerland.

76. Pyle RL (2019) *Advanced Technical Diving Mesophotic Coral Ecosystems*. pp. 959–972. Springer, Cham, Switzerland.

1974	Report of National Estate Committee of Inquiry. AIMS opened at Cape Ferguson, North Queensland.
1975	The Great Barrier Reef Marine Park Authority (GBRMPA) established.
1977	Consortium of Reef Island Research Stations formed. Australian Marine Sciences and Technologies Advisory Committee established.
1978	*A Coral Reef Handbook: A Guide to the Flora, Fauna and Geology of Heron Island and Adjacent Reefs and Cays*. Publication by GBRC. GBRC consults to the GBRMPA on geology, geomorphology and biology of the Capricorn-Bunker Group.
1982	Assets of HIRS transferred to UQ. GBRC changed name to Australian Coral Reef Society (ACRS) and became an incorporated entity. GBR declared a World Heritage site. First GBR zoning plan comes into effect (Capricorn-Bunker group).
1983	Far northern GBR section zoning plan declared.
1987	Elizabeth and Middleton Reef Marine Nature Reserve established.
1991	Shark Bay declared a World Heritage Site.
1993	Elevated Nutrients on Coral Reef Experiment (27 research groups).
1993	Australia's National Resource Assessment Commission Coastal Zone Inquiry.
1994	A 25-year strategic plan for the GBR World Heritage site.
1997	75th Anniversary of ACRS – Heron Island Conference.
1998	Queensland East Coast Trawl Fishery mandates the use of turtle exclusion devices to reduce by-catch.
1999	The Whitsundays Plan of Management enacted. Dead coral limestone extraction in Moreton Bay, Queensland banned. Cairns Area Plan of Management enacted. Solitary Islands Marine Park, New South Wales established.
2000	Lord Howe Island Marine Park declared.
2001	Far northern GBR section zoning plan in operation.
2003	Jurien Bay Marine Park, Western Australia established. GBRMPA Representative Areas Program for the GBR Marine Park Zoning Plan. Review of fishing on the GBR, buy-back of licences. The GBRMPA released the Reef Water Quality Protection Plan: For Catchments Adjacent to the GBR World Heritage site. The GBRMPA accredited the first Traditional Use of Marine Resources Agreement (TUMRA) to a conglomerate of six Girringun Traditional Owner groups (Djiru, Gulnay, Girramay, Bandjin, Warragamay and Nywaigi).
2007	Lord Howe Island group declared a World Heritage site.
2008	*The Great Barrier Reef: Biology, Environment and Management* book – publication supported by ACRS with proceeds going to ACRS students.
2009	Two oil spills of significant threat to tropical marine environments. First GBR Outlook Report released by the GBRMPA (subsequently released every 5 years).
2011	The Ningaloo Coast declared a World Heritage site.
2012	Commonwealth Marine Reserve Network established.
2013	The Coral Sea Marine Park established.
2013	Abbot Point port expansion plan approved, marine scientists' Port Development letter.
2014	Queensland Ports Development Strategy released. Reef 2050 Long-Term Sustainability Plan. Senate Inquiry into the Management of the GBR.
2015	Abbot Point Report from ACRS.
2016	ACRS science-based policy plan for Australia's coral reefs – *Benefits of Mangroves* – science brief to Hon. Mark Butler MP. Great Kimberley Marine Park declared.
2018	GBR Partnership established ($444 million for the GBR Foundation). Dampier Marine Park established.
2019	Houtman Abrolhos Islands National Park established.
2020	Draft policy on the GBR interventions under review.
2021	Great Barrier Reef black teatfish export discontinued, guidelines developed for sustainable domestic harvesting.
2022	'Making Waves: A Century of Australian Coral Reef Science' Centenary conference and exhibition at the Queensland Museum and University of Queensland, Brisbane.

Fig. 6.1. The diversity of life on a coral reef. (Image credit: Matt Curnock)

The spectre of climate change arises

The 1990s marked a turning point, as the mass coral bleaching event of 1998 turned the focus of Australian coral reef scientists increasingly towards understanding the implications of climate change for coral reefs. The recently formed consortium of reef island stations coordinated an approach that brought together university researchers to make observations and carry out experiments using field-based facilities in order to elucidate how some of the environmental changes associated with climate change would play out on coral reefs (see 'Tropical marine network of stations' in Chapter 3). Two avenues of research in particular opened up to study: (1) how increasingly warmer seas and (2) acidifying oceans had an effect on corals (see Chapter 8).

As well as transitioning from puzzling out the fundamentals of reefs to better understanding the new emergent threat of climate change, coral reef scientists were increasingly communicating their findings outside of the realm of research journals. The science community found a new role for itself, sounding the alarm among the general public and with politicians about this new threat to Australia's coral reefs and the livelihoods that depended on them (see Table 6.2 below).

Scientists speak out for Australia's coral reefs

The dialogue for successful coral reef management does not take place entirely in scientific papers. This is a public conversation that takes place over decades, through the work of many hands and

Table 6.2.   Primary developments relating to climate change in Australian coral reef science

1928-29	First observations of coral bleaching on the GBR made at Low Isles during The Great Barrier Reef Expedition of 1928–29
1980s & '90s	Mild mass coral bleaching events affecting small areas of the GBR (e.g. 1987).
1998	Mass coral bleaching event, first global event recorded on GBR.
2002	Mass coral bleaching event (regional). The Townsville Declaration on Coral Reef Research and Management developed at the International Forum on Threats to Coral Reefs.
2005	The GBRMPA established GBR Coral Bleaching Response Plan. ARC Centre of Excellence for Coral Reef Studies established.
2006	Mass coral bleaching event (regional).
2009	Coral Reef Crisis Working Group and resulting statement released to media and ministers.
2011	UNESCO visit to discuss listing the GBR as 'in danger'.
2016	Global mass coral bleaching and mortality event. ACRS science-based policy plan for Australia's coral reefs – comments on reef policy plans in the 2016 election. *Discussion: Advancing Climate Action in Queensland.*
2017	Global mass coral bleaching and mortality event. ACRS letter to Prime Minister Malcolm Turnbull requesting immediate action to reduce carbon emissions to protect the GBR.
2020	Global mass coral bleaching and mortality event.

minds. To this end, the Australian coral reef science community has worked through university research groups, museums, environmental consultancies, and state and federal government departments to translate biological and geological scientific principles into timely, practical advice for Australia's coastal and marine environment.

Here, we outline some examples of scientists providing expert advice on coastal development, planning marine reserves and managing oil spills. A collective picture emerges of a dedicated community of scientists with diverse expertise, whose work goes beyond collecting data, doing laboratory experiments and reporting findings in scientific journals to amplify their voices to influence environmental governance of Australia's coral reefs.

Dredging for coastal development

Marine biologists and water quality experts have evaluated the environmental implications of coastal development proposals, notably the development or expansion of ports and marinas around Australia (Fig. 6.2). A particular concern for coral reefs is the impact of dredging and dumping of dredged material on benthic communities, such as seagrasses and coral reefs. Patches of coral can be directly smothered by slurry, or impacted by remobilised dredge spoil after it has been dumped. This can result in long-term stress from the prolonged supply of fine sediments, increased turbidity and particulate nutrients and pollutants to reefs.

Within the Great Barrier Reef Marine Park, development proposals for tourism infrastructure and

Fig. 6.2.   (A) A coal terminal at Hay Point on the central GBR. (B) Looking south towards Abbot Point from the mouth of the Burdekin River, anchored ships wait to dock and load up at Abbot Point. (C) Dredging activities at Onslow, WA. (D) The *Brisbane*, a cutter suction dredge barge, dumping dredge material from the Port of Cairns in the Great Barrier Reef World Heritage site. (Image supplied with kind permission by WWF-Australia; image credits: (A) Susan Curnock, (B, C) Matt Curnock, (D) Xanthe Rivett)

marinas have exposed corals to the impacts of dredging. However, it was the expansion of ports in response to the Queensland mining boom (2010–12) that pitted the environment against coal exports for ports at Gladstone, Rockhampton, Mackay, Bowen and Hay. The impacts of dredging activities on corals within the Great Barrier Reef Marine Park were becoming a matter of public concern, prompting the GBRMPA to develop a Port Strategy. Matters reached a climax in 2013 when, a week before submissions on the Port Strategy closed, it was announced that plans to expand the huge Abbot Point coal port in northern Queensland had been approved.

The Abbot Port expansion was needed to service the Carmichael mine, where premium coal would be extracted from in the Galilee Basin in rural Queensland and transported 400 km by train to the coastal port at Abbot Point, for export by ship through the GBR. This controversial port expansion proposed to dredge 3 million m^3 of sediment to make way for coal export ships, and then dump the dredged spoil inside the Great Barrier Reef World Heritage site. A 2013 report commissioned by the GBRMPA highlighted the potential for dredge material to migrate over larger spatial scales and longer timescales than had previously been appreciated [5]. The large volumes of sediment proposed to be dredged would likely smother seafloor life, and become remobilised to chronically impact nearby corals and exacerbate impacts upon an ecosystem already in decline.

The 'Fight for the Reef' campaign coordinated jointly between the Australian Marine Conservation Society and the Australian World Wildlife Fund was a large, successful public engagement exercise. Over 10 000 people gathered in Brisbane to rally against the Abbot Point dredging. The ACRS coordinated an open letter, signed by 280 concerned marine scientists to outline their concerns, which garnered international press attention and shifted the tone of the debate from one of 'Greens and activists fighting for the reef' to scientists sharing their expertise. Many members wrote submissions and reports in what became known as the 'coal versus coral war' (see 'Science meets the public, policy and management practitioners' in Chapter 8).

The GBRMPA revised its initial approval of the plans after several prominent scientists spoke out about the implications for marine benthic communities. This attention prompted AIMS and the GBRMPA to convene a 19-member expert panel to review the impacts of dredging, which were synthesised in a report that suggested a review of dredging policy in areas that had been designated for conservation [6]. The issue became influential within the 2015 Queensland state election when an unpopular proposal supported by the Queensland Liberal National Party (LNP) to move the dredge spoil dump site from the Marine Park to nearby Caley wetlands received 79 000 submissions largely opposing the idea. After one term, the LNP lost the Queensland seat to the Australian Labour Party, who supported a less controversial land-based dump site. The Reef 2050 Long-Term Sustainability Plan was released later on that same year, banning the disposal of capital dredge spoil within the Great Barrier Reef Marine Park and World Heritage site.

Designing marine protected areas

Much of Australia's marine estate is protected through a national network of protected areas (Fig. 6.3). The selection of locations for marine reserves typically combines physical, biological and socio-economic criteria to account for Traditional Owners, as well as the fishing and tourism industries. The size and placement of reserves can often be a controversial issue (see 'The story of the Coral Sea Marine Park: science, policy and advocacy' in Chapter 7). The success of marine protected areas requires that management objectives for biodiversity within the area be clearly specified and well understood. A manager may wish to protect a range of representative habitats in a region, or prioritise a threatened or commercially valuable species. Targeted research to encourage sustainable fisheries within Australia's marine protected areas may seek to estimate the size of area that it is necessary to protect in order to retain viable habitats, evaluate stocks of exploited species, identify vulnerable life stages and map out connectivity among reserves and broader links among ecosystems [7].

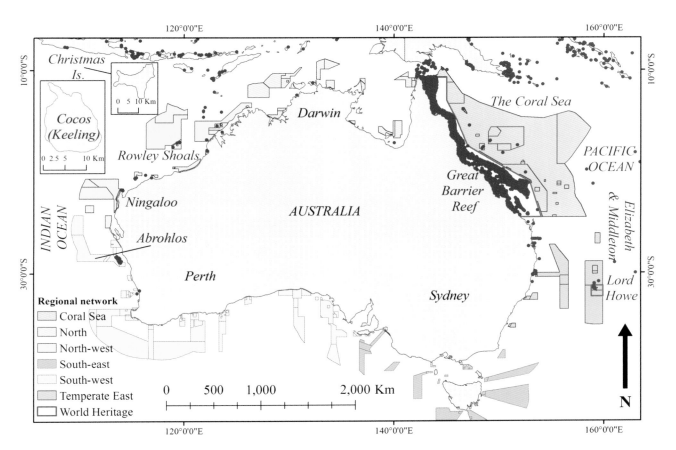

Fig. 6.3. Network of marine parks around Australia. The black outlines indicate parks that contain coral reefs and the red outline indicates World Heritage sites. (Image credit: Sarah Hamylton)

Marine biologists and conservation scientists have helped to design many of Australia's marine protected areas across a broad range of geographical scales, including Jurien Bay Marine Park (1851 km²) and Rowley Shoals (400 km²) and regional examples of larger, networked parks across the GBR (344 000 km²), the Coral Sea Marine Park (989 836 km²) and the Great Kimberley Marine Park (74 469 km²). At the national scale, conservation scientists have designed the Australian Marine Parks Network to include a representative range of different coral reef environments and, where possible, promote connections between protected reefs that allow for larval transfer of juvenile corals and spill-over of fish stock to occur between reefs.

The Ningaloo and Muiron Islands Marine Park

The World Heritage-listed Ningaloo coast is home to the largest fringing reef in the world along the North West Cape of Western Australia, ~1200 km north of Perth (Fig. 6.4). Ningaloo Reef sits at the convergence of regional temperate and tropical currents, resulting in diverse coral, algae and filter-feeding communities and abundant fishes. It also attracts several characteristic species with special conservation significance such as turtles, whale sharks, dugongs, whales and dolphins.

A management plan developed over 5 years for the ~2919 km² covering Ningaloo Marine Park and Muiron Islands Marine Management Area illustrates how different scientific experts can contribute to marine protected area management plans.

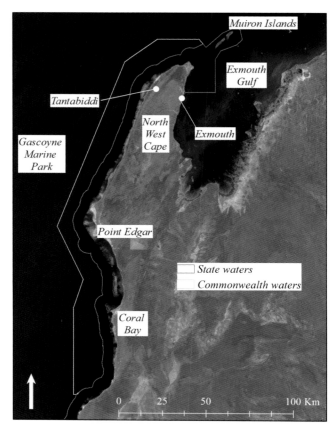

Fig. 6.4. Ningaloo Marine Park, Western Australia. (Image credit: Sarah Hamylton)

Fig. 6.5. A whale shark swimming with pilot fish. (Image credit: Kenny Wolfe)

Only by comparing against a meaningful benchmark of biodiversity within undisturbed benthic communities could any threats to existing communities be determined, alongside their potential for recovery and the adequacy of protection for any given area.

The hydrodynamic processes governing the circulation of seawater in and around Ningaloo were also identified as a critical knowledge gap. A major socio-economic objective of the park managers was to support tourism. One of the world's largest known seasonal aggregations of some 300 whale sharks (Fig. 6.5) occurs from March to June each year within 10 km of the reef crest inside the Ningaloo Marine Park. A nature tourism industry of 15 licensed tour groups takes visitors snorkelling with whale sharks, and is valued at $12 million annually to the region. The majority of interactions occur between Tantabiddi and Point Edgar in the north of the park and north-west of Point Maud in the central area of the Park. It is understood that the whale sharks aggregate where nutrient-rich waters upwell from the deep ocean. Biologists and oceanographers have worked together within the park to better understand spatial and temporal patterns in related biophysical variables (e.g. temperature, nutrients and zooplankton) that may govern whale shark aggregation and migration.

The wide-ranging consultation for drafting the Ningaloo Marine Park and Muiron Islands Marine Management Plan illustrates how conservation scientists, hydrologists, oceanographers, water-quality

Released in draft form in 2004, the plan received 5771 submissions during the 3 month consultation period, including advice from the Australian Museum, AIMS, the ACRS and several university research groups. This scientific advice spanned a range of themes, including the need to monitor undisturbed biodiversity inside and outside fishing sanctuaries, to better understand water circulation and provide for a full range of habitats within the protected areas. Importantly, the factors governing seasonal whale shark aggregations within the park needed to be better resolved.

Benthic ecologists emphasised the paucity of information on the undisturbed marine life in and around Exmouth Gulf, particularly filter-feeding communities on the inner reefs (many of which had been impacted by trawling), soft-sediment communities within and outside proposed fishing sanctuaries, and the inshore reefs that had been heavily fished.

experts and marine biologists can work together to design a comprehensive marine protected area that balances both biophysical and socio-economic needs along Australia's coastline.

The Montara oil spill

An extensive offshore oil and gas drilling industry has developed inside Australia's Exclusive Economic Zone, which extends some 200 nautical miles offshore from the coastline to cover a total area of 8.2 million km². Offshore exploration has increased in Commonwealth waters steadily since the 1970s, regulated by the Australian Government, which issues permits, leases and licences and oversees the environmental management and oil well integrity of international oil and gas companies.

In the aftermath of several high-profile oil spills, including from the SS *Torrey Canyon* (England,

1967), the Santa Barbara oil well (USA, 1969) and the *Oceanic Grandeur* in the Torres Strait (1970), a royal commission inquiry was established into petroleum drilling in the area of the GBR (1970–74). Its main recommendation was that all drilling for oil within the GBR province be postponed and only permitted once the impact of a potential oil spill on a coral reef was known. This effectively ended drilling for oil around the GBR, although oil was still transported on ships through the region. Exploration subsequently focused on petroleum basins in deeper waters around the north-west coast, offshore from the Kimberley region. Since drilling began in the early 1900s, with major expansion of the industry in the 1970s, there have been several major oil spills from both oil wells and ships around the Australian coastline. Many of these have occurred close to fringing reefs (Fig. 6.6).

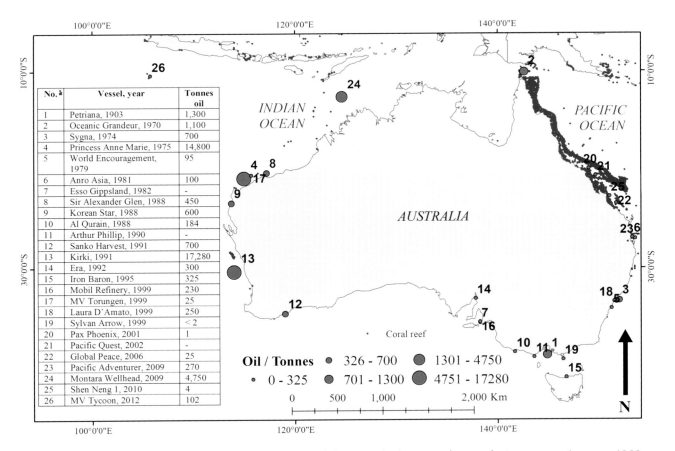

Fig. 6.6. Major historical oil spill incidents from wells and ships inside the Australian exclusive economic zone, 1903–2012. (Data source: Australian Maritime Safety Authority; image credit: Sarah Hamylton)

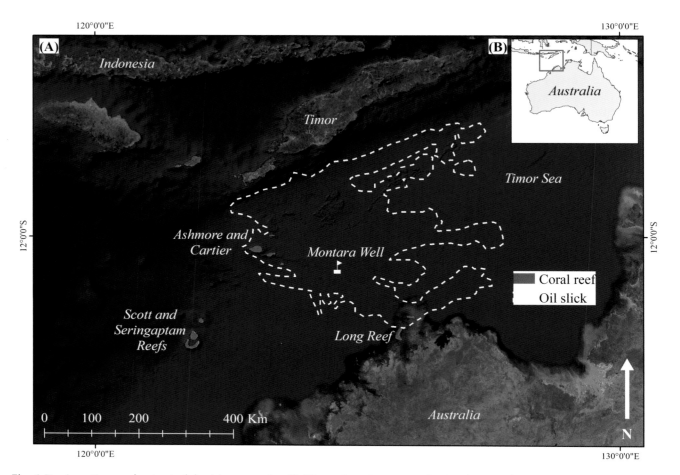

Fig. 6.7.   Location and extent of the Montara oil spill, Timor Sea. (Image credit: Sarah Hamylton)

On 21 August 2009, a blowout occurred on an oil well platform operated by PTTEP Australasia on the Montara Shelf between Australia and Indonesia. The well sat around 200 km off the Kimberley coastline with several reefs and islands nearby, including the Ashmore and Cartier reefs around 100 km to the west, Scott and Seringopatam reefs 250 km to the south-west and Long Reef around 150 km to the south-east, some 500 km west of Darwin along the mainland Australian coastline (Fig. 6.7). Crude oil flowed into the Timor Sea for 11 weeks before the well could be capped. The Australian Maritime Safety Authority (AMSA) estimated a daily flow of ~64 tonnes of oil into the adjacent waters, in what was Australia's biggest offshore petroleum industry oil spill. During this time, marine scientists mapped large volumes of

oil spreading across the sea surface from publicly available radar and MODIS satellite images of the region and called for greater efforts to manage the impacts of the spill on the regional environment and marine life, particularly nearby coral reefs. The delayed response was due to the lack of preparation for a spill. Chemicals to disperse the oil and a replacement drill rig equipment needed to be shipped from Singapore and Geelong, Victoria. AMSA undertook 130 aerial surveys to map the extent of the spill for planning containment operations, including spraying dispersant from above to degrade the oil and containing and recovering the oil via a boom operated by a pair of boats (Fig. 6.2C).

The Montana Commission of Inquiry was launched 2 days after the well was capped, resulting in significant reform of Australia's offshore oil and gas industry. In a submission to the Inquiry, the ACRS noted that the foreseeable delay in the response to the spill highlighted the need for greater pre-emptive safety measures to manage the risk of, and adequately respond to, spills to provide greater protection for coastal and offshore marine biodiversity. The need for a tighter approval process for offshore petroleum drilling, including the involvement of independent scientific bodies and greater transparency in the environmental assessment of activities of this industry, was clear. In response, the National Offshore Petroleum Safety and Environmental Management Authority was established to regulate Commonwealth waters for safety, well integrity, environmental management and day-to-day operations of petroleum activities. Under the recommendation of scientists from AIMS and CSIRO, who worked with ACRS to develop a series of reform recommendations, the authority also took on responsibility for monitoring the distribution and fate of oil, as well as impacts on marine wildlife, shoreline ecology and fish catches in nearby coastal and offshore marine environments.

The need to consider the transboundary effects of spills was not considered in the inquiry. Over 15 000 Indonesian seaweed farmers who operated along the coast of Rote Island, Indonesia, sought, and later received, compensation for cultivated seaweeds that had been killed or severely affected, with estimated losses of A\$2.4 billion because of the Montara incident [8].

Australia's role in international coral reef science and management

Pat Hutchings

Australia has been well placed to become a global leader in coral reef science and management, with vast areas of reef inside its exclusive economic zone and a relatively wealthy economy. There is a long history of Australian coral reef science being communicated internationally; indeed, the early collaborations between the University of Queensland,

the GBRC and the Royal Society led to detailed and extensive discussion and publication of Australia's earliest reef science at the Royal Society in London. The International Coral Reef Symposia became the main international meeting for communicating coral reef science, at which Australian reefs and scientists were well represented.

The 1st International Coral Reef Symposium was held in Mandapam Camp, India, in January 1969. This was convened by Dr David Stoddart and Sir Maurice Yonge, the latter having been involved in the Great Barrier Reef Expedition to Low Isles in 1928–29 (see 'The Great Barrier Reef Expedition of 1928–29' in Chapter 3). As Chair of the Continuing Committee for Coral Reef Symposia, Stoddart suggested to the GBRC that it would welcome an invitation from Australia to host the next meeting. In May 1969, the GBRC approved this proposal to host the 2nd International Coral Reef Symposium in 1973, which was subsequently held on board the 'Marco Polo', with a combination of research papers presented and field trips to Lizard Island, Low Isles and Heron Island. Held from 22 June to 2 July and convened by Dr R. G. Orme with Dr Patricia Mather, the Honorary Secretary of the GBRC who was instrumental in the organisation and planning, the symposium (supported by both the Australian and Queensland governments) attracted 264 researchers from 17 countries, with 120 papers presented. At the opening of the conference, the Federal Minister for Science conveyed to delegates that, beyond hoping for a successful meeting, he was looking to the recently established AIMS to make significant progress on increasing our knowledge and understanding of the Reef.

An enduring part of the symposium was the meeting of the Coral Reef Working Group of the Scientific Committee on Oceanic Research (SCOR), chaired by Stoddart and with a group of Australian and international scientists. Over a period of 8 days at the Heron Island Research Station, sampling techniques for studying coral reefs were developed and tested, both at Heron Island and at One Tree

Island. Their overall objective was to standardise coral reef research techniques to aid comparative studies between different reef areas. Their findings were published in a monograph of oceanographic methodology as the UNESCO Handbook on *Coral Reefs: Research Methods* [9]. For a long time, this was the prevailing international guide for coral reef fieldwork, accounting for recent advances in the study of functional coral reef ecology [10].

As well as considerable growth in the number of delegates and presentations at subsequent symposia, there has been an increasing focus on climate change and ocean acidification, reef management and governance issues [11]. In particular, the Coral Triangle Initiative on Coral Reefs, Fisheries and Food Security (CTI) brought together six South-East Asian nations to help coordinate research and management objectives. Similarly, it brought the governance of many Indo-Pacific reefs to the attention of Australia's coral reef science community, widely recognising for the first time the connectivity and fundamental importance of this area to the Australian coral reefs.

Translating science into management

At the 9th International Coral Reef Symposium held in Bali in October 2000, the Chair of the GBRMPA, Virginia Chadwick along with Alison Green, presented a paper entitled 'Managing the Great Barrier Reef Marine Park and World Heritage site through critical issues management: science and management'. It stressed both the need for science and the importance of ensuring that management decisions were based on the best scientific information available. This paper introduced an entire session on managing the world's largest coral reef ecosystem, the GBR, which had been declared a World Heritage site in 1981.

Subsequent papers dealt with the Representatives Areas Program (RAP) for protecting biodiversity of the GBR, which formed the basis of the 2004 rezoning plan. The RAP provided a clear scientific basis for the zoning of this multi-use marine park to allow a range of uses, fisheries, tourism and recreation while also reducing land-based impacts on the reef (see Chapter 5). The rezoning plan was the result of continual interaction between the researchers, managers and users of the reef over several years as the GBRMPA hosted committees on biodiversity and World Heritage sites, fisheries, tourism and water quality. This participatory approach ensured that the zoning plan was well accepted and facilitated its implementation; indeed, the RAP has since been widely recognised both in Australia and worldwide as a benchmark for zoning in marine parks [12].

Regional capacity building for understanding and managing coral reefs

The capacity of Australia to manage its reefs is partly due to its domestic wealth, which supplies resources and expertise for management, combined with lower human populations and a correspondingly limited potential impact from pollution, as well as lower reliance on subsistence fisheries relative to many developing countries of the world where coral reefs are found. In developing countries, small-scale fisheries provide critical nutrients and protein through the livelihoods of millions of women and men, for the food security of around 4 billion consumers globally, and over 1 billion low-income consumers in the developing world [13]. However, the science and management strategies developed in Australia have been applied, at least in part, in such countries.

Australian researchers have long worked in coral reef systems elsewhere in the world, especially in the South Pacific and South-East Asia. Immediately to Australia's north, the Coral Triangle represents a global epicentre of marine biodiversity. Recognising the need to safeguard the region's marine and coastal biological resources, the CTI was formed in 2007 as a multilateral agreement between the governments of Indonesia, Malaysia, Papua New Guinea, the Philippines, Solomon Islands and Timor-Leste. As a partner to the CTI, the Australian Government provided financial aid of $13.2 million to help establish multi-use, integrated planning of large marine areas across the Coral Triangle region. Under the Seascapes

Goal of the CTI Regional Plan of Action, the Federal Government has funded Conservation International to build an understanding of, and a capacity for, integrated marine planning and action [14]. The Australian Commonwealth Scientific and Industrial Research Organisation (CSIRO) has collated ecological and socio-economic information to identify priority seascapes for protection, assist with production of the Coral Triangle Atlas, and inform regional policy and planning through improved knowledge of the marine assets of the Coral Triangle and the threats to those assets. Similarly, support has been provided to encourage sustainable fisheries for Pacific Island nations. This includes protecting coral reefs through the international CTI and awarding scholarships for students from Pacific countries to undertake studies in fisheries, marine science, climate change and environmental management at Australian institutions [15].

The influence of Australian coral reef science and management: an Indonesian perspective
Nurjannah Nurdin

Indonesia has thousands of islands and reefs. In many ways, the Indo-Pacific community, particularly Indonesia, looks to Australia for guidance on best practice in reef science and management. Here, we compare the management frameworks and challenges on the GBR with those of a similar system of reefs and islands in Spermonde, Indonesia. Just as the GBR formed along the coast of Queensland, the reefs and islands of Spermonde have formed along the south-west coast of Sulawesi. The Spermonde shelf sits 40 km offshore, supporting 137 reefs and 54 islands which are comparable to those seen on the GBR, including mobile sand cays, vegetated cays and low wooded islands [16]. Geologists have dated some of the corals and found that, like the GBR, the Spermonde shelf was flooded as sea levels rose, so that the reefs we now see have formed over the last 6000–6500 years over a shallow continental shelf some 30–50 km wide [17].

National, regional and local coastal management partnerships

In Australia, responsibility for managing the GBR falls to several government offices. The Federal Office of Environment oversees World Heritage sites and the GBRMPA, while Queensland State Government agencies, such as the Queensland Parks and Wildlife Service, are responsible for national parks, the Queensland Department of Fisheries regulates fishing within the park, and each agency works in partnership with Traditional Owners and local industries, while coastal ports are managed by the State Government. The integrated coastal zone management frameworks developed in Australia in the 1970–80s in the years leading up to and following declaration of the Great Barrier Reef World Heritage site are a regional blueprint for coastal management. Indonesia has developed similar cross-jurisdictional partnerships to provide for the complex mix of marine, coastal and island issues and find ways to provide for the needs of a range of possible users in marine parks.

In Indonesia, at the national level, the Ministry of Marine Affairs and Fisheries was established in 1999 to encourage sustainable use of Indonesian coastal resources. The Coral Reef Management and Rehabilitation Program (COREMAP) started in 2001 to facilitate communities in marine conservation with the broader aim of supporting fisheries and raising public awareness of the need for, while fostering community participation in, coral reef management. This program has encouraged alternative income to improve the welfare of coastal communities, combining both top-down and bottom-up approaches to create healthy coral reef resources, abundant fish and a prosperous community. Coral reefs are protected and conserved through natural rehabilitation efforts, while the community participates by maintaining and utilising resources wisely. The COREMAP program empowers communities to protect and conserve coral reef resources to improve the welfare of coastal communities and small islands. One of the efforts to manage resources in marine waters is the establishment of marine conservation areas and marine protected areas that function as fish savings.

In Sulawesi, the marine environment is divided up into geographical areas for management. The Spermonde reefs are located in the administrative area of Liukang Tupabiring district, which falls within the Pangkep Regency of the South Sulawesi Province. The people living here depend fundamentally on the richness of marine resources for their livelihoods. Public understanding of the functional importance of coral reefs, for example as spawning ground for fish and offshore buffers to protect beaches, is critical and forms the basis for the formulation of coastal management policies.

Managing reefs for coastal livelihoods

Despite their similarities, there are different challenges for managing the islands in the Spermonde. Unlike the islands of the GBR, people live on the Spermonde islands (Fig. 6.8). Since the 1940s, the population of Makassar has expanded out to the reef islands, meaning that 50 out of the 54 islands now have houses built on them. Some individual islands have become crowded with large populations. The most heavily populated island is Barrang Lompo, an island ~300 m wide that sits around 12 km offshore from Makassar City, which now has around 4000 people living on it.

As with Australia, there is often a trade-off between regional economic growth and the

Fig. 6.8. Salemo Island in the Spermonde archipelago, with housing infrastructure, jetties and a mosque in the centre of the island. (Image credit: Sarah Hamylton)

condition of the reef environment. In Sulawesi, there is a pressing need to incorporate people's livelihoods into coastal environmental management, as is the case across much of South-East Asia. All the islands have formed on top of coral reef platforms, which are commonly covered with live coral and seagrass that support people to make a living from coral reef fisheries, seaweed farming and collecting crabs, squid, octopus and shellfish directly from the reef flats. Each community depends on coastal resources for their livelihoods in their own unique way (Fig. 6.9). On Sanane Island (population of 1198 people, 229 houses) most of the people are purse seine fishers, although other gear (fishing rod, longline and trawlers) is used. At Podang-Podang Lompo Island (population of 880 people, 189 houses), rod and net fishers catch shrimp and crab, while others are teachers, repairers and boat builders. Finally, on Saugi Island (population of 427 people, 102 houses) the main occupation is crab fishing, while other are craftsmen of wood and boats, and some make a living through crab peeling, fish and lobster cultivation.

The lack of sustainable methods in shrimp aquaculture and the use of destructive practices in the fisheries industry are two key challenges for managing reefs on which peoples' livelihoods depend in South-East Asia. In the 1980s, much of Sulawesi's mangrove forests were converted into brackish water ponds in a period of focused and sustained aquaculture expansion known as the 'blue revolution'. Government policies instructed provincial offices to intensify shrimp cultivation, which was then a profitable export commodity. In the intervening years, limited access to electricity on offshore islands has meant that aquaculture ponds cannot be properly aerated, so they have become anoxic and redundant.

The use of destructive fishing techniques, including dynamite and cyanide fishing, is illegal yet still practised in the Spermonde islands. In 2001, the local government established Pangkep District Regulation No. 10, which prohibited destruction of coral reefs in the waters of Pangkep Regencies, including extracting organisms

Fig. 7.5. Aerial photograph looking north across Coral Bay, Ningaloo Coast. (Image credit: Mike Van Keulen)

Fig. 7.4. The 2002 'Save Ningaloo' march in Fremantle, where 15 000 people gathered as part of a campaign in which the community indicated strong opposition to a major proposed resort on the Ningaloo Coast. (Image credit: Roel Loopers)

vetoed the construction of a white-shoe resort and marina near Coral Bay. This signified a major reprieve for the remote reef and its pristine shores, because it didn't just stop one misbegotten project; it also forestalled the stampede of opportunistic developments that would have followed in its wake. So, it was a momentous announcement. And a very rare choice for a West Australian premier to make. Forced to choose between a private business venture and an obscure stretch of coral, Gallop ruled in favour of the ecosystem. And given the quarry mentality that persists here on the western frontier, it remains a decision as historic as it was unlikely.

HOW THE REEF WAS WON. That headline ran nationwide, trumpeting the news that Ningaloo was safe. To some degree it reflected the story we told ourselves, even those of us who knew better, those who understood that the price of victory is eternal vigilance. But that's how it felt. Ningaloo was safer. Saved, even.

Gallop wasn't punished politically for sparing the reef. To the contrary. His Labour government rode a surge of popular support into a second term. And his decision unleashed a decade of progress in terms of how our second great coral ecosystem is conserved and managed.

In 2005 the Ningaloo Marine Park was enlarged. Sanctuary protection was extended from 10 to 34 per cent. Overdue planning controls finally secured the reef coast for low-impact nature-based tourism. And in 2011 the Ningaloo–Cape Range area was added to the list of World Heritage sites. It looked as if one of the planet's last healthy coral reefs was finally being given the status and protection it deserved.

The gains of that pivotal period weren't just environmental. They produced a major social and economic dividend, especially in Exmouth, the community closest to these World Heritage assets. The town's sustainable tourism industry began to boom. People came from all over the world to see what all the fuss was about. Suddenly everyone wanted to swim with a whale shark.

In the wake of these turning points, most decision-makers responsible for the region's future seemed to understand the link between the health of the reef and the fortunes of the local community. And as a result, Ningaloo looked like being that rare thing, an enduring good news story. Since then, the reef's global prestige has continued to grow, and Exmouth's tourism reputation has risen steadily along with it. Ningaloo's tourism brand is globally recognised. It's also becoming a marine science research hub and could one day become a specialised regional educational centre as well.

Compared to the GBR, the Ningaloo Reef is tiny, a mere 260 km long. But because it's so remote and because it's adjacent to arid rangelands that produce no agricultural runoff, it's in much better shape than its beleaguered eastern cousin. It's also much closer to shore. At many points you can snorkel at Ningaloo without needing a boat. You just wade out, put your face in the water and marvel. Wow, a turtle. Far out, a dugong. Visitors continue to come from all over the world to swim in the reef's clear waters, to see its vivid corals, sponges, sharks, rays and fish. For thousands of domestic and international tourists, the opportunity to swim with whale sharks, manta rays and turtles in a single excursion is unrivalled anywhere in the world. For the past 4 years, tourists have also been able to enjoy guided swims with humpback whales. Hundreds of people have told me that a visit to Ningaloo is 'life-changing', and I know how they feel. I've been swimming and snorkelling and surfing there for 30 years and the sense of awe and privilege it inspires never wanes.

But here's the thing: Ningaloo isn't saved. Worse than that, its future is now in jeopardy. Because during all those years of celebration and consolidation, despite the success of ecotourism and the research attention and the prestige that came with World Heritage listing, the fossil fuel industry was moving in. And by moving in, I mean territorially, culturally, and politically. While the reef still has passionate local defenders and hardworking government agencies looking to its welfare, the sense of stewardship once evident among local decision-makers has faded to little more than lip service. Civic leaders still tout the virtues of 'World Heritage values' and 'sustainable tourism', but those buzz-words sound pretty hollow while those who use them court the oil and gas giants and promote industrial projects deeply incompatible with the values they signify.

Some Australians will be surprised to learn how influential Woodside, Rio Tinto and Chevron have become in a nature-based tourism town like Exmouth. Many more would be shocked to see what a map of offshore oil and gas tenements in the Ningaloo region looks like today. In some places our historic 'line in the sand' looks faint indeed. Few Australians understand how close the rigs are to Ningaloo Reef already. Or how hard it's been for conservationists and regulators to maintain the slim buffer that currently exists between the drills and pipes and the World Heritage property. At night the rig flares are visible from the beaches and lagoons. The monstrous flame of Chevron's Wheatstone gas facility at Onslow lights up the sky like an endlessly rising moon. The flare is visible from Exmouth – 60 km away.

Chevron's massive and contentious Gorgon LNG operation at Barrow Island was granted permission to operate under strict terms. Expected to emit 6.1 million tonnes of CO_2 per annum, the operator committed to sequestering 3.4 million tonnes of that pollution. But Chevron has never complied. Millions of tonnes of methane and CO_2 continue to spew from this project every year. The LNG industry has grown to become one of Australia's most egregious polluters. And don't fall for the furphy about gas bolstering our nation's energy needs; most of it is exported.

Ningaloo Reef is encircled by oil and gas operations. When tourists catch a glimpse of those sinister flames on the horizon, it is unexpected and unsettling. But the jarring reality is minimised by local civic leaders. Nobody wants to talk about possible leaks or explosions. And the prospect of a spill as catastrophic as the one at the Montara platform is quickly dismissed. That great oil disaster of 2009 happened further north, in the Timor Sea.

Somewhere safely foreign-sounding. The industry's local defenders assert that something so dreadful could never happen in the oilfields off Ningaloo. But even without a spill or a blowout, the oil and gas industry remains the biggest threat to the reef's survival. Which makes it the greatest threat to Exmouth's future. Because the most significant danger to the world's coral reefs is the unchecked emission of carbon dioxide. Some of Australia's biggest carbon polluters are right on Ningaloo's doorstep. Their flares are visible to the naked eye. Their emissions are not. But they are real and proximate. And the danger they present must not be ignored or downplayed. The current gas rush on the North-West Shelf is no small phenomenon. But like all resource booms, it will soon pass. The benefits of the industry are not only overstated and ephemeral – they also come at an unacceptably high cost. Because their negative consequences will endure forever. Long after the LNG industry's demise, the dangerous CO_2 released by extraction and combustion will remain in our atmosphere. Heating the oceans, turning them acid, killing coral. Degrading the conditions of life on Earth. So when you consider Ningaloo's fate in those terms, it's reasonable to feel more than just a passing moment of discord at the idea of this industry continuing to explore, operate and expand. The Big Fossils are a real and present danger. Not just to Ningaloo. And not only to coral reefs globally, but to human survival.

But now to a bit of good news. Although the fossil fuel industry has colonised and degraded the coastal landscapes of the Pilbara to the near north, it's never managed to establish a beach head ashore in the Ningaloo region. Until recently, such a prospect was inconceivable. Because in that part of WA, Ningaloo has been the exception. But, in recent years, regulatory oversight has been slipping and, even as its social licence falters and fades, the industry's political influence seems to increase. And as cheaper, cleaner renewables present better alternatives, the gas rush only intensifies. Big Fossil operators seek opportunities to increase returns, enhance their influence and enlarge their opportunities. Some local decision-makers are all-too-eager to oblige them. Sadly, this includes the idea of bringing industrial projects ashore at Ningaloo.

In 2017, news came to light about the ambitions of a Norwegian company called Subsea 7 to build a 500 ha pipeline assembly and launch facility at Heron Point, deep in Exmouth Gulf. This proposal threatened considerable impacts to both the terrestrial and marine environments. Just to build the two 10 km rail lines needed to haul pipelines to the water, a lot of native vegetation needed to be bulldozed. A 380 m launch-way was to be cut through the dunes and laid across the untouched beach and over the corals and sponges of the intertidal zone so tugs could drag these enormous steel juggernauts into the water. Then, in lengths of up to 10 km each, and stabilised by 500 massive pendant chains, they'd be towed 40 km north through the gulf and out through the Ningaloo Marine Park to gas fields across the horizon.

This was always a terrible, reckless idea. The attractions for Subsea 7 included access to cheap land, suitable terrain and a very compliant local shire. But Exmouth Gulf isn't just another scarred bit of Pilbara real estate. It's Ningaloo's nursery. A major rest area and nursery for the world's most successfully recovering humpback whale cohort. It's also the foraging ground for one of the world's last stable dugong populations. Home to rare species of dolphins, sawfish and sea snakes, it is important habitat for several species of endangered turtles. So this is a manifestly inappropriate site for such a large and intrusive industrial operation.

Plans for this project were well advanced before the tiny local environment outfit, the Cape Conservation Group, got wind of them in October 2017. At that time, the Shire of Exmouth was without an elected council and under administration following adverse findings at WA's Crime and Corruption Commission. Without the alertness of a few volunteers, this project might have sailed through without assessment by the Environmental Protection Authority (EPA). Which shows how close the unthinkable can come to being quietly inevitable.

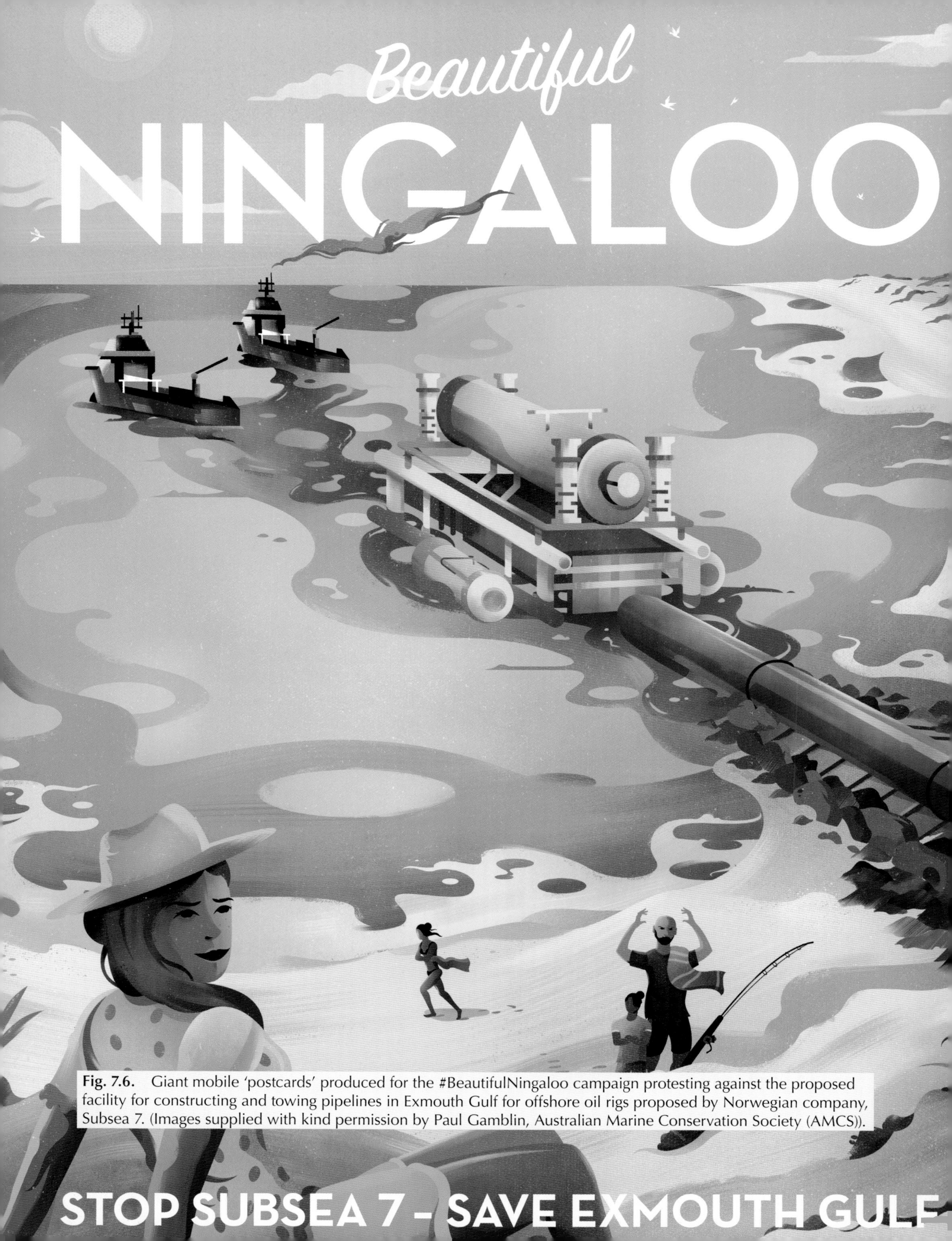

Fig. 7.6. Giant mobile 'postcards' produced for the #BeautifulNingaloo campaign protesting against the proposed facility for constructing and towing pipelines in Exmouth Gulf for offshore oil rigs proposed by Norwegian company, Subsea 7. (Images supplied with kind permission by Paul Gamblin, Australian Marine Conservation Society (AMCS)).

The local chamber of commerce was hotly in favour of Subsea 7's project. And the shire was so eager to oblige the proponent, it was determined to wind back local zoning protections to smooth the way.

The Subsea 7 project and the plight of Ningaloo's nursery in Exmouth Gulf quickly spilled beyond the confines of Exmouth. And the reasons for that are straightforward. It may be a long way from the centres of power, but this is an ecosystem of national and international significance. With every passing year it attracts more research interest.

Exmouth Gulf is one of the last intact arid-zone estuaries left in the world. Two thousand six hundred square kilometres of unique and biodiverse waterway, it has massive, untouched mangrove forests, tidal flats, seagrass meadows and heat-resistant corals. These habitats are where billions of small fry begin life before graduating to open waters. Exmouth Gulf supports 800 species of teleost fish alone. That's twice as many as Ningaloo Reef. The International Union for the Conservation of Nature (IUCN) has long acknowledged the gulf's World Heritage values and recommended it be included in the Ningaloo World Heritage estate. But to date these recommendations have fallen on deaf ears in WA. And now more than ever, this remarkable and unrepeatable ecosystem warrants increased conservation, not industrial-scale pressure. Because any degradation of the gulf is likely to have implications for the integrity of Ningaloo Reef.

In 2017 conservationists and researchers feared that if by some perverse circumstance Subsea 7 got approval, much of the protection Ningaloo has garnered in the past 20 years would exist only on paper. The security so many Australians battled to win for this world treasure would evaporate because this proposal would beget others along the shores of Exmouth Gulf. It loomed as a gateway project. Approval would signal that the gulf was now open to the oil and gas industry. And that meant that Ningaloo's nature-based tourism hub at Exmouth could face the kind of industrial development that's already disfigured coastal communities like Onslow, Dampier and Port Hedland to the near north. Should that happen, the low-impact ecotour industry at Ningaloo would be hammered. The health of a major natural icon would be compromised. And WA's excellent reputation as a custodian of World Heritage properties would be in jeopardy.

The status Ningaloo currently enjoys was not achieved by accident. It's the fruit of many years of scientific attention and unstinting conservation advocacy. It's been the site of generational social and political change. For me, it's a beacon of hope. I'm proud to have been a part of the Save Ningaloo campaign in the early 2000s that helped stimulate all this. During that period 100 000 Australians stood up for the protection of Ningaloo Reef. In 2002, 15 000 citizens marched through the streets of Fremantle to save it (Fig. 7.4).

I'm also proud to be one of the founders of the Protect Ningaloo campaign that sprang up in in 2017 to defend Exmouth Gulf from industrialisation. Our efforts follow, quite literally, in the footsteps of that earlier enterprise. The original defenders of Ningaloo are still with us – in their thousands. In the years since 2002, many have become parents and grandparents with an even deeper stake in the future. They've brought family, neighbours and friends into the cause and together they've demonstrated a mainstream determination to hold the line in the sand at Ningaloo.

I'm glad to report that Ningaloo's nursery has become the focus of increased national attention the way its sister-system was nearly two decades ago. Protect Ningaloo's national advocacy ensured that the Subsea 7 proposal received proper scrutiny and faced rigorous assessment by the EPA. In 2019, 54 000 people made submissions to the EPA against that proposal. In December 2020, Subsea 7 finally withdrew its application.

So, once again, there was good news for Ningaloo. But Exmouth Gulf isn't out of the woods yet. Just as the waterway begins to get the scientific attention and public interest it deserves, and just as our extinction crisis and climate emergency begin to hit home, industrialisation looms once more.

In late 2020, a company called Gascoyne Gateway Ltd announced its ambition to build a major deep-water port for oil and gas, mining, military and cruise-ship uses on the gulf's western shore. And K+S Salt is now seeking to build a large industrial salt works on the untouched tidal flats of the eastern shore. To celebrate in these circumstances would be foolish.

If Exmouth Gulf becomes a noisy, crowded industrial waterway subject to increased pollution, heavy shipping movement, dredging and clearing of sensitive habitat, a World Heritage-level ecosystem will be put in jeopardy and the sustainable economy that depends on it will be undermined.

Ningaloo is an exceptional place (Fig. 7.5). It supports exceptional creatures and exceptional biodiversity. It attracts exceptional people. It has enjoyed a period of exceptional management. But to endure and to flourish it requires continued stewardship of an exceptional standard. That means bolstering protection, not weakening it.

Australia is such a big island, expensive to get to, costly to travel across, and its distances and spaces are daunting. Even so, there are a handful of places tourists will make epic journeys to visit. Ningaloo is one of them. Humans yearn for places of respite from the ugly madness of the corporate world. We need to know there are still some exceptional sites left intact, ecosystems whose richness and enduring health afford us hope for the future, even if we never get the chance to visit them ourselves. For our own sanity and honour we want to believe there are some places so special they'll never be offered up to industrialisation.

Not now. Not ever. I believe Ningaloo is one of those places. And I'm not the only one who thinks so.

That's why I think this can still be a good news story, why I'm determined to make sure it will be. Because that historic line in the sand remains. The era of cavalier exploitation must remain behind us. Ordinary Australians simply won't put up with it anymore. They understand that it's wasteful and dishonourable and that it threatens our future. That's why they'll defend an ecosystem like Ningaloo. For love of the place, out of hope for their children, and for the enduring principle of the common good against the interests of a powerful few.

The story of the Coral Sea Marine Park: science, policy and advocacy
Imogen Zethoven

The Coral Sea stretches from outside Australia's GBR to New Caledonia, Vanuatu, the Solomon Islands and Papua New Guinea. The Australian section spans nearly 1 million km² of tropical Pacific Ocean and contains a great diversity of habitats and wildlife, including 34 coral reefs and 56 sandy islets. Within its depths, seamounts and canyons punctuate one of the world's largest continental margin plateaus. The region is renowned for its ocean giants, including whales, sharks, tuna, marlin and swordfish, as well as seabirds, turtles and *Nautilus* (Fig. 7.7A). Some species here are found nowhere else in the world [13].

The Coral Sea was first charted by Matthew Flinders in the early 19th century and, until recently, it received very little attention. The region is of great interest to maritime archaeologists as many ships have been grounded on its perilous reefs. The Battle of the Coral Sea was a major air and sea battle in World War II, when the combined forces of Australia and the United States of America halted the Japanese naval advance across the Pacific Ocean. In 1982, two fully protected National Nature Reserves were established at Lihou Reef and Coringa-Herald Cays, which are important breeding and foraging sites for seabirds (Fig. 7.7B).

Japanese longliners started fishing the Coral Sea for black marlin and tuna in the early 1950s. An Australian longline fishery has operated out of Cairns since 1987, harvesting large catches of yellowfin and bigeye tuna in the 1990s. Recreational charter fishers operate here, alongside hand and trap collectors and trawlers targeting redfish, rock lobster, trochus, sea cucumber and aquarium

Fig. 7.7.   (A) A masked booby, *Sula dactylatra*, and (B) South-west Herald Cay, a fully protected national nature reserve in the centre of the Coral Sea. (Image credits: Daniela Ceccarelli (A); image supplied with kind permission by the Australian Department of Home Affairs, photographer unknown (B))

fishes. Catches are brought into the Queensland coastal ports of Mooloolaba, Townsville and Cairns to be sold in domestic Australian and global markets.

Australia's vast marine protected area: a square peg in a round hole

In 2005, The Pew Charitable Trusts embarked on the Global Ocean Legacy project, which sought to identify national sites around the world that could be permanently and fully protected [14]. The organisation was concerned that protection of the sea lagged far behind protection on land. Even as recently as 2008, only 0.2 per cent of the total marine area under national jurisdiction was fully protected [15]. Pew identified a few oceanic-scale, biologically important and unspoiled marine ecosystems, including Australia's Coral Sea, where very large marine parks were feasible. These areas could afford to offer partial refuge to severely depleted pelagic (open-water) fish and serve as reference sites for studying the effects of climate change and ocean pollution.

The Australian Government had already established a comprehensive and representative national network of protected 'multi-use' zones inside which a range of activities were permitted. In Australia, Pew's goal of establishing a large, fully

protected zone was the equivalent of fitting a square peg into the round hole of the existing national policy framework. Yet the two contrasting approaches had the potential to complement each other by achieving separate conservation goals and it was therefore critical to build the scientific case for pursuing both.

In 2010, Pew released the 'Scientific Rationale for the Designation of Very Large Marine Reserves' global statement. Signed by 245 marine scientists from 35 countries, it called for the scope of marine parks to match the scale of important ecosystem processes of dispersal and migration that would ensure the long-term recovery of highly mobile species. It substantiated the case for large, fully protected reserves.

As very little was known about the Coral Sea's biodiversity, scientists and conservationists also had to justify why the Coral Sea warranted protection. In 2008, a further statement outlined the unique natural values of the Coral Sea, including providing critical habitat for threatened turtles, whales, dolphins and seabirds [16]. It emphasised that the Coral Sea was one of the few places remaining on Earth where large pelagic fishes (tuna, billfish and sharks) have not been severely depleted and also described the region's unique military and civic heritage [17, 18].

Battle of the Coral Sea campaigns: Barry Wrasse vs the 'lock-up' scaremongers

The campaign to establish the Coral Sea Marine Park was launched at the Australian National Maritime Museum in Sydney in September 2008. Two retired Chiefs of the Royal Australian Navy were recruited to publicise the wartime heritage value of the area and advocate in Canberra, alongside environmental non-governmental organisations (NGOs) under the banner 'Protect our Coral Sea'. Delegations of environmental NGOs, scientists and retired navy chiefs met with government Ministers, Shadow Ministers, Opposition backbenchers and the Australian Greens to present the case for the marine park.

On 19 May 2009, the federal Environment Minister Peter Garrett declared a Coral Sea Conservation Zone. Under the *Environment Protection and Biodiversity Conservation Act 1999* (EPBC Act), Conservation zones provide temporary protection for an area while its status as a potential marine park is under discussion. The campaign had achieved a milestone in securing recognition of the Coral Sea, which now had its own identity as a distinct marine treasure.

The announcement of the Conservation Zone drew fierce opposition from the charter fishing sector. A major scare campaign was mounted, focused mainly on Cairns, but also spreading the word along the Reef coastline that the Coral Sea was undergoing a 'lock-up' and claiming that a complete fishing closure of the Great Barrier Reef Marine Park would be next. In the Australian Parliament, the Coalition Opposition attempted to disallow the conservation zone in the Senate. Although the move was narrowly defeated, it revealed the extent to which the fate of the Coral Sea and the wildlife it supported hung in the balance of both public and political opinion.

The alliance of conservation groups launched a high-profile public communications campaign soon after the announcement of the conservation zone to engage and motivate a wide audience to support protection of this remote yet important region off east Australia. Campaigners understood

that protecting such a large area was ambitious, and that success depended on a combination of political championing and widespread public support, critically in local areas.

Barry Wrasse was the star of the public campaign: a memorable and friendly fish who was broadcast in an advertisement on commercial TV, in cinemas and on the inflight entertainment of a major airline (Fig. 7.8). The campaign also produced a short film about the Battle of the Coral Sea, including interviews with two war veterans in which they conveyed their strong support for protecting the Coral Sea's natural and cultural heritage values for future generations. The film proved to be an important and persuasive part of communication with the then federal Minister for the Environment, Tony Burke of the Labour Party.

Fig. 7.8. Barry Wrasse, the memorable and friendly face of the Protect Our Coral Sea campaign.

The Protect our Coral Sea campaign held conversations with people from all walks of life through community outreach efforts in Cairns, Brisbane and Sydney to raise awareness of the Coral Sea, enlist their support for the protected area, and allay fears of an inshore 'lock-up'. To keep building the case for protection, The Pew Charitable Trusts commissioned a detailed, independent biophysical study of the region, which confirmed the biological importance and relatively intact status of the Coral Sea [19]. It found that the Coral Sea was probably the only tropical oceanic environment not markedly impacted by fishing, and where a very large scale protected area could be established and effectively managed.

Protected areas shaped by lobbying: a backward step for the Coral Sea

After more than 2 years of political advocacy, public communications, local outreach and online mobilisation, Minister Burke announced a proposal to establish the world's largest marine park in the Coral Sea. In a November 2011 media release, the minister said: 'There is no other part of Australia's territory where so much comes together – pristine oceans, magnificent coral, a military history which has helped define us and now a clear proposal for permanent protection.'

The government proposed one single fully protected marine zone in the eastern half of the reserve, covering 51 per cent of the proposed 989 842 km^2 marine park (Fig. 7.9A). In the remainder of the park, other forms of fishing, including charter fishing and longlining, were permitted. While there was popular support for the marine park, the fishing limits were not embraced by everybody. The minority Labour Government was extremely cautious and wary of jeopardising support from their recreational fishing voters. The Protect our Coral Sea alliance had mobilised hundreds of thousands of supporters to back the proposal. In December 2012, the Coral Sea Marine Reserve was declared. The next step was to develop a management plan, which would introduce a zoning system and reduce fishing effort.

The EPBC Act required two phases of public consultation for a management plan. Progress was slow, with stakeholders and supporters experiencing consultation fatigue. On 12 March 2013, the final plan was tabled in Parliament, along with management plans for four other Commonwealth marine bioregions: the Temperate East, the South West, the North West and the North. Fifteen Parliamentary sitting days were required to pass in both houses before the plans could come into effect. In early June, the Liberal Opposition sought to disallow the plans in the House of Representatives, but failed. However, before 15 sitting days could pass in the Senate, Labour Prime Minister Kevin Rudd called an election, which was subsequently won by the opposing Liberal Party.

Shortly after the election, the Liberal Government, under the leadership of Tony Abbott, announced a review of the proposed management plans, establishing a scientific expert panel and, for each region, a stakeholder review panel. Commercial and recreational fishers were invited to join these stakeholder panels, but not conservation groups. Concerned at the direction things were going, the independent Ocean Science Council of Australia released a statement signed by well over a thousand Australian and international scientists calling on the government to expand the fully protected zones throughout the Commonwealth marine estate.

In January 2018, the Liberal Government, now under the leadership of Malcolm Turnbull, introduced a radically different Coral Sea management plan to Parliament (Fig. 7.9B). The single, large, fully protected zone was replaced by several much smaller zones. Full protection had been cut by 53 per cent, from 502 626 km^2 to 238 400 km^2. A trawling zone of 4300 km^2 in Labour's management plan was expanded to 66 480 km^2 in the Liberal Government's plan. Over the entire Commonwealth marine estate, fully protected zones were cut by almost 400 000 km^2 – the deepest cut to protected areas anywhere in the world.

The Labour Government had set aside $100 million for fisheries structural adjustment to

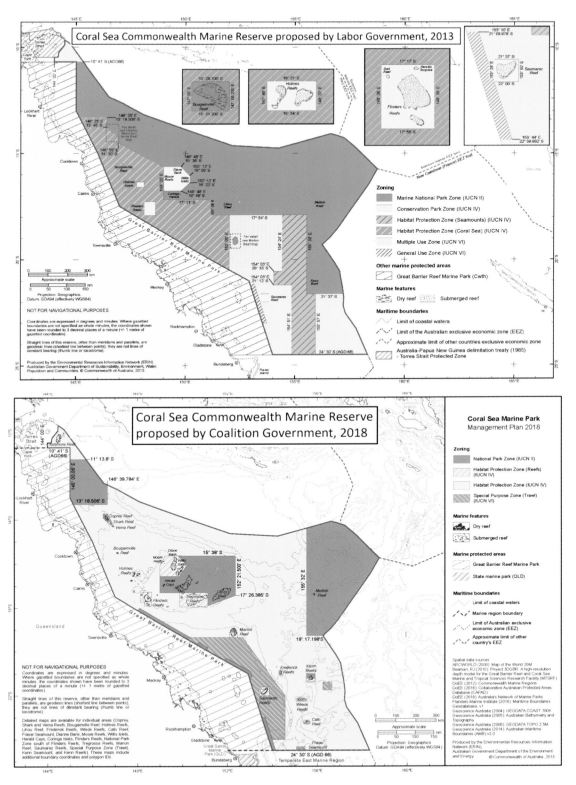

Fig. 7.9.   Maps outlining the Coral Sea areas proposed for protection by (A) the Labor Government in 2013 and (B) the Coalition Government in 2018, led by Malcolm Turnbull. (Image credits: Environmental Resources Information Network)

offset impacts from its proposed marine parks across the entire Commonwealth marine estate. In contrast, in 2016 the Coalition Government announced a $56 million package which included funding for fisheries adjustment, community engagement and science. In the fullness of time, only $9.5 million would become available through Fishing Business Assistance Grants, with some funding for vessel-monitoring systems. This figure is dwarfed by the package of $200 million, which was released under the fisheries structural adjustment policy following the 2004 rezoning of the Great Barrier Reef Marine Park.

In a show of support for better understanding the Coral Sea marine life, the government funded James Cook University to carry out a 3-year investigation of the health of corals and fish communities in the Coral Sea Coral Reef Health Project. This revealed unique coral and reef fish communities that are distinct from those of the adjacent Great Barrier Reef Marine Park and similar to temperate rocky reefs in the Tasman Sea to the south and New Caledonia, Vanuatu and the Solomon Islands to the east [20].

In 2010, the Australian Government signed a declaration with France–New Caledonia on the management of the Coral Sea. Two years later, New Caledonia declared it would establish a 1.3 million km² multiple-use marine park and full protection was to follow for all the coral reefs inside the park, an area covering 28 000 km² – a stunning achievement.

Perhaps the take-home message from the story of the Coral Sea Marine Park is that Australia's marine environment is most comprehensively protected when conservation groups and coral reef scientists work together alongside local stakeholders. Climate change is a real and present danger to the corals and biodiversity of the Coral Sea, as it is to coral reefs globally. Fully protected areas are more resilient to these changes than fished areas [14]. It follows, therefore, that substantial protection is needed to maintain the values of Australia's Coral Sea Marine Park. Despite the drastic loss of legal protection in recent years, there is reason for

people who have an interest in protecting the reefs of the Coral Sea to be hopeful. When our understanding of the biology and ecology of outstanding marine treasures is expanded, alongside critical community support for their permanent and full protection, these unique natural wonders are best placed to receive the responsible environmental stewardship they deserve.

Australian coral reefs on the World Heritage list
Jon C. Day

Australia's World Heritage sites (WHSs) include the GBR, the Lord Howe Island Group, Shark Bay and the Ningaloo Coast. Each WHS has been formally recognised by UNESCO's World Heritage Committee and inscribed on the List of World Heritage. Such a listing may only occur after the values within an area have been assessed as being exceptional, superlative or universally outstanding.

These four Australian WHSs are some of the most important coral reefs in the world. For example, when the GBR was listed as World Heritage in 1981, the IUCN evaluation stated: 'if only one coral reef in the world were to be chosen for the World Heritage List, the Great Barrier Reef is the site to be chosen.' Today the GBR is believed to be the most biodiverse WHS on the planet due to its extensive cross-shelf and latitudinal extent [21].

Worldwide there are 29 WHSs containing coral reefs [22], including internationally significant areas such as the GBR, the Belize Barrier Reef, Aldabra Atoll (Seychelles), Papahānaumokuākea (USA), Tubbataha Reefs (Philippines) and Phoenix Islands (Kiribati). Collectively, these 29 WHSs are considered globally to be among 'the best of the best' for coral reefs.

The basis for World Heritage listing
A World Heritage designation recognises the global significance of a place based on its natural and/or cultural values. Provided those values meet one or

more of the 10 criteria in the World Heritage Operational Guidelines [23], the area may be considered to be of outstanding universal value (OUV), and then listed as a WHS.

The term 'OUV' is a fundamental cornerstone for World Heritage and is central to an area's nomination, and then its subsequent listing, monitoring and periodic reporting as a WHS. OUV explicitly requires values that are globally significant. While other values within a WHS may be recognised as being nationally, regionally or locally significant, only those values that have been assessed as being universally outstanding and globally significant comprise its OUV.

Six of the 10 overall criteria for OUV are regarded as 'cultural criteria' (numbered (i)–(vi); see [23] for further details), with the remaining four (vii)–(x) considered as the 'natural criteria' as shown below [23]:

(vii) contain superlative natural phenomena or areas of exceptional natural beauty and aesthetic importance

(viii) outstanding examples representing major stages of earth's history, including significant on-going geological processes, or significant geomorphic or physiographic features

(ix) outstanding examples representing significant on-going ecological and biological processes in the evolution and development of terrestrial, fresh water, coastal and marine ecosystems and communities of plants and animals

(x) contain significant natural habitats for in-situ conservation of biological diversity, including threatened species of outstanding universal value from the point of view of science or conservation.

The 1972 World Heritage Convention calls for the identification, protection, conservation, presentation, and transmission to future generations of cultural and natural heritage that is 'so exceptional as to transcend national boundaries and to be of common importance for present and future generations of all humanity' [23]. In 2021, there were 218 natural, 897 cultural and 39 mixed properties on the World Heritage list, which equates to a total of 1154 WHSs. The World Heritage Committee consists of representatives from 21 of the States Parties that have signed the convention, and is elected by the UN General Assembly, usually for a 4-year term. The committee meets annually and is responsible for the implementation of the convention.

Statement of Outstanding Universal Value

Each WHS has a Statement of Outstanding Universal Value (SOUV) that outlines the relevant criteria, the key values and why the WHS is considered to be universally significant. The SOUV is adopted by the World Heritage Committee at the time of a property's inscription and includes statements about the *integrity* and/or *authenticity* of the property, along with an outline of the protection and management system to ensure the values endure (integrity requires the WHS to be of adequate size, protected from threats, and including all the elements necessary to express its OUV; authenticity is only applicable to WHSs inscribed for their cultural values). The requirement for a SOUV was introduced in 2005, with its fundamental purpose being 'the basis for the future protection and management of the property' [23].

Australian coral reefs designated as World Heritage

The four WHSs containing corals within Australian waters are listed in Table 7.1 and their locations around Australia are shown in Fig. 7.10.

Once a WHS has been inscribed on the World Heritage List, there are usually some potential benefits. These include increased tourist visitation, employment opportunities and income for local communities, as well as a greater focus on management and protection of the area. Listing may also be accompanied by greater scrutiny, particularly due to the area's internationally acknowledged importance.

Table 7.1.   Australian World Heritage (WH) properties containing coral reefs

	Australian WH property	Year inscribed on WH list	WH criteria for listing (refer to the text)	Importance of coral to the WH property
1	Great Barrier Reef	1981	(vii), (viii), (ix), (x)	'the world's most extensive coral reef ecosystem … Includes coral reefs, shoals, sand banks and coral cays and over 600 types of coral …'
2	Lord Howe Island Group	1982	(vii), (x)	'the most southerly coral reef in the world, … a rare example of a zone of transition between algal and coral reefs'
3	Shark Bay	1991	(vii), (viii), (ix), (x)	'corals occur in Shark Bay (over 80 coral species identified), but it is not a significant area for coral'
4	Ningaloo Coast	2011	(vii), (x)	'the world's largest fringing reef … more than 300 documented coral species'

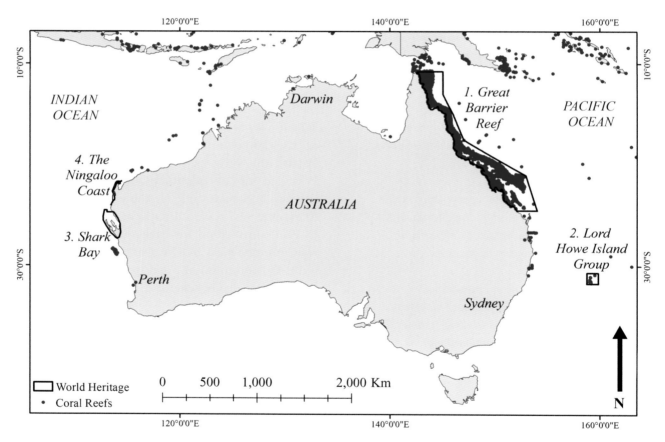

Fig. 7.10.   The four Australian World Heritage sites containing coral – numbers refer to Table 7.1. (Image credit: Sarah Hamylton)

World Heritage reporting

At each annual meeting, the World Heritage Committee examines reports addressing the state of conservation of a select number of WHSs. These reports are prepared by the World Heritage Centre and/or the Advisory Bodies and may be based on information collated from the Advisory Bodies, the relevant State Party, an NGO, or from an appropriate publication [e.g. 24, 30]. The committee may also request a State Party to prepare a specific report addressing the threats to the OUV of a particular WHS. Over the years, Australia has been requested to provide State Party reports on the GBR, Kakadu, Wet Tropics, Tasmanian Wilderness, Shark Bay and Macquarie Island.

The committee may also request a State Party to undertake specific actions when the values of a WHS are being adversely impacted. For example, in 2012, the committee requested several actions for the GBR, including a strategic assessment and a long-term plan for the sustainable development of the WHS, an independent review of Gladstone Harbour management, a request to 'ensure that development is not permitted if it would impact individually or cumulatively on the OUV of the [GBR]', and 'an explicit assessment of the OUV in future GBR Outlook Reports'. The committee also warned that 'in the absence of substantial progress, the possible inscription of the [GBR] on the List of World Heritage in Danger' [25].

'In danger' listing

The World Heritage Committee may decide to inscribe a WHS on the List of World Heritage in Danger if the OUV is faced with 'specific and proven imminent danger' or 'potential major threats that may have deleterious effects on its OUV'.

The factor(s) that are threatening the values and /or the integrity of the WHS must be amenable to correction by human action, and some States Parties therefore argue that climate change is not applicable for 'in danger' listing. Some legal experts and NGOs, however, have mounted a cogent argument that in danger due to climate change is consistent with the operational guidelines [26].

Significant threats impacting Australian WHSs have prompted the World Heritage Committee to indicate a possible 'in danger' listing on at least two occasions. The first instance was over uranium mining impacting the values of Kakadu National Park in 1999. More recently, concerns over coastal developments impacting the GBR were raised in

several State of Conservation reports. In both instances, Australia lobbied hard and managed to avert an 'in danger' listing [27].

Despite strong lobbying by certain NGOs who maintain that the GBR should be listed as in danger, to date this has not occurred. However, reports about the severe mass-bleaching events (2016, 2017 and 2020) have ensured the GBR has remained in the international spotlight.

The UNESCO List of World Heritage in Danger: a lever for reform
Tiffany H. Morrison

The GBR is one of ~250 ecosystems to hold UNESCO World Heritage status based on its OUV to humanity, alongside Vietnam's Halong Bay, Florida's Everglades and Ecuador's Galápagos Islands. Such ecosystems are accorded World Heritage status under the 1972 UNESCO World Heritage Convention, which is widely regarded as one of the world's most powerful global environmental regimes. This convention is ratified by 192 countries and generates discernible environmental impact due to direct and specific application at the ecosystem level, wielded through national government accountability.

The declining state of the GBR has led many concerned actors to look for additional national and international levers for reform to encourage better environmental stewardship of this site. UNESCO's List of World Heritage in Danger identifies sites facing major problems that threaten the very characteristics for which they were initially designated as World Heritage. The 'in danger' list is explicitly designed to bring global attention to a problem ecosystem, encourage emergency action, and mobilise national and international assistance, both financial and technical (see the previous section) [28, 29].

Experience in Belize demonstrates that the 'in danger' list can generate meaningful environmental benefits. The Belize Barrier Reef Reserve System was placed on the list in 2009 because of uncontrolled coastal development and mangrove loss. The World Heritage Fund provided technical and financial assistance for restoration. By 2018, mangrove coverage was back to 95 per cent and the entire maritime zone was under a moratorium on oil and gas production. Restoration work is ongoing, but the Belize reef is no longer on the list.

Environmental NGOs first requested that the World Heritage Committee consider the GBR for inclusion on the 'in danger' list in 1997, noting multiple threats such as ground transport infrastructure, major visitor accommodation and associated tourism infrastructure, and mining. At the time, the Australian Government successfully opposed the suggestion and reasons for inclusion on the 'in danger' list. In 2005, environmental NGOs again unsuccessfully petitioned the committee, noting increasing impacts of climate change on the GBR.

In 2012 the World Heritage committee began to seriously consider listing the GBR as in danger, in response to multiple port developments along the coast, including a plan to dump dredged sediment from the Abbot Point coal port inside the World Heritage site, as well as continuing poor water quality and impacts from climate change. Since then, UNESCO has repeatedly warned that it could list the GBR as in danger, on the grounds of poorly managed coastal development and continued declines in water quality in 2012, 2013 and 2014, and severe coral bleaching and inaction on climate change in 2021.

In 2015, the Australian Government narrowly avoided an 'in danger' listing through political and industrial lobbying as well as the development of an integrated ports strategy and a long-term sustainability plan that addressed proximate threats, yet failed to address climate change (see 'Science meets the public, policy and management practitioners' in Chapter 8). A subsequent national audit and Senate inquiry found that a substantial portion of finance for the long-term sustainability plan was delivered in a non-compliant (non-competitive and non-transparent) process to a private organisation with limited capacity and expertise, thereby

jeopardising the ability to achieve the key actions and outcomes set out in the agreed plan [30]. The following year also saw the worst coral bleaching in the world's history. Since then, recurrent and severe coral bleaching, including the global marine heatwaves of 2014–17, has devastated parts of the reef. The gap between consecutive bleaching events has now shrunk drastically – from an average of 25 years in the 1980s to just 6 years since 2010 (Fig. 7.11).

By 2021, climate change and poor water quality were continuing to place the GBR under serious threat. The GBR's endangered state had been confirmed in independent reports by scientists, the IUCN, the Australian Government's own reporting, and the WHC's own threat intensity coefficient, which indicates that the GBR's reported threats are equal to or higher in intensity than other sites that are already on the 'in danger' list [29, 31]. However, lobbying again in 2021 delayed a decision on the World Heritage 'in danger' listing. The Australian Government, under pressure from the mining and agricultural industries, lobbied the World Heritage Committee to argue that an 'in danger' listing would decrease economic revenues, that the process is unfair, that Australia should not be held responsible for global change, and that UNESCO should not supersede national sovereignty on climate change policy.

Growing numbers of Australian voters, industries and politicians are now calling for stronger action on the thorny problem of climate change. Accepting an 'in danger' listing could be just the strategy to tip the balance in the right direction and provide the Australian Government with a unique opportunity to pass from gridlock to action on climate change. After back-to-back mass coral bleaching and mortality in 2016–17, and the devastating bushfires in early 2020, even the Queensland Tourism Industry Council has emphasised the extreme importance of maintaining coral reefs under the spotlight in a call to the world to do more on climate change.

Other countries, like the United States of America, have made comparable political calculations in

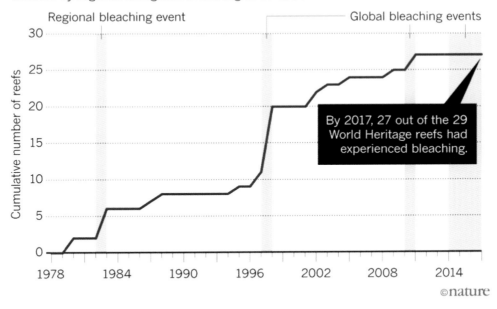

Fig. 7.11.   About 93 per cent of World Heritage listed coral reefs, including the Great Barrier Reef, have been affected by regional and global bleaching since 1980. (Image credit: Tiffany Morrison, reproduced from [33])

relation to similar challenging environmental and developmental conflicts. In 1993, for example, the US administration under Bill Clinton requested that UNESCO certify the Everglades as in danger. This helped bring industry opponents on board to better manage coastal development in Florida. Had the GBR been similarly listed as in danger in 2015, proposals to mine fossil fuels in the wider Queensland catchment may have struggled to get approval, and Australia and the world may have moved closer and faster towards meaningful action on climate change.

UNESCO and concerned scientists and observers are also working hard to increase the usefulness of the 'in danger' list as a policy tool. They are worried that increased politicisation of decision-making by the World Heritage Committee, particularly the influential lobbying by governments to determine whether or not sites are to be classified as 'in danger', is undermining both the integrity and capacity of World Heritage

governance frameworks to protect outstanding universal values [28, 29]. In February 2020, a consortium of 76 organisations and individuals petitioned UNESCO to consider climate change in its World Heritage decisions and policies are moving toward including climate-related degradation of a World Heritage Area as the basis for 'in danger' listing. This will shine a harsh spotlight on the intensifying geopolitics of climate change. Advanced economies, like Australia, with high *per capita* emissions but limited climate action will need to find more sophisticated ways of protecting ecosystems and jobs [32].

The politics governing survival for the Great Barrier Reef

David Ritter

The cynical behaviour of an individual may blight a friendship, but the institutionalisation of cynicism within a polity through the unconstrained politics of vested interests can poison trust across whole societies, with profound consequences for the world at large. In July 2017 perhaps Australia's most iconic anti-corruption figure, Tony Fitzgerald, noted that partisan interest had fundamentally undermined certain norms that are essential to good governance in our country. As Fitzgerald explained, Australia's system of government 'takes it for granted that political power will be exercised by people who know how to behave properly and can be trusted to do so' [34]. Unfortunately, the former judge observed, contemporary standards suggested that winning was now all that seemed to matter:

> [M]isleading and deceptive political conduct is regarded as clever … Important information is withheld, distorted and manipulated and falsehoods and propaganda are euphemistically misdescribed as 'spin'. It is almost impossible for voters to distinguish information and rational opinion from nonsense and deceit. [34]

A self-interested approach to governance means that what are meant to be impartial rules are instead interpreted to material advantage, and information is presented to achieve a vested agenda, rather than to transparently facilitate greater understanding. Under these circumstances, structures and resources that are intended for public good become hollowed out, or worse, redirected to unscrupulous ends.

In the opinion of the author, the dynamics invoked by Fitzgerald are on full display in the context of the Australian Government's ongoing failure to protect the GBR. While science can account for what is happening to the GBR in terms of physics and chemistry, it is politics and political economy that explain why inadequate measures are being taken by Australian governments to maximise efforts to abate the risk. The political imperative to control the narrative through spin, and the favouring of certain vested economic interests over others, has taken precedence over the health of natural ecosystems.

As numerous scientific studies have made clear, the greatest threat to the survival of the GBR is global warming (see Chapter 8) [35]. For Australia to maximise the future chances of survival for the GBR, the very minimum course of action would involve a credible national plan to achieve domestic greenhouse gas emissions reductions at a speed consistent with the survival of the ecosystem, coupled with a moratorium on new coal, oil and gas – the fossil fuels that are the greatest driver of global warming. Although responsible Australian ministers have continued to assert that their sincere intention is to create the conditions for the GBR's survival and recovery, neither of the above steps has yet been taken and Australia continues to experience notoriety for its obstructive stance in international climate negotiations [36].

Without effective action on climate change, expending resources on other priorities cannot secure the long-term health of the GBR. It is a simple matter of the cascading order of threats: unless greenhouse gas emissions are very rapidly abated, the Reef will not survive in anything like its current form. In bureaucratic and legal terms, siloing of 'the environment' from climate and energy policy has all too often led to the decoupling of 'environmental threats' from energy policy – as if, in a carbon-constrained world, the latter did not hold the ultimate key to the success of the former.

Ongoing public political controversy over whether to place the GBR on the List of World Heritage in Danger has been marked by political lobbying that puts political appearances ahead of physical realities. The government's own 2019 report shows many of the values for which the reef was inscribed on the World Heritage list in 1981 have declined and that the condition of the ecosystem had deteriorated from poor to very poor [31].

Around this time, the Australian Academy of Science told a Senate inquiry into farming practices that impact water quality on the GBR, that it was 'greatly concerned about a recent tendency to "cherry pick", dismiss, misrepresent or obscure scientific evidence' on the GBR to suit an agenda of misinformation [37]. Throughout the arduous, year-long inquiry process, which was described by leading reef experts in mid-2020 as a 'politically motivated charade', increasing numbers of scientists spoke of a refusal from politicians to accept evidence in efforts to avoid admitting the deteriorating state of the Reef's health [38].

The UNESCO World Heritage Centre and the IUCN jointly submitted that the GBR was facing imminent danger based on its declining condition due to climate change, and met the criteria to be listed as such. Nevertheless, a succession of Australian environment ministers have argued that the listing was not appropriate – and these political views have ultimately been supported by the World Heritage Committee.

In his analysis of contemporary standards of governance and political decision-making, Tony Fitzgerald diagnosed the source of the problem as being political parties with vested interests in and of themselves, as well as third parties who gain special, influential access [34]. Harvard Law School Professor Lawrence Lessig has defined the malformation of public entities for the benefit of sectional interests as a form of 'institutional corruption':

when there is a systemic and strategic influence which is legal, or even currently ethical, that undermines the institution's effectiveness by diverting it from its purpose or weakening its ability to achieve its purpose, including, to the extent relevant to its purpose, weakening either the public's trust in that institution or the institution's inherent trustworthiness. [39]

Lessig's thinking is apt to the case of the GBR. It is a matter of record that elements within political parties have combined with segments of the commercial media, along with the coal, oil and gas sector and other big polluters (aided and abetted by some peak industry representative bodies, service sectors like lawyers and accountants, and others with an interest), to effectively negate the Australian Government's ability to effectively tackle climate

change –- and to give the best chance to the GBR [36]. Law and policy have been effectively diverted to the great detriment of the public good.

We have now reached the 40th anniversary of Australia's promise to the global community that we would protect the GBR through the precepts of the World Heritage Convention. Under the terms of the treaty, each state signatory recognises and accepts a duty to ensure the identification, protection, conservation, presentation and transmission to future generations of world heritage and 'will do all it can to this end, to the utmost of its own resources' [40]. Although largely untested in Australia, the common law also recognises a foundational obligation upon the sovereign to hold in trust for enduring public use, and benefit, certain common resources: an ancient obligation that should surely not least be enlivened by the recognition of world heritage values in the GBR [41].

References

1. Clare P (1971) *The Struggle for the Great Barrier Reef,* Collins, London and Sydney.
2. Gray K (2021) *The Artful Activist. He Saved the Barrier Reef.* Ho007 VI.4. p. 12. Mission Beach Historical Society, Mission Beach.
3. Wright J (1977) *The Coral Battleground.* Spinifex, Melbourne.
4. McCalman I (2013) *The Reef – A Passionate History.* pp. 276–301. Viking, Melbourne.
5. Gunn M, Brouwer C (2016) Ninney Rise and John Busst Memorial, Bingil Bay, North Qld Conservation Management Plan. Biotropica Australia.
6. WWF Australia (2018) Backyard barometer. WWF Australia, <https://www.wwf.org.au/knowledge-centre/resource-library/resources/backyard-barometer#gs.1j4vte>.
7. Albrecht R, Cook CN, Andrews O, Roberts KE, Taylor MF, Mascia MB, Kroner RE (2021) Protected area downgrading, downsizing, and degazettement (PADDD) in marine protected areas. *Marine Policy* **129**, 104437. doi:10.1016/j.marpol.2021.104437

8.   Roberts KE, Hill O, Cook CN (2020) Evaluating perceptions of marine protection in Australia: does policy match public expectation? *Marine Policy* **112**, 103766. doi:10.1016/j.marpol.2019.103766

9.   The Great Barrier Reef Marine Park Authority (2015) Reef 2050 Long-Term Sustainability Plan, <http://www.environment.gov.au/marine/gbr/publications/reef-2050-long-term-sustainability-plan-2018>.

10.   The Great Barrier Reef Marine Park Authority (2019) Position Statement Climate Change, <https://elibrary.gbrmpa.gov.au/jspui/bitstream/11017/3460/5/v1-Climate-Change-Position-Statement-for-eLibrary.pdf>.

11.   Australian Academy of Science (2021) The risks to Australia of a 3°C warmer world, <https://www.science.org.au/supporting-science/science-policy-and-analysis/reports-and-publications/risks-Australia-three-degrees-c-warmer-world>.

12.   International Panel of Climate Change (2019) 'Special report on the ocean and cryosphere in a changing climate', <https://www.ipcc.ch/srocc/>.

13.   Director of National Parks (2018) *Coral Sea Marine Park Management Plan 2018*. Director of National Parks, Canberra.

14.   Nelson J, Bradner H (2010) The case for establishing ecosystem-scale marine reserves. *Marine Pollution Bulletin* **60**, 635–637. doi:10.1016/j.marpolbul.2010.04.009

15.   Wood LJ, Fish L, Laughren J, Pauly D (2008) Assessing progress towards global marine protection targets: shortfalls in information and action. *Oryx* **42**, 340–351. doi:10.1017/S003060530800046X

16.   Hoegh-Guldberg O, Hughes T, Marsh H, Pandolfi J, Possingham H (2008) Urgent need for improved protection of the Coral Sea. In *An Australian Coral Sea Heritage Park*. (Ed. I Zethoven) p. 12. The Pew Charitable Trusts, Sydney.

17.   Stevens DM (2008) The military history and heritage of the Coral Sea. In *An Australian Coral Sea Heritage Park*. (Ed. I Zethoven) pp. 34–45. The Pew Charitable Trusts, Sydney.

18.   Hundley P (2008) The Civic Maritime History and Heritage of the Coral Sea. In *An Australian Coral Sea Heritage Park*. (Ed. I Zethoven) pp. 46–56. The Pew Charitable Trusts, Sydney.

19.   Ceccarelli D (2011) 'Australia's Coral Sea: a biophysical profile'. Report for the Protect our Coral Sea Coalition, <https://www.pewtrusts.org/-/media/legacy/uploadedfiles/peg/publications/report/pewcoralseabiophysicalprofileaug2011pdf.pdf>.

20.   Hoey A (2020) *Coral Reef Health in the Coral Sea Marine Park*. Parks Australia, Canberra.

21.   Day JC (2016) The Great Barrier Reef Marine Park – the grandfather of modern MPAs. In *Big, Bold and Blue: Lessons from Australia's Marine Protected Areas*. (Eds J Fitzsimmons and GC Wescott) pp. 65–98. CSIRO Publishing, Melbourne.

22.   Heron SF, Eakin CM, Douvere F, Anderson KL, Day JC, Geiger E, Hoegh-Guldberg O, Van Hooidonk R, Hughes T, Marshall P, Obura DO (2017) *Impacts of Climate Change on World Heritage Coral Reefs: A First Scientific Assessment*. UNESCO World Heritage Centre, Paris.

23.   UNESCO (2019) *The Operational Guidelines for the Implementation of the World Heritage Convention*, <https://whc.unesco.org/document/178167>.

24.   Day JC, Heron SF (2019) The Great Barrier Reef is in trouble. There are a whopping 45 reasons why. *The Conversation*,<https://theconversation.com/the-great-barrier-reef-is-in-trouble-there-are-a-whopping-45-reasons-why-122930>.

25.   World Heritage Committee (2012) Decision: 36 COM 7B.8, Great Barrier Reef (Australia) (N154). UNESCO, Paris, <https://whc.unesco.org/en/decisions/4657>.

26.   Earth Justice, Environmental Justice Australia (2017) World Heritage and climate change: the legal responsibility of states to reduce their contributions to climate change – a Great Barrier Reef case study, <https://earthjustice.org/sites/default/files/files/World-Heritage-Climate-Change_March-2017.pdf>

27.   Day JC, Heron SF, Hughes T (2021) Not declaring the Great Barrier Reef as 'in danger' only postpones the inevitable. *The Conversation*, <https://theconversation.com/not-declaring-the-great-barrier-

reef-as-in-danger-only-postpones-the-inevitable-164867>.

28. Cameron C, Rössler M (2013) *Many Voices, One Vision: The Early Years of the World Heritage Convention*. Routledge, Oxford, UK.

29. Morrison TH, Adger WN, Brown K, Hettiarachchi M, Huchery C, Lemos MC, Hughes TP (2020) Political dynamics and governance of World Heritage ecosystems. *Nature Sustainability* **3**, 947–955. doi:10.1038/s41893-020-0568-8

30. Morrison TH (2017) Evolving polycentric governance of the Great Barrier Reef. *Proceedings of the National Academy of Sciences of the United States of America* **114**, 3013–3021. doi:10.1073/pnas.1620830114

31. GBRMPA (2019) 'Great Barrier Reef Outlook Report. GBRPMA, Townsville.

32. Morrison TH (2021) Great Barrier Reef: accept 'in danger' status. *Nature* **596**, 319. doi:10.1038/d41586-021-02220-3

33. Morrison TH, Hughes TP, Adger WN, Brown K, Barnett J, Lemos MC (2019) Save reefs to rescue all ecosystems. *Nature* **573**, 333–336. doi:10.1038/d41586-019-02737-8

34. Fitzgerald T (2017) The only hope for democracy is for politicians to stand up to political parties. *ABC News*, 13 July (online).

35. Hughes T, Kerry J, Alvarez-Noriega M, *et al.* (2017) Global warming and recurrent mass bleaching of corals. *Nature* **543**, 373–377. doi:10.1038/nature21707

36. Ritter D (2018) *The Coal Truth: The Fight to Stop Adani, Defeat the Big Polluters and Reclaim Our Democracy*. UWA Publishing, Crawley.

37. Australian Academy of Science (2019) Australian Academy of Science Submission to the Inquiry into the identification of leading practices in ensuring evidence-based regulation of farm practices that impact water quality outcomes in the Great Barrier Reef. Parliament of Australia: Submission 31, <https://www.aph.gov.au/Parliamentary_Business/Committees/Senate/Rural_and_Regional_Affairs_and_Transport/GreatBarrierReef/Submissions?main_0_content_1_RadGrid1ChangePage=2_20>.

38. Garret, G; Hoegh-Gulderg, O; Chubb, I. (2020) A response to written questions on notice from Senator Rennick by Dr Geoff Garrett AO, Professor Ove Hoegh-Gulderg and Professor Ian Chubb AC FAA after a public hearing in Canberra on 28 August 2020. Parliament of Australia: Answers to Questions on Notice 13, <https://www.aph.gov.au/Parliamentary_Business/Committees/Senate/Rural_and_Regional_Affairs_and_Transport/GreatBarrierReef/Additional_Documents?docType=Answer%20to%20Question%20on%20Notice>.

39. Lessig L (2013) Foreword: 'Institutional corruption'. *The Journal of Law, Medicine & Ethics* **41**(3), 553–555. doi:10.1111/jlme.12063

40. United Nations Educational, Scientific and Cultural Organisation (1972) *Convention Concerning the Protection of the World Cultural and Natural Heritage*. UNESCO, Paris.

41. Wood M (2013) *Nature's Trust: Environmental Law for an Ecological Age*. Cambridge University Press, Cambridge, New York, NY.

8

A changing climate for Australian reefs

We are now living in the Anthropocene, a geological era in which humans exert the dominant influence on the Earth's climate and environment. This chapter reviews the implications of climate change for Australia's coral reefs, with a focus on issues such as mass coral bleaching and mortality, the poleward movements of coral species, the longer-term record of human influence on Australia's reefs over the last thousand years, and how scientists can predict the influence of future ocean acidification on reefs – a busy but crucial agenda.

Innovative scientific tools for enhancing reef resilience are also discussed, alongside some of the challenges that face their application. The chapter concludes with perspectives on how to govern reefs in the Anthropocene, through science-based national policy as well as via the Intergovernmental Panel on Climate Change (IPCC) and in an era of shifting national and international governance paradigms. Multiple disciplines and perspectives are adopted as we discuss one of the most serious issues facing coral reefs in Australia and globally.

Climate change and Australia's coral reefs
Ove Hoegh-Guldberg

Coral reefs have been a prominent feature of Australia's coastal regions for millions of years. Up until the European colonisation of Australia, coral reefs like the Great Barrier Reef (GBR) are likely to have been vibrant coral assemblages that persisted for thousands of years under a relatively stable environment. In the past, coral reefs were characterised by high levels of coral cover and diversity, which provided habitat, in turn, for tens of thousands of species of mammals, fish, invertebrates, plants and many others. These abundant reefs also provided resources for Indigenous Australians, who harvested marine species for food, and drew inspiration from the Reef and its surrounding coastal areas.

The majesty and resilience of Australia's coral reefs

The size and complexity of Australia's coral reefs convinced many that they were invulnerable to human activities. As recent history attests, however, this was not true. Some of the earliest impacts of coastal land use practices on coral reefs by European settlers (e.g. clearing catchment forests for intensive agriculture) became evident as early as the 1870s with sediments in reef waters increasing five-to-ten fold in a few decades [1]. As these activities increased, mostly clear coastal waters turned brown and green as sediments spread further offshore and excess nutrients fuelled algal blooms that began to impact coral reefs and associated ecosystems.

The pace of change increased steadily over the early 20th century, with the post-war boom triggering a dramatic increase in people, consumption, coastal land use, and industrial activities. The near-complete agricultural development of the Queensland coastline for cattle and sugarcane was seen as inevitable, as was the heavy fishing pressure placed on the GBR by recreational and commercial fishing. The potential for oil and gas reserves underneath the GBR led to a fierce political battle in the 1960s and 1970s about whether the GBR should be mined. This culminated in the establishment of the Great Barrier Reef Marine Park by the Labour Federal Government in 1975, with oil and gas exploration and exploitation being banned within the park boundaries. The GBR would later be listed by the United Nations Educational, Scientific and Cultural Organization (UNESCO) as a World Heritage site in 1981, further consolidating the ban on drilling for fossil fuels.

Global climate change: insult to injury through local impacts

While there was much tussling over issues such as increasing nutrient runoff, crown-of-thorns starfish outbreaks and overfishing, the GBR sailed through to 1980 with many believing it was so large and untouchable that almost anything could be done to it, and it would either resist long-term damage or bounce back smartly. The idea that rising carbon dioxide (CO_2) from the burning of fossil fuels would ultimately be the greatest threat to the Reef was not on the radar, although researchers elsewhere such as Peter Glynn from the University of Miami were beginning to speculate about the significance of coral reef responses to surface ocean warming in the Eastern Pacific and Caribbean following the exceptionally strong El Niño event of 1982–83 [2].

Around this time, there were also hints of changes afoot that would have tremendous significance for the future of coral reefs, including the GBR. Substantial areas of reef-building corals were turning white (hence referred to as 'coral bleaching') over a few weeks as they lost their all-important brown *Symbiodinium* symbionts, or simply died. While elevated levels of light, UV radiation and temperature could trigger mass coral bleaching and mortality in experiments, the main cause was eventually tracked down to small increases in the maximum sea temperature (1–2°C) at the height of summer [3, 4], with ecological studies in the 1990s revealing increasingly significant impacts [5, 6]. A small contingent of Australian-based scientists focused on causal mechanisms as well as the implications of mass coral bleaching and mortality [7]. During the same period, NOAA's satellites increasingly revealed that mass coral bleaching and mortality could be successfully predicted from sea surface temperature anomalies alone. These thermal anomalies enabled the prediction of when and where mass coral bleaching would occur, adding to a global picture of how rising global temperatures were affecting coral reefs.

The first global mass coral bleaching event

While localised observations of coral bleaching on the GBR go back as far as the 1928–29 Great Barrier Reef Expedition, reports of mass coral bleaching began to emerge with greater frequency and intensity from the mid-1980s, some of which were part of global mass coral bleaching events. To the best of our knowledge, there have been eight mass coral bleaching events in total on the GBR (1987, 1998, 2002, 2006, 2016, 2017, 2020 and 2022), six of which were associated with global mass coral bleaching events (1998, 2002, 2016, 2017, 2020 and 2022). The link to sea temperature was increasingly undeniable, with predictions of future sea temperatures from rising greenhouse gases strongly suggesting that they would exceed safe levels by mid-century [9].

On the GBR, coral bleaching took a steep change in 1998 with exceptionally warm (record) conditions across the tropics, which also impacted Scott Reef in north-west Australia. These warmer than normal conditions resulted in 11–83 per cent of shallow-water coral colonies on the GBR undergoing sustained bleaching, with many corals dying [8]. Soon afterwards, temperatures began to increase in the western Indian Ocean, which severely affected reefs in places like the Maldives. As the northern summer developed, South-East Asia, northern Asia and the Caribbean began to show high levels of mass coral bleaching that resulted in considerable coral mortality. Generally, these impacts on reefs from across the world also coincided with anomalously high sea surface temperatures [9].

The mechanistic linkage between elevated sea temperature and the bleaching and mortality of corals also implied how the world's coral reefs might change if the climate models were correct. This led to the question: if atmospheric CO_2 doubles, would coral reefs exceed their temperature tolerance? Working with climate scientists from the European Union and CSIRO, it was apparent that doubling of CO_2 would lead to serious and unprecedented levels of mass coral bleaching and mortality globally by mid-century (Fig. 8.1, [9]). These estimates ('every second year' by 2020–30) appear conservative given the increased frequency and intensity of recent devastating mass coral bleaching and mortality events in response to changes in sea temperature [10]. This led to considerable discussion about the implications for people and

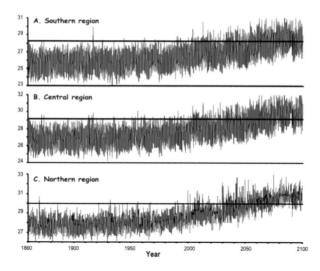

Fig. 8.1. Sea temperature data from 1900 to 2100, generated by a global model forced by a doubling of atmospheric CO_2, shows the proximity of the bleaching and mortality threshold for three regions of the Great Barrier Reef. (Image credit: Ove Hoegh-Guldberg, reproduced from [9])

ecosystems within a wide array of scientific, political, and public forums.

The global warming hiatus

One of the challenges in projecting the demise or not of coral reefs in Australia and globally to climate change is that the implications go well beyond that of coral reefs [9]. In this regard, the push by many special interest groups (e.g. Institute for Public Affairs; large coal, oil and gas mining interests) as well as conservative governments to ignore or misreport the science was daunting to many scientists. This was especially challenging given the acrimony and mistruths involved, and the fact that few scientists expected, or were trained in the art of, public debate [11].

To make things even more challenging, global warming appeared to 'stabilise' in many parts of the world, including Australia (from 1998–99 to 2013). This was due to variability in the climate system with this period being referred to as the 'hiatus'. This period of 'stability' meant that much of the world's tropical oceans went for a decade without a mass coral bleaching and mortality event. For those

trying to ignore the evidence of climate change, this supposed trend was much welcome news. Unfortunately, it was also used to undermine the link between the future of coral reefs and climate change, with the putative 'hiatus' being used as 'evidence' that oceans weren't warming, and that the science of climate change regarding coral reefs must be wrong. Recent events show that this was far from the truth, revealing that 10–15 years of data are not sufficient to capture the overall trend in sea surface temperature given other sources of variability. Depending on the decade in which data were obtained, one could get positive through to negative trends.

These socio-political dynamics, however, falsely undermined many scientists and ultimately, it can be argued, slowed the global response of society to the implications of climate change for coral reefs and much more.

It gets worse: ocean acidification comes to town

In 1999, Joanie Kleypas and colleagues wrote a critically important paper on the impact of rising atmospheric CO_2 levels on the chemistry of the ocean [12]. The central premise was that adding CO_2 from the burning fossil fuels to the ocean (via the atmosphere) would influence the pH, as well as the carbonate ion concentration, in the upper ocean (0–1000 m). Most importantly, Kleypas and colleagues argued that these changes were significant enough to interfere with the ability of corals and other reef organisms to lay down and maintain their calcium carbonate ($CaCO_3$) skeletons, which is a critical feature of the carbonate stocks of coral reefs.

As the ability to maintain carbonate structures decreases, the decline in the three-dimensional structure of reefs reduces the habitat for thousands of species, and potentially exposes millions of people in tropical coastal regions worldwide to greater impacts of waves in the absence of a protective reef crest barrier. Dove and colleagues explored the impacts of ocean acidification on coral reef ecosystems in large measure mesocosms at Heron Island over the last 10 years [13] (see 'Predicting coral reef futures' section, this chapter). The changes are profound and will be with us for the next 10 000 years, the time it will take to bring the acidity of the ocean to the pre-industrial state through continental weathering and other processes.

Climate change on the Great Barrier Reef today

Over the past 20 years, our understanding of climate change, particularly in the ocean, has advanced dramatically. The latest United Nations IPCC special report on the cryosphere, oceans and climate change shows that the ocean is changing at a rate that probably exceeds anything seen in at least the last 50 million years [14]. Indeed, over the past 50 years rapid changes in sea temperature, alkalinity, sea level and storm intensity have been seen across the world. Critically, these changes exceed the pace that many coral reef organisms take to evolve given the long generation times (years to decades).

Over the past 5 years, we have also seen four of the largest mass coral bleaching and mortality events in recent Australian history (2016, 2017, 2020 and probably 2022; see next section), with estimated losses of as much as 50 per cent of the GBR's shallow-water (< 5 m) reef-building corals, some of which were previously described as pristine (e.g. the northern section of the GBR) due to the low impact of local factors such as pollution and overfishing (Fig. 8.2). At the same time, Australia has become hotter and drier, experiencing record mega-fires in its forest ecosystems.

What if we had acted on climate change in 1980?

Here is a simple question: how much easier would it have been to tackle climate change if we had acted on it when we first linked its impacts to natural and human systems? Table 8.1 illustrates how dealing with climate change would have been a simpler problem if we had listened to the science available in the early 1980s. For example, CO_2 emissions of greenhouse gases would only need to be reduced by two-thirds of the amount required today to return to zero. Oceans would have stabilised more rapidly and probably avoided some of the 30 per cent increase in surface ocean acidity.

Table 8.1. The state of climate change in 1980 and today (numbers are approximate)

Climate variable	1980	Today
CO_2 emissions ($GtCO_2$/year)	20	32
Atmospheric CO_2 concentrations (ppt)	339	412
Global surface temperature (°C)	14.18	15.25

Similarly, rates of sea-level rise (SLR; not shown in Table 8.1) would be more manageable relative to today, while the increased intensity of storm systems would have been less. This analysis gives pause for thought and brings into sharp focus why trusting the science is so important when it comes to climate change. In the present case, the scientists working on Australia's coral reefs have been trying to tell us that we are headed in an ultimately disastrous direction. Ideally, we will get back on track and double down on our efforts to tackle climate change today rather leaving it even longer before we act.

Responses of coral assemblages to recurrent mass bleaching

Terry Hughes

As documented throughout this book, climate change and runoff of pollutants from land are the major ongoing threats to GBR ecosystems. Even before the creation of the Great Barrier Reef Marine Park in 1975, many coastal locations were already struggling with elevated rates of sedimentation and inflow of nutrients caused by the expansion of land clearing and agriculture (see the discussion in other sections). Similarly, stocks of many marine species, including sharks, crocodiles, turtles, dugong, pearl oysters, *Trochus* and sea cucumbers, were severely depleted long before the advent of systematic monitoring programs. Reef-wide monitoring of coral cover on offshore reefs began in the mid-1980s, after the first two recorded outbreaks of crown-of-thorns starfish, obscuring ecological baselines. In 2014, the Australian government reported to UNESCO that 24 out of 41 values that collectively comprise the outstanding universal value of the GBR World Heritage site had deteriorated since its inscription in

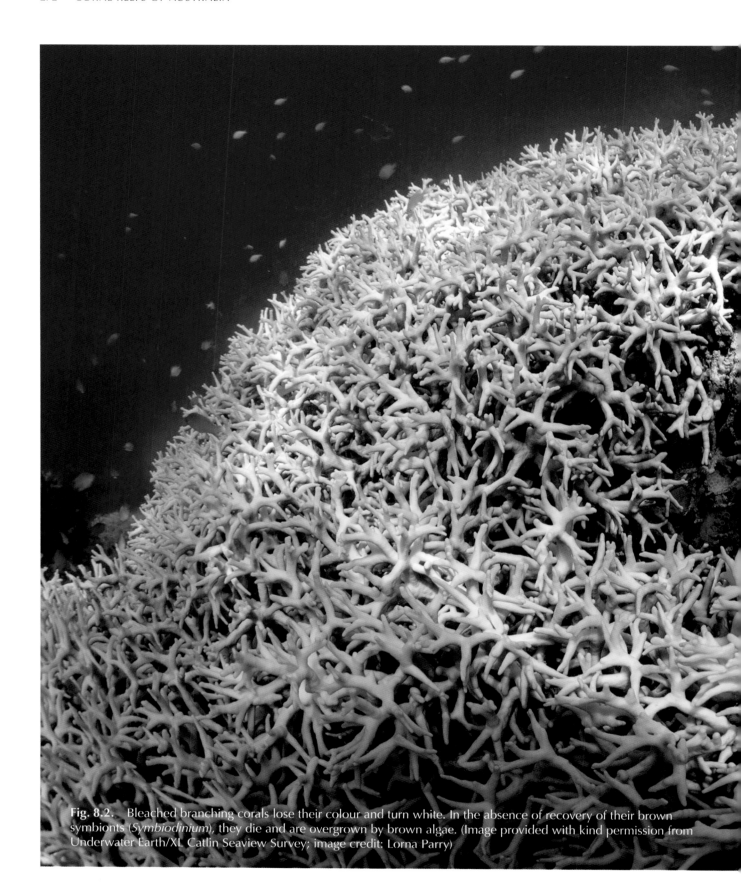

Fig. 8.2.   Bleached branching corals lose their colour and turn white. In the absence of recovery of their brown symbionts (*Symbiodinium*), they die and are overgrown by brown algae. (Image provided with kind permission from Underwater Earth/XL Catlin Seaview Survey; image credit: Lorna Parry)

Fig. 8.5. Collage of tropical species that have made it to Sydney, including butterflyfish, damselfish, Moorish idols, cleaner wrasse, bicolour angel fish and lineolate surgeonfish. (Image credits: Jack Hannan, David Booth, Gigi Beretta)

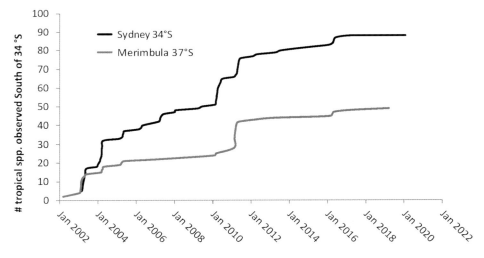

Fig. 8.6. Cumulative species richness of coral reef fishes in Sydney and Merimbula. (Image credit: David Booth, redrawn from [22])

when it comes to the minimum water temperature in winter, we expect a rapid increase in overwintering in the coming decades. It's not just about the EAC and sea surface temperatures. Tropical organisms are also vulnerable to 'east coast low' storms, which are becoming more intense under climate change. We have shown a negative effect of these severe storms (direct loss of vagrants in the turbid seas) but later positive effects (kelp beds ripped out to reveal urchin barrens, which attract new settler vagrants).

How do the vagrants fit in within the local fish assemblages, given they can account for over 20 per cent of the fish numbers at some sites? Some form mixed schools with local species, and this can have benefits for both species [23]. Below ~18°C, few vagrants survive. They become sluggish and are easy prey for local predators. However, species that successfully overwinter are becoming more common in Sydney, with some species even spawning. Establishment may become imminent.

New corals and associated fauna

An exciting recent development has been the expansion of subtropical hard corals into Sydney [24, 25]. Once established, the brooding coral species *Pocillopora aliciae* will spread locally. This has been the case in Sydney, where it now occurs in sizeable beds on otherwise barren rock.

Newly settled Sydney corals support a suite of coral reef species we have not recorded previously on our 20-year surveys on nearby rocky reefs, including several fish species that are only seen on coral beds (Fig. 8.7). While some species, such as the domino humbug damselfish, have been

Fig. 8.7.   New hard corals in Sydney and associated species. (A) *Pocillopora aliciae*, Sydney. (B) Humbug damselfish, *Dascyllus reticulatus*, on *Pocillopora aliciae*, Sydney. (C) Dick's damselfish *Plectroglyphidodon dickii*. (D) Trapeziid crab on *Pocillopora aliciae*, Sydney. (Image credits: John Sear).

spotted in the long-term surveys on rocky reefs, they are in much higher numbers on the new corals. Other taxa, such as trapeziid crabs, which can have a commensal relationship with pocilloporid corals, have appeared at more southern sites.

What does the future hold?

From two decades of surveys and associated experiments, the influx of coral reef fish vagrants appears to be on the increase and partly linked to EAC patterns. The likelihood of overwintering increases with water temperature, as a direct result of both performance drops at lower (under 18°C) winter temperatures and the arrival of new coral habitats. We predict a possible threshold increase in overwintering and local reproduction in the coming decades. This increase would be tempered by the increased intensity of east coast low storm events, predicted under climate change scenarios, which can wipe out vagrants [26]. Ocean acidification may also counteract temperature effects by slowing urchin movement southwards, and so inhibiting rocky barren formation [27].

Historical degradation and shifting baselines on Australian coral reefs

John M. Pandolfi

The concept of shifting baselines

Most Australians are unaware of the magnitude of change that has occurred since European invasion of the continent. What did the first Europeans experience when they arrived and how was that different from the thousands of years leading up to this event when First Nations Peoples alone interacted with nature? To answer these questions, it is essential to acknowledge the 'shifting baseline syndrome', where each new generation views the condition of nature as pristine, even though it has been long preceded by generations of human impact and degradation [28]. The shifting baseline syndrome can lead scientists, managers and other stakeholders to set inappropriate and too

optimistic management goals based on their own short-term experience of an environment degraded long before their time.

One way to account for the effects of the shifting baseline syndrome is to incorporate historical records contained within archival sources into ecological assessments of ecosystem condition. Historical archives include newspaper articles, popular media accounts, naturalist accounts, ship's logbooks, early navigational and Admiralty charts, early government fishery reports, recreational fishing club records and competitions, and photographs housed in local, state and national archives. Ecological data that can be extracted from these sources include, but are not limited to, the number of fish caught, catch rates, reef area, species composition, diversity and abundance, size, fish landed by location, and species targeted. Data can also be extracted from oral histories derived from interviews with stakeholders such as commercial and recreational fishers [29], and extension of this approach to traditional knowledge from Indigenous communities helps describe changes to species and coastal environments over even longer timeframes. These records represent an untold account of the scale and effects of human activity on past marine ecosystems, but remain largely untapped by ecologists and fisheries scientists.

Human pressures on the Great Barrier Reef

Globally coral reefs have suffered huge losses over historical timeframes, such that no single reef can now be referred to as 'pristine' [30] (Fig. 8.8). This applies equally to the coral reefs of Australia. The GBR has historically been affected by changes in water quality, depletion of fisheries, and climate change. The GBR has a long history of intermittent growth along the Queensland coastline, from 650 thousand years ago to the present. Since the coldest period of the last ice age some 20 000 years ago, Aboriginal people have been displaced inland by rising sea levels. From 18 000 to 6000 years BP, sea level rise averaged 14 cm per decade, or an order of magnitude faster than today. As the sea level stabilised, modern configurations of coastal reefs,

Fig. 8.8.   The slippery slope of coral reef decline through time. Degradation on coral reefs globally has resulted in a trajectory of decline from which recovery is not guaranteed. (Image credit: Mary Parrish, reproduced from [31])

seagrass meadows, oyster beds and mangrove habitats developed alongside the Aboriginal societies inhabiting adjacent terrestrial ecosystems.

The arrival of Asian and European people in Australia dramatically escalated environmental exploitation with the emergence of new extractive industries (e.g. logging of coastal forests, mining of guano, hunting of dugong, and export fisheries for sea cucumbers, turtles, and pearl oysters) that opened access to regional and global markets [32]. Land clearing, agriculture and the introduction of livestock around the turn of the 19th century led to dramatic increases in terrestrial runoff of sediment, suspended solids, nutrients, fertilisers and pesticides into the GBR catchment [33]. This increase in land-based runoff since European invasion led decades later to depletion and even collapse of coral populations (Fig. 8.9) and continues to degrade Queensland's coastal reefs [34].

The extraction of fossil fuels, coastal development and dredging for ports have added to the diversity and extent of human impacts, reducing water quality in nearshore reefs. Simultaneous exploitation of eastern Australian finfish and reef fish since the European invasion of Australia has depleted many fisheries along the GBR, including Spanish mackerel [35] (Fig. 8.10). Finally, climate change has resulted in several mass bleaching and mortality events.

Managing for a shifted ecological baseline

Historical data tell us what was natural in the sea, thereby enabling us to determine whether contemporary systems are acting within the historical range of variability exhibited before large-scale human impacts [36]. This guides collective social decisions about how far and in what way to facilitate recovery in coral reef ecosystems. Historical data provide context for understanding the magnitude of loss of individual species, including fisheries. Reconstructing historical fisheries (e.g. [37]) can help improve the reliability of stock assessment models by providing an understanding of the earliest unfished biomass available – which helps us to evaluate and regulate reef fisheries. Historical data also provide clues on where evolutionary novelty might be concentrated [38], and where genetic diversity in the sea might best be conserved.

Clear and objective baselines derived from knowing how different today's oceans are from their past condition, and their trajectory of change through time, are instrumental in formulating effective management strategies for the recovery of coral reef communities. These strategies help us to characterise the status and trends of coral reef ecosystems, identify drivers of change and provide data on past ecosystem states. This type of information is key for evaluating the conservation effects of specific management actions on living

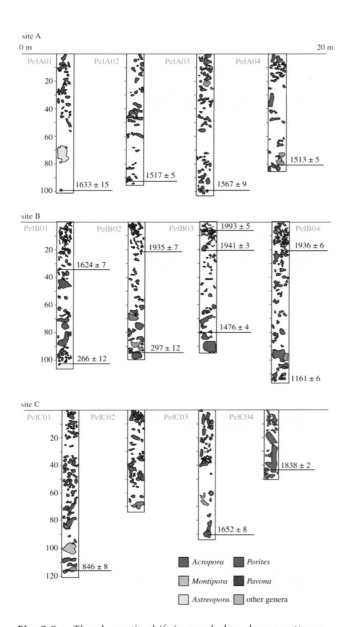

Fig. 8.9.   The dramatic shift in coral abundance patterns associated with land-use changes since the European invasion of the Queensland coastline. Cores of the reef sedimentary matrix at Pelorus Island, Great Barrier Reef, show either continuous accretion of *Acropora*-dominated coral assemblages (red) or a shift to *Pavona* (blue). Core depth (cm), with U-series ages (CE ± errors). (Image credit: George Roff, reproduced from [34])

reefs – that is, for measuring the success of those actions. Historical knowledge of ecosystems provides an essential bridge between science and management goals and outcomes.

Predicting coral reef futures
Sophie Dove

The climate change threat to coral reefs took decades to be appreciated by both scientists and the public. Alarm bells were initially rung in the early 1980s when unprecedented and widespread whitening of coral (bleaching) was observed. As late as 2004, most scientists still perceived overfishing or eutrophication to be the greatest threat to reefs, downplaying the potential impacts of ocean warming and ocean acidification [39].

As a scientific community, we have now observed enough damage to reefs to reach the consensus that most coral reefs are doomed unless we meet the Paris Agreement objective of limiting global warming to well below 2°C above pre-industrial temperatures. Reefs are doomed because repeated bouts of elevated sea surface temperatures will impede recovery and reduce coral diversity. On the other hand, ocean acidification is less well understood. Indeed, scientists have different perspectives on whether ocean acidification is an additional threat to the future of coral reefs.

It is often remarked that corals build coral reefs; however, the processes that dismantle reefs, other than cyclones, are barely mentioned. Acidification is an external factor that facilitates the dismantling of reefs as much as it reduces their growth rate through reductions in coral calcification. Carbonate reefs gain $CaCO_3$ faster than it is lost, permitting them to keep up with SLR or replace carbonate stock lost to erosion.

The carbonate balance of a reef is not open to direct observation. The relatively scant fossil and chemo-stratigraphic record used by geologists preferentially documents carbonate gains rather than losses. These records prime us to believe that erosion and the loss of carbonate platforms are too slow to be significant, while the build-up of carbonates rapidly responds to ameliorating conditions [40]. Geological perspectives have a tremendous influence on how we distinguish present-day carbonate reefs from non-carbonate or marginal reefs. For practical purposes, carbonate reefs are coral

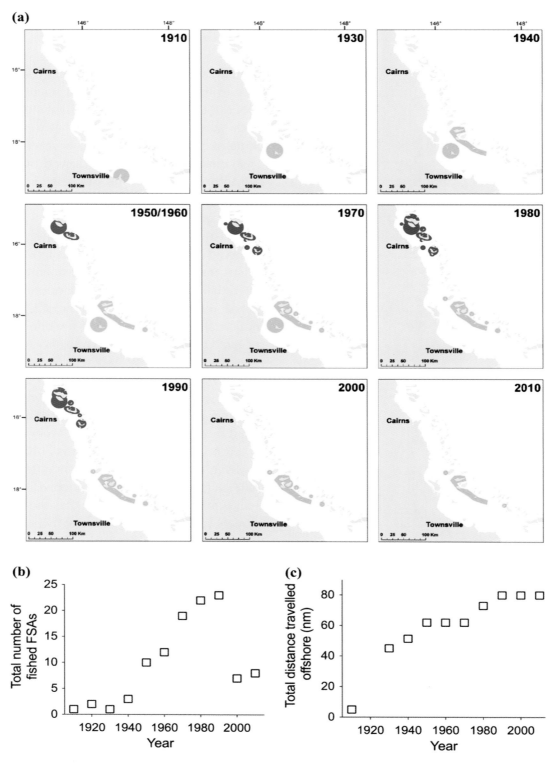

Fig. 8.10. The Great Barrier Reef's Spanish mackerel spawning-aggregation fishery (FSA) expanded from Townsville to Cairns in the 1950s; then overfishing caused contraction of the fishery back to Townsville in the 2000s. (a) Geographical distribution of effort by decade, (b) total number of fished FSAs by decade, and (c) total distance fishers travelled offshore by decade (nm, nautical miles). (Image credits: Sarah Buckley, reproduced from [35])

communities with a significant carbonate base. Marginal reefs lack the carbonate base but maintain the potential to produce one [41]. This perspective leaves little space to determine how acidification drives carbonate reef loss other than via its impact on hard corals.

How much of a threat is ocean acidification to coral calcification?

Coral calcification is a complex process that proceeds by extension and densification. As corals grow, they extend by building a thin $CaCO_3$ scaffold, which is subsequently infilled with further deposits of $CaCO_3$ (densification). Multiple studies have demonstrated that ocean acidification undermines densification more than extension, resulting in corals with weakened skeletons that otherwise appear normal [42].

Coral calcification occurs in a pocket-like space separated by a layer of cells from the ocean (Fig. 8.11). Inside this pocket, the primary carbon source for calcification is respired CO_2, rather than ocean carbonate ions (CO_3^{2+}) [43]. Corals can calcify because they use energy to pump protons (H^+) from the pocket (site of calcification) into their tissues allowing CO_2 to be converted to CO_3^{2+}. However, ocean acidification impedes the downstream removal of these protons, creating potentially toxic consequences for corals that continue to calcify.

The observation that corals can internally regulate the pH at the site of calcification is therefore not a justification for omitting acidification as a significant predictor of reef futures. Nonetheless, apart from a few thought-provoking exceptions where long-term exposure to acidification (~pH 7.4) led to the complete dissolution of previously deposited $CaCO_3$ contained within living coral [44], studies performed at lower acidification levels show only reductions in calcification. In this sense, warming is more impactful than acidification because heat kills corals (and dead corals don't calcify). On the other hand, acidification appropriate

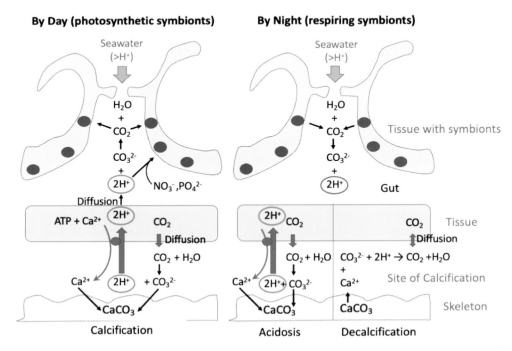

Fig. 8.11. Coral calcification when the proton buffering capacity of seawater is impeded by ocean acidification (> H^+). By day, acidification is mostly solved in corals hosting actively photosynthetic symbionts. Protons (H^+) diffusing into the gut are dissipated by CO_2 and nutrient uptake (nitrate, NO_3^-). By night, symbiont respiration does the opposite, potentially leading to acidosis or decalcification. (Image credit: Sophie Dove)

to the 21st century only reduces their rate of calcification.

Acidification-driven decreases in the rate of coral or algal calcification are also rated less impactful than those associated with, for example, overfishing or eutrophication. Maximum potential rates of coral extension are often seen as surplus to requirements and only limited by space. This assumption leads to the false conclusion that hard corals recover $CaCO_3$ stocks at exponential rates [41]. Reductions in an exponential recovery trajectory are perceived as less significant than limitations in recovery associated with algal overgrowth or a failed supply of recruits. The view that algal overgrowth (even triggered by acidification) might be contained by an appropriate and thriving cohort of herbivores also appears to diminish the potential impact of acidification on future reef community dynamics.

How much of a threat is ocean acidification to the loss of previously deposited $CaCO_3$?

Models of seawater chemistry have estimated that a 10-fold increase in atmospheric partial pressure of CO_2 (pCO_2) may dissolve all $CaCO_3$ in open seawater. In this regard, we are unlikely to observe such acidification levels by 2100, but such levels are projected for scenarios post-2200 in which we do little to limit CO_2 emissions over the next decade [45]. Acidification is particularly serious given that the return of the ocean to the present alkaline state is projected to take at least 10 000 years. As we approach the end of the century, metabolically assisted acidification will begin to undermine the net carbonate balance of reefs [46]. Some surface communities may even continue to look relatively healthy as they shift towards slow-growing coral communities that favour tissue biomass over skeletal density. At the same time, foundational carbonate structures may be destabilised by decalcification and bioerosion driven by ocean acidification.

Previously deposited $CaCO_3$ can be undermined by metabolically assisted acidification either via bioerosion when an organism invades dead corals or via the localised production of CO_2 from the metabolism of an upstream or proximate community.

Bioerosion is presented as the counterpart of calcification on reefs. However, organisms erode $CaCO_3$ via mechanical methods, as much as they erode through the production of acids that dissolve $CaCO_3$ into its component ions. Below the visible surface communities of reefs are dark crevices and caverns formed by bioerosion and long past events of rapid coral extension. These crevices in the carbonate base are lined with heterotrophic filter-feeders whose respiration rates, unbalanced by photosynthetic activity, locally deplete O_2 and enrich the water column with CO_2 [47]. Metabolically produced CO_2 can lower seawater $CaCO_3$ saturation state to levels that lead to carbonate dissolution. For example, CO_2 produced by decomposing microbes acidifies the water that surrounds reef sediment particles. Acidification of the open ocean then amplifies this effect, leading to sediment dissolution. Likewise, by stimulating respiration, warming can increase the CO_2 concentration of reef waters percolating through sediments, especially at night or in the deeper recesses of the reef where photosynthesis is inactive.

Bioerosion and $CaCO_3$ dissolution are natural processes on reefs. Still, mathematical estimates of their holistic impact on the carbonate balance of a reef are not computable from field surveys. The use of disparate proxies for measurements, the hidden nature of many localities for $CaCO_3$ dissolution, and the role of ecological succession and localised conditions on the involved biological processes are significant impediments to accurate quantification.

In coastal nations like Australia, it is imperative to understand the fate of carbonate platforms. Will they erode and disappear over the present millennium? Or will they persist as degraded navigational hazards and coastal barriers that support a subset of current services with appropriate management?

Scientists have needed to see the negative impact of underwater heatwaves on reefs to shift to the consensus perspective that global warming is highly detrimental. However, no survey of a coral reef community can predict the net carbonate status of a future reef and reveal the true scale of the threat posed by acidification.

Predicting coral reef futures: the role of mesocosms

The most realistic mesocosm experiments include a diverse community of organisms that fulfil the broad spectrum of known functional groups deemed essential to coral reef ecosystem dynamics, and a natural supply of reef water. Such ecosystems can reveal community succession and transition to alternate states. Still, they can also be used to directly monitor ecosystem calcification rates without the need to sum often disparate measurements of the individual carbonate components [48].

Biophere2 was the first substantive mesocosm experiment monitoring the effects of calcium ion concentration $[Ca^{2+}]$ and carbonate ion concentration $[CO_3^{2-}]$ on the calcification rate of coral reefs at a community scale [49]. The experiment demonstrated a 40 per cent reduction in net daytime calcification associated with doubling atmospheric pCO_2. More recent experiments manipulate temperature and pCO_2 in replicated mesocosms with net ecosystem calcification measured by day and night (Fig. 8.12). These experiments characterised the long-term response of a subtropical reef in the

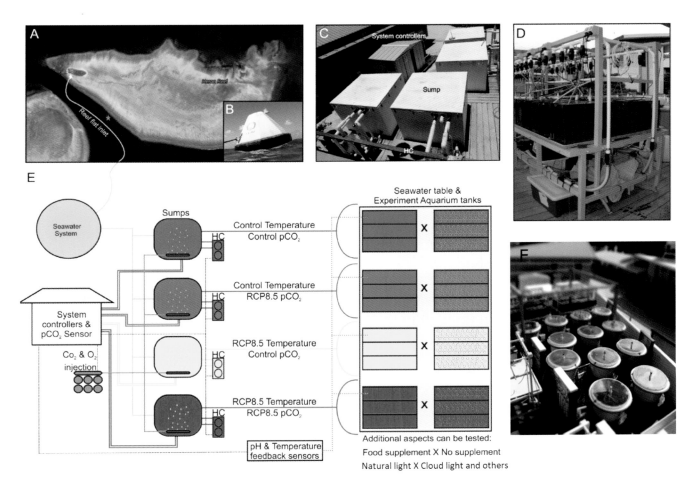

Fig. 8.12. Mesocosm experimental set-up at Heron Island (Southern Great Barrier Reef). (A) Reef seawater input. (B) CSIRO buoy measuring bi-hourly temperature and pCO_2 used for present day. (C) Large upstream sumps to control water flow-through rates (D) Small downstream tanks for performing experiments on specific organisms. (E) Schematic of an experimental design. (F) Larger downstream mesocosm tanks fitted with marine blue filters to provide light quality and intensity appropriate to 5–8 m water depth at Heron Island. (Image credits: Matheus A. Mello-Athayde)

GBR. They showed that under unmitigated pCO$_2$ scenarios, the expected phase transitions to algal reefs dominated by cyanobacteria occurred (Fig. 8.13A, B). Critically, the experiment showed that even if you assume that a high density of reef calcifiers remains, and no violent waves are present, the combination of elevated temperatures and high levels of acidification will drive significant net reef decalcification by day and by night (see the blue regions in Fig. 8.13C, D).

Mesocosms and other experiments have a vital role to play in understanding the future by experimentally manipulating environmental parameters in a manner that closely approximates natural ecosystems, with well-constructed and replicated mesocosms offering a glimpse of a future that is otherwise unobservable (Fig. 8.14). The outcomes of mesocosm experiments should alert both the scientific community and the broader public about this largely invisible yet insidious, chronic threat to coral reefs.

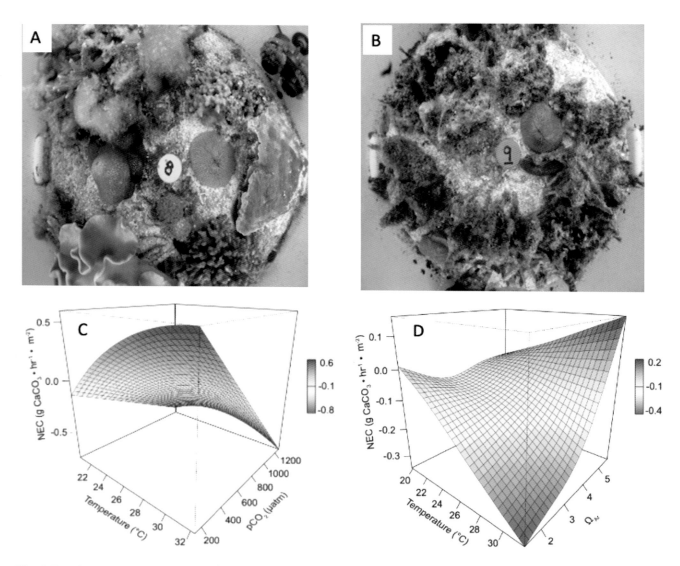

Fig. 8.13. Mesocosm status 17 months into experiment. (A) Present day. (B) RCP 8.5 treatment. Effects of the interaction between temperature and acidification (pCO$_2$ or aragonite saturation (Ω_{Ar})) for a reef with an average calcifier cover of 30 per cent: (C) by day, (D) by night. (Image credits: Sophie Dove, redrawn from [48])

Fig. 8.14. The Free Ocean CO_2 Enrichment (FOCE) experiment to test the effects of ocean acidification at Heron Island. (Image credit: David Kline)

Adapting for Australia's reefs of tomorrow: the complex landscape of reef restoration and interventions

Gergely Torda

As coral populations decline globally due to anthropogenic impacts, particularly climate change, the need for active reef restoration has come to the forefront of discussions. While it may be possible for corals to keep pace with the current trajectory of the changing climate in which they find themselves through natural recovery and adaptation, interventions may be required to assist natural processes that aid reef recovery or bolster resilience. The debate on whether or not interventions are required to safeguard coral reefs focuses on the ecological, socio-economic, cultural and ethical appropriateness, risks, costs and benefits of human interventions. The roadmap for managers and scientists is also complicated by the fact that what is appropriate for one reef may not be appropriate for another, so whether and what type of intervention should be applied requires consideration in a unique decision-making framework.

If all emissions were halted immediately, it would take time for the ocean to re-equilibrate to favourable environmental conditions for coral reefs. This rationale has motivated some scientists to explore ways in which they can 'buy time' for corals to survive the various manifestations of climate change by applying a toolbox of possible interventions. The practicality of intervening will be challenging at large spatial scales like entire reef systems. This is a new research frontier largely because in the past there was no demand for restoration on large spatial scales. To date, coral reef restoration has targeted individual reef sites, typically with high tourism value, with a median size of 100 m² [50], and it has become clear that the same techniques cannot be cost-effectively upscaled to entire reef systems. Addressing the problem of scalability therefore demands technological innovations, for example, in developing the tools and techniques to efficiently deploy corals that will survive across broad spatial scales using gardening approaches, or locating and harvesting from wild coral spawn slicks (see the following sections).

Another challenge is increasing the resilience of coral propagules to adverse conditions (e.g. heatwaves), so that restored sites do not simply succumb to the next disturbance event. Here, the goal is to produce propagules for restoration at vast spatial scales that are 'climate-proofed' [51] by stress-hardening, microbiome engineering [52], or fast-tracking the evolution of the coral host via selective breeding [53]. Because complex interactions are present at all tiers of biological organisation, concerns exist about how the manipulation of one element (one gene, one trait or one species) may have unforeseen and irreversible cascading effects on whole ecosystems. This raises fundamental questions about complex ecosystem behaviour that will need to be addressed before any intervention can responsibly proceed. While the race to fill these critical knowledge gaps continues to focus on reliable predictions of intervention outcomes, assessing risks, costs and benefits, some novel restoration approaches are already being tested in laboratories and in the wild. The plethora of interventions being considered each carry their own levels of risk that need to be prudently evaluated, and in some cases, where natural recovery processes are intact, the best action will be to proceed without intervention.

Ultimately, rapidly curbing greenhouse gas emissions will safeguard coral reefs. The potential of active interventions should not shift focus away from the fundamental need to address climate change, but aid in their resilience while proximal and distal stressors are reduced.

Reef rehabilitation and restoration

David Suggett and Kate Quigley

Australian reef researchers, managers, stakeholders and policy makers met in Townsville in 2017 to discuss a critical emerging question: could 'reef restoration' approaches provide effective management aids – beyond existing protection and mitigation measures – to combat the declining health of Australian coral reefs? [54]. Less than a year later, the first in-water restoration programs were established on the GBR, capitalising on the growing global movement for localised community-led coral propagation, commonly referred to as 'coral gardening' [50]. Like plants on land, coral propagation refers to the process by which new corals are produced from fragments taken from an adult coral colony. This is often achieved by regenerating part of one parent plant into a new plant. Because the process occurs asexually (i.e. through grafting), it produces a genetically identical coral to the parent. Coral nurseries are often used to grow coral species of interest, which are subsequently planted back onto the reef.

Propagation and outplanting have increasingly been adopted on the GBR, using a variety of methods, such as from 'Caribbean Coral Trees™' on Fitzroy Island, floating nursery platforms (e.g. Opal Reef [55] and the Whitsundays; see Fig. 8.15), land-based coral raceways at Daydream Island and on-reef stabilisation structures at Moore Reef and Green Island (Seastars [56], Moore Reef, Green Island).

These approaches to restoration typically yield relatively high survival rates of outplants (> 65 per cent; [50]), but are often restricted to localised re-planting because of high costs and labour

Fig. 8.15. Outplanted corals floating on nursery platforms as part of a restoration initiative at Opal Reef, offshore from Port Douglas, northern GBR. (Image credit: Johnny Gaskill).

associated with maintaining propagation infrastructure and the speed of re-planting. While such factors ultimately limit 'scalability' and restrict activity to specific sites, stakeholder groups can form collectives. For examples, GBR tourism pooled efforts across multiple 'high value' tourism sites in 2019 under the Coral Nurture Program [57] and have since planted tens of thousands of individual corals. This program was a world-first partnership between the tourism industry and researchers to validate the effectiveness of new propagation approaches integrated into existing tourism operations [55, 58], including low-cost and low-effort nurseries [55] and outplanting devices (Coralclip® [59]) that could re-plant coral faster than before (hundreds per hour compared to tens per hour). Such approaches have provided industry operators with new stewardship responsibilities including crown-of-thorns starfish removal, particularly at high-value tourism reefs where relatively 'healthy' areas can rehabilitate more impacted areas. Using local corals is likely to maximises the effect of local adaptation, where individuals perform best in their local ranges.

New challenges now arise as in-water, asexual-propagation-based restoration practices in Australia are increasingly adopted and become socially accepted to support local site management. To what extent are such practices able to retain (or rebuild) reef ecological communities that are inherently complex? Can more targeted propagation also be achieved, as in the Caribbean where propagation has focused on a few species [60], to ensure that re-planted populations have enhanced resilience to accelerating environmental stressors? Such challenges will partly shape the future Australian reef research and management landscape.

Moving beyond coral gardening

Restoration methods that take advantage of sexual reproduction provide an opportunity to recombine genetic material of species, thereby increasing overall genetic diversity and the potential for rapid adaptation for novel or desired traits, like enhanced heat tolerance or faster growth [61]. These methods are rapidly emerging given their potential for broad-scale deployment both within and between sites [50] (see the next section). For corals on the GBR, sexual reproduction is particularly amenable for restoration at scale because most species are hermaphrodites (i.e. both male and female [62]), thereby allowing the physical separation of eggs and sperm from a single colony in the laboratory and the downstream manipulation of genetic contributions sourced from different coral colonies. Moreover, these colonies can be sourced from a variety of habitats, reefs or even species for the directed enhancement of genetic traits (see 'Enhancing corals using assisted evolution' later in this chapter). In brief, bundles consisting of eggs and sperm packaged together are collected, and separated into eggs and sperm to be mixed in specific combinations [63]. Sexual propagation can occur in laboratory settings, or in the field in

nurseries that float directly above reefs, or in aquaculture tanks set up on boats.

Field tests of floating larval nurseries on reefs have been established in the northern and southern GBR (Moore Reef and Heron Island; see Fig. 8.18). Both laboratory and field-based methods for sexual propagation can produce millions, if not billions, of individual corals with unique individual genotypes that can, if they survive, be planted onto the reef. These corals would be likely to have higher levels of genetic diversity compared to corals produced via asexual fragmentation, and could be placed onto damaged reefs to enhance recovery potential or to facilitate the addition of specific genetic material (e.g. heat tolerant alleles [53] to high-value tourist sites, or those at risk of damage such as heat sensitive reefs).

The use of sexually produced offspring provides an opportunity for new genetic combinations to be made and for genetics to be mixed across multiple reef environments. Sexual propagation requires a detailed understanding of the reproductive life cycle of each coral species and population of interest. Although huge strides have been made to document and predict reproduction in corals [64], a full understanding and therefore ease of manipulation of the diverse aspects of coral reproduction is still critical to the success of these methods across many coral species [reviewed in 61].

Harvesting coral spawn slicks for reef restoration
Christopher Doropoulos

Following their early work on coral reproduction at Heron Island on the southern GBR, Kojis and Quinn in 1982 hypothesised that the synchronous release of eggs and sperm was likely to be the most common method of reproduction in scleractinian, reef-building corals [65]. This was quickly confirmed by the scientific discovery of mass spawning of entire coral communities on the central and northern GBR [66]. The prospect was raised of harvesting coral larvae that concentrate *en masse* in coral spawn slicks and spread across the sea surface for kilometres following mass coral spawning events for large-scale coral restoration [67].

With the discovery of mass coral spawning, coral spawn slicks were first harvested for reef restoration in Australia in 1997 at a tiny beach town called Coral Bay, at Ningaloo Reef. A small fraction – but still millions – of gametes was collected from wild spawn slicks and transferred into floating ponds on the reef (with a holding capacity of 6000 L) for development into larvae [68]. Once the larvae were ready to metamorphose from their swimming and planktonic stage into newly settled coral polyps on the reef, differential water pressure was used to push the larvae from the floating pond into enclosed tents on the reef. The Ningaloo trial showed that while enhancing the number of new coral settlers on the reef was successful, boosting longer-term coral recovery as compared to control areas was not.

In the Philippines, larvae were directly transported from the laboratory and injected into net enclosures on the reef [69]. This work showed that direct seeding of larvae onto the reef can boost recovery compared to control reefs, resulting in coral colonies that matured into reproductive and spawning adults. In Okinawa, Japan, harvested wild coral spawn slicks were reared in large ponds with a total capacity of 22 000 L situated in ports, rather than in floating ponds on the reef itself [70]. Larvae that developed from captured wild spawn slicks were then settled onto devices such as 'coral pegs', where they grew into mature, spawning adults that were then out planted onto reefs.

A resurgence of work that harvests wild coral spawn slicks to produce coral larvae for reef restoration on the GBR began in 2017 at Heron Island, the Whitsundays, Magnetic Island and Moore Reef (Fig. 8.18), using new technologies to optimise the various steps involved in the approach, which remain the major hurdles to achieving large-scale applicability (Fig. 8.16). Passive boom systems and automated technologies have been used to detect,

concentrate and collect slicks, replicating and increasing the size of floating ponds to scale culturing capabilities, and releasing larvae using automated surface and subsurface vehicles to scale larval delivery to reefs [71]. Capturing and transferring wild coral spawn slicks using oil booms and surface skimmers for culturing on vessels, for translocation and mass delivery to reefs, is also being developed through scientist and industry partnerships. Using demographic modelling, comparisons have been conducted between scaling coral reef restoration by (1) capturing from wild coral spawn slicks or (2) relocating gravid coral colonies, incorporating industrial scale techniques borrowed from other marine engineering applications such as oil spill remediation, dredging operations and land-based aquaculture [72]. Harvesting from wild spawn slicks provided the greatest opportunities for scaled reef restoration, and trials outside Heron reef harvested and cultured 29 million embryos, producing > 5 million competent larvae in a 50 000 L experimental aquaculture facility that was built on a tugboat (Fig. 8.17). The next stage of the work is to deliver the larvae to reefs *en masse*.

Scientists from throughout Australia are continuing to develop techniques for harvesting from wild coral spawn slicks for the purpose of reef restoration. Harvesting wild coral spawn slicks remains one of the most promising approaches as this technique retains natural genetic and species diversity, has a low impact on natural communities, and has promising potential for reef scale application and long-distance transportation such as from the northern to southern GBR. There are some major hurdles to overcome, particularly mass delivery onto reefs, before the approach can become a routine application administered by practitioners. Partnerships among ecologists, biological oceanographers, and marine engineers will, we hope, provide the innovation required to fulfil the dream of harvesting naturally produced coral larvae following mass spawning events for the benefit of large-scale coral reef restoration.

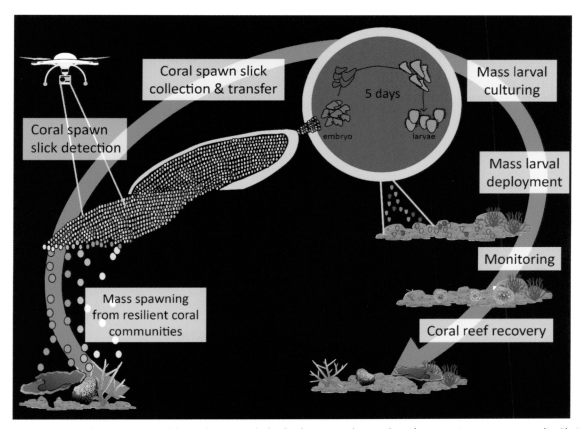

Fig. 8.16. Workflow of harvesting wild coral spawn slicks for larger scale coral reef restoration. (Image credit Christopher Doropoulos)

Fig. 8.17. Scientists from CSIRO and Van Oord harvested tens of millions of gametes from coral spawn slicks on the water's surface outside Heron Island to be cultured in a facility built on a tugboat. (Image credit: Remment van Hofstede)

Enhancing corals using assisted evolution

Madeleine van Oppen and Kate Quigley

The biggest challenge faced by today's coral reefs is rapid climate warming and the associated increase in intensity, duration and frequency of summer heat waves. Further, the impacts of rising temperatures can be exacerbated by other stressors, such as ocean acidification and nitrification. Over half of the shallow-water corals (< 5 m water depth) on the GBR died during the summer heat waves of 2016, 2017 and 2020, and most of the world's reefs are predicted to disappear within this century under the current CO_2 emission trajectory [73].

While there is evidence of some coral adaptation to warming temperatures [74], there is concern that natural rates of adaptation to rising temperatures are too slow to prevent further coral loss. The approach of assisted evolution (AE) has been proposed for corals, inspired by examples from agriculture, aquaculture, livestock farming and the conservation of iconic endangered species [75]. In these fields, selective breeding, hybridisation and microbial manipulation have successfully

Fig. 8.18. Harvested coral spawn slick developing into larvae in an open-bottomed floating pool at Moore Reef. (Image credit: Juergen Freund)

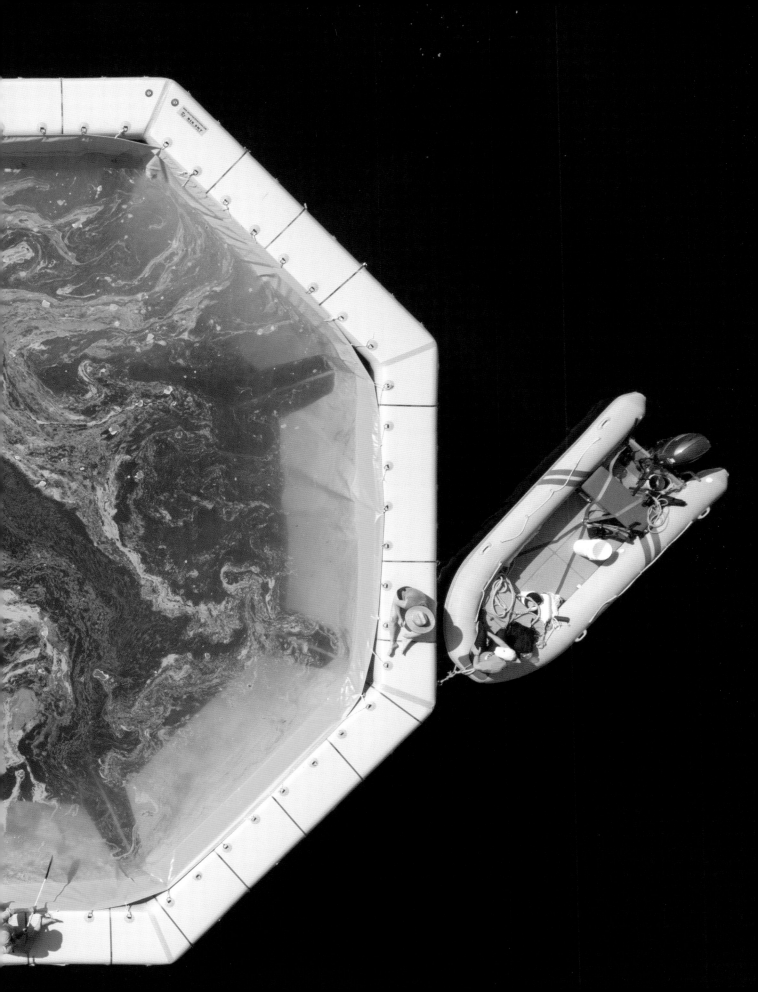

enhanced species traits of interest to humans. 'Assisted evolution' is a collective term for several approaches aimed at enhancing the climate resilience of corals via the acceleration of naturally occurring evolutionary processes. Here, we outline four AE approaches that hold promise for application to corals: (1) the elicitation of a non-genetic (epigenetic) adaptive response via pre-exposure of corals to sublethal levels of stress; (2) the mixing of gene pools via translocation (assisted gene flow); (3) the active cross-breeding between colonies from different populations or species (intra- and interspecific hybridisation), or selective breeding of individuals from the same population; and (4) modifying the composition of coral-associated microbial communities via inoculation with natural microbes, and introducing 'enhanced' microbes into the coral microbiome, where the microbes are enhanced via experimental evolution.

Although the field of AE is still young; however, it has been demonstrated that all of these interventions have the capacity to help corals adapt to rapid climate warming.

Pre-exposure to sublethal stress levels

Short-term exposure to low levels of stress can prime organisms for future, potentially more extreme bouts of exposure to heat, salinity or UV radiation. In corals, this may occur through mechanisms such as the early expression of a particular suite of genes ('front loading'; [76]) probably driven by epigenetic changes [77]. Epigenetic changes do not alter the DNA sequence, but can affect gene expression via other mechanisms such as the modification of the methylation state of the DNA, the histone proteins that help condense the DNA or the expression of small non-coding RNAs. Pre-exposure to stress during early development or throughout the life of the coral has in some (but not all) cases led to acquired tolerance that allows corals to withstand large fluctuations or increases in temperature [78]. Pre-exposure may also induce changes to microbial communities that can sometimes be passed on to the next generation [79], possibly

contributing to enhanced tolerance across generations. This approach may be used to enhance environmental tolerance of coral larvae during the transport of spawning slicks or the *ex situ* rearing of coral larvae and recruits for subsequent deployment on damaged reefs. The placement of coral nurseries in naturally warming locations, such as reef lagoons, could also be considered as a means of pre-conditioning adult coral fragments. Note that our understanding of the efficacy of coral pre-conditioning and epigenetics in general is scant.

Selective breeding, hybridisation and assisted gene flow

Selective breeding as a conservation tool has been used for a diverse range of species, although it has generally focused on increasing genetic diversity and has not necessarily been aimed at enhancing particular traits. Hybridisation refers to the process of breeding individuals from genetically distinct populations of the same species (intraspecific hybridisation), or of different species (interspecific hybridisation), to produce hybrid offspring. Hybridisation, as well as the selective breeding of corals from the same population, has in some instances improved heat tolerance. Assisted gene flow is the translocation of corals within their natural distribution range. It is anticipated cross-breeding between transplants and native corals of the same species will occur naturally following transplantation, producing intraspecific hybrids as a result. Intraspecific hybridisation between colonies from a warm as well as a cooler reef has been shown to increase heat tolerance of hybrid offspring 10 to 26 times, depending on the life stage and symbiotic state measured [80]. While intraspecific hybridisation in corals has focused on breeding heat tolerant gene variants into the receiving populations, the crossing of different species increases genetic diversity and creates new combinations of gene variants through the mixing of divergent genomes. When tested under laboratory conditions, this resulted in hybrid offspring that exhibited 14–34 per cent higher survival compared with pure bred offspring under elevated temperature and pCO$_2$ depending on the parental species [81]. Field deployment of both intra- and interspecific hybrids has commenced and assessment of their performance is underway (Fig. 8.19).

Modifying the composition of coral-associated microbial communities

Corals associate with diverse bacterial communities that are critical for their health and functioning. The development of bacterial probiotics for the purpose of enhancing particular coral traits is receiving increasing attention, and some early findings are promising. For example, the exposure of coral early life stages to whole microbiomes isolated from the mucus of several coral species resulted in distinct bacterial communities whose composition correlated with that of the inoculation with microbes

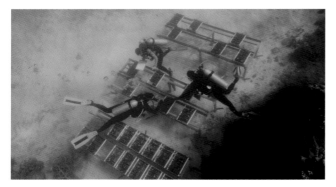

Fig. 8.19. Corals produced using assisted evolution methods being out-planted on tables. (Photo supplied with kind permission from the Australian Institute of Marine Science; image credit: Kate Green)

when measured 4 months post-dosing [82]. In another study, inoculation of adult corals with a cocktail of cultured bacteria reduced the negative impact of elevated temperature [83].

Similarly, the provisioning of algal endosymbionts (Symbiodiniaceae) at both early life history and adult stages is also being explored (Fig. 8.20). The provisioning of Symbiodiniaceae in the genus *Durusdinium* resulted in increased survival under heat exposure relative to provisioning with either *Symbiodinium* or *Cladocopium* [74]. When these same taxa were provided to adult corals at the height of experimental temperature increase, *Cladocopium*-exposed corals showed significantly better survival compared to those without symbionts or those with *Durusdinium*, although the algae were not incorporated into the coral's microbiome [84]. Adjustments to coral-associated bacterial and algal communities in some instances benefit corals at least in the short term, and may be retained over time. However, it is not well understood which microbial species are likely to provide benefits under climate warming, how they can be best administered, and whether they will be incorporated into the coral microbiome in a stable manner.

'Enhanced' microbes into the coral microbiome

The enhancement of microbial symbionts of corals through experimental evolution has so far only focused on the photosymbionts that live inside the coral gastrodermal cells. A recent study demonstrated that about one-third of the heat-evolved

Fig. 8.20. The experimental evolution of symbiotic algae. (Photos supplied with kind permission from the Australian Institute of Marine Science; image credit: Marie Roman)

Symbiodiniaceae strains tested showed increased resistance to coral bleaching after reintroduction into coral larvae [85]. Mechanisms underpinning this enhanced trait seem to include reduced rates of photosynthesis in combination with increased rates of carbon fixation. The heat-evolved symbionts also affected the expression level of coral host genes, resulting in the early activation of the coral heat-stress response.

Studies on experimental evolution of coral-associated bacteria are lacking. The short generation times of most bacteria (compared to Symbiodiniaceae) and the evidence for rapid adaptive responses in experimental evolution studies with other bacteria provide confidence this approach will be able to shift trait values of culturable coral-associated bacteria. Whether and how laboratory evolved bacteria affect coral traits or whether they can enhance coral thermal bleaching tolerance is unknown.

In summary, targeting multiple members of the coral holobiont using AE, for example through the strategic pairing of heat tolerant corals hosts (via hybridisation) with heat-tolerant symbionts (via experimental evolution), can result in significant increases in heat tolerance. Moving forward, these approaches need to be (1) scaled up by testing them for a larger number of coral and microbial symbiont species, (2) tested in the field, and (3) assessed for their biological, ethical, and socio-cultural risks.

Natural extreme reefs as potential coral resilience hotspots

Emma Camp and Verena Schoepf

Coral reef organisms flourish under certain optimal environmental conditions (e.g. temperature and light, see 'The basics of coral biology' in Chapter 4). However, we continue to discover locations where corals survive in seemingly hostile conditions, some of which share similarities with predicted future climate scenarios. These suboptimal habitats house so-called 'extreme' or 'marginal' corals that extend our understanding of the

broad range of environmental conditions under which corals can persist (see 'Marginal reefs: distinct ecosystems of extraordinarily high conservation value' section in Chapter 4). Marginal or extreme habitats can act as a natural laboratory to study how corals can survive under suboptimal environmental conditions, and what, if any, are the costs of trade-offs to such survival [86]. Furthermore, these reef environments may provide additional refuge in the form of habitats that maintain favourable environmental conditions which are being lost elsewhere [87], or by acting as resilience hotspots that provide a source of stress-hardened corals [86] that may become targeted species in coral nurseries [88]. Growing acknowledgement of the value of extreme and marginal corals has increased the research focus of these systems on both sides of the continent.

Along Australia's iconic GBR lie shallow reef flat and lagoon systems, turbid reefs and deep reefs (> 30 m depth) that exemplify Australia's extreme and marginal coral environments. Shallow reef and lagoon habitats are thought to precondition corals to environmental extremes through often highly variable and acute environmental conditions. For example, mangrove lagoons on the GBR house over 30 species of corals living under conditions that are warmer, more acidic, and lower in oxygen than the average value predicted for the GBR in the year 2100 [89] (Fig. 8.21). Strategies behind corals' survival in mangroves are species-specific and include altered physiology as well as unique symbiotic partners, but often come at the cost of reduced calcification [89]. Despite low light and muddy waters, inshore turbid reefs on the GBR also house a surprising diversity and cover of corals, which have been reported to be less impacted by coral bleaching [90]. Consequently, turbid reefs have been proposed as potential coral refugia from climate change. Similarly, deep reefs may act as potential refuge environments as they typically experience less disturbances than shallow reefs. However, such refuge capacity remains unclear, with some deep corals reported to bleach on the GBR during 2016 due to local oceanographic conditions [91].

Fig. 8.21. An *Acropora* coral growing among mangrove roots at Low Isles on the Great Barrier Reef. Such corals survive in warmer, more acidic and lower oxygen conditions than are typical for coral reef environments. (Image credit: Emma Camp)

High-latitude, cooler water (subtropical) environments where corals exist at the edge of their environmental limits in the Solitary Islands (30°S), Lord Howe Island (31.5°S), and Sydney Harbour (34°S) also provide important new habitat for tropical corals in a rapidly warming ocean and may act as refugia from climate change.

Some of the world's most unique – and arguably most extreme – coral reefs are, however, found in the remote Kimberley region in north-west Australia. There, abundant and highly diverse coral reefs exist despite the world's largest tropical tides (up to 12 m), which create muddy waters and extreme fluctuations in environmental conditions. Furthermore, shallow reef areas regularly dry at low tide for several hours (see 'Kimberley corals

exposed' in Chapter 1). The resulting environmental extremes have made the corals of the Kimberley remarkably stress-resistant – but does this mean that they will also cope better with ocean warming and acidification? The answer is potentially 'yes', although caveats exist. Some species of corals and coralline algae from the Kimberley were found to be resistant to future acidification levels [92], suggesting that their reef-building capacity could remain unaffected. Intertidal corals also have enhanced heat tolerance and bleaching resilience [93], but many reefs in the Kimberley nevertheless bleached severely during the 2016 global bleaching event. The ability of these reefs to cope with the ocean warming predicted for the next few decades may be limited [94], highlighting the vulnerability of even stress-hardened corals to climate change. But there is a silver lining: some corals of the Kimberley can rapidly adjust their physiology to cooler temperatures without losing their enhanced heat tolerance [94]. Although this would have to be confirmed outside the laboratory, this makes them promising candidates for heat-resistant coral nurseries [88] or AE approaches [75].

Since many extreme and marginal reef systems do not possess the aesthetic and functional values associated with clear-water coral reefs, they have often been overlooked and understudied. However, they could facilitate the persistence of coral reefs due to their potential to serve as resilience hotspots, refugia and natural analogues for future ocean conditions [86]. As we continue to learn more about these fascinating systems, this knowledge may inform and expand available management options, including new interventions.

Science meets the public, policy and management practitioners
Russell Reichelt

When the Great Barrier Reef Marine Park Authority (GBRMPA) was newly formed in the 1970s, it employed scientists at a time when the science-management relationship was relatively weak. It was unusual for the leadership of the Australian Institute of Marine Science to talk to the Chair of the GBRMPA or, for that matter, any politician.

This section describes how scientists have worked with the GBRMPA to shift its original goal from fostering protection and wise use of the Reef towards restoring and maintaining a damaged reef facing the greatest management challenge yet, that of climate change (Fig. 8.22).

Reef politics and the role of science

Reef politics have been active and often heated since the 1960s. Scientific reviews of the Reef's status and future risks are readily drawn into public debates on impacts from industries such as

Fig. 8.22. 'Coral Greenhouse' installation at the Museum of Underwater Art, John Brewer Reef. The installation represents a pivot in management policy for the Great Barrier Reef World Heritage site, and the structure serves to educate the public about reef restoration. (Image credit: Matt Curnock)

tourism, agriculture, ports and shipping, and the use of fossil fuel for energy generation causing climate change.

Political leadership has been the key driver of reef protection since the 1970s. Since the Emerald Agreement in 1979, the joint management arrangement between Queensland State and the Commonwealth governments has been informed by science and responded comprehensively to a range of risks to the Reef, including changes to Reef zoning plans, moderation of fishing impacts, improvements for water quality and control of crown-of-thorns starfish. Governance frameworks to achieve this have been supported by the strategic assessment of the GBR by the GBRMPA and the Queensland Government. More recently, efforts have been consolidated in the Reef Trust Partnership coordinated by the Great Barrier Reef Foundation and guided by science experts in partnership with the GBRMPA, the Australian Institute of Marine Science, CSIRO, James Cook University, the University of Queensland and more.

Coral bleaching: the stark reality of climate change

Until the late 1990s, the impacts of climate change on coral reefs received little attention from scientists, policy-makers and reef managers. The looming crisis for the GBR from global warming was foreshadowed in a reef science paper by Ove Hoegh-Guldberg in 1999 [9]. Charlie Veron's explicit book *The Life and Death of the Great Barrier Reef* in 2006 was another major milestone in explaining the problem. While this was unfolding, for most reef scientists the impending dramatic consequences of global warming were not top of mind and public outcry was muted. Even the 1998 coral bleaching event came and went without much attention from the broader Reef science community and even less from the government. The first Kyoto Agreement within the United Nations Framework Convention on Climate Change (UNFCCC) was later ratified in 2007, at which point Australia argued it should be able to increase its greenhouse gas emissions.

The GBRMPA Outlook Report in 2009 put the case strongly that climate change far outweighed all risks to the GBR. But it was the port developments associated with the mining boom from 2009 to 2017, especially the Gladstone Harbour gas facilities and the expansion of all the major ports along the GBR coast, that attracted public attention and encouraged more scientific cooperation to assess the risks associated with exporting fossil fuels.

Reef politics and science of fossil fuels and coral bleaching: the last 10 years

The GBR's first well-documented mass coral bleaching in 1998 caused extensive areas of bleached coral and ~4–6 per cent of the GBR corals died from heat stress. At the time, relatively few reef scientists would have been aware of Peter Glynn's 1984 reports of the mass bleaching in Panama associated with the 1982–83 El Niño [2] and very few science teams were working actively on the drivers of mass coral mortality.

The Queensland mining boom raised the profile of climate change and clouded the options for how to respond. In 2010–11 the southern end of the GBR experienced the expansion of multiple liquefied natural gas refineries on Curtis Island in Gladstone Harbour. This was an area outside the Great Barrier Reef Marine Park, but within the boundaries of the Great Barrier Reef World Heritage site. By 2013–14, there were prospects of expansion of all the five ports along the Reef Coast, stretching 1000 km from Gladstone in the south to Cairns in the north, and the creation of multiple new ports in areas at the time not developed as industrial sites. The first major development involving the Reef directly was a proposal to add additional loading berths at the Abbot Point coal export terminal near Bowen. This was part of broader planned increases in coal mining in central Queensland, which would be exported overseas through multiple port facilities along the coast – referred to as 'the mining boom'. During the mining boom, many varying perspectives and interests became intertwined: the multiple views by scientific experts; the local, state and Commonwealth governments; the environmental movement; the mining industry; and the public's interest.

Before the port expansion proposals, scientists had advised that the impacts of ports on water quality were localised and should not be confused with the greater risk of large-scale runoff of sediments, fertiliser and chemicals from disturbed rivers and coastal catchments. The scientific advice on the risks posed at Abbot Point departed dramatically from this viewpoint.

The Queensland Government owned North Queensland Bulk Ports and sought to expand the capacity of the Abbot Point coal terminal near Bowen in 2011. This expansion would involve dredging channels for coal ships to access the port. North Queensland Bulk Ports requested permission from the GBRMPA to dispose of clean dredged material ~30 km to the east of the port in the deep sandy shipping channel between the coast and shallow coral reefs.

The GBRMPA issued a permit in 2014 when it appeared that the project could be implemented with minimal environmental impacts if there were strong safeguards to stop work if circumstances arose that might damage the environment. A year later in 2015, the 2014 permit was revoked before work had started on port expansion and before the contractors had demonstrated they could meet all the environmental safeguards. The revocation was part of a larger review by the GBRMPA of all port development along the entire coast.

Strong advocacy from the concerned public and the environmental movement had a major impact on the political situation developing not only at Abbot Point, but also further along the coast. There were calls for the practice of disposing of dredged material to be banned within the GBR Region. The debate about Abbot Point, the port site where coal was to be loaded onto bulk carrier ships that would then transport it through the Reef, pitted the environment against coal extraction at both a local and an international scale (Fig. 8.23). The underlying driver of intense and much broader concern was that the cargo to be carried on ships through these dredged port channels was fossil fuel being sent

Fig. 8.23. A coal ship at Abbot Point terminal in Bowen, central Great Barrier Reef. (Image credit: Jumbo Aerial Photography)

for burning overseas, leading to the release of massive quantities of greenhouse gases. This was the very activity that would lead to further global warming that was killing the GBR and, indeed, coral reefs globally. The environmental laws then, and still today, do not assess future risks posed by the practice of burning coal.

The Abbot Point concern was all about the coal and with good justification. This controversial national political issue became known as the 'coal versus coral war'. Collective action to curb carbon emissions would be needed on a massive scale to remedy this and the GBRMPA was now operating within a national debate about energy policy and the need to curb carbon emissions that had become increasingly polarised. To reach the goal of limiting global warming to 1.5°C would require all of Queensland's abundant coal reserves to remain in the ground and unburned. A popular saying in the GBRMPA at the time was: 'The stone age did not end because humans ran out of stone. The coal age will end similarly.'

A single table prepared by Jon Brodie, a water quality expert at James Cook University, changed the course of events. He showed the full extent of coastal engineering and dredging that would occur if all the planned proposals were approved. Most had not reached formal proposal stage but included more coal loading terminals near

Rockhampton, Mackay and Bowen. The total volume of material to be dredged was close to 150 million tonnes, which put the original Abbot Point proposal of 3 million tonnes into a different light. Abbot Point was not a low-impact project if it became a precedent for the much larger set of projects. This demanded a new, collective way of looking at the environmental impacts of port development across several sites. Unless the laws were changed, each project could be potentially approved because the environmental regulations are about individual actions, not the accumulated effect of them all.

Armed with this picture of the cumulative impacts and working with the ports and the political leaders, the GBRMPA proposed regulation amendments to the *Great Barrier Reef Marine Park Act* to ban any disposal of 'capital' (i.e. dredge material associated with new ports inside the Great Barrier Reef Marine Park), an amendment that would revoke the earlier approval for Abbot Point. The Environment Minister, then Greg Hunt, approved and the change was enacted in Federal Parliament in 2015. The Queensland Government declared that there would be no new ports allowed.

What happened at Abbot Point offers lessons on how best to integrate high quality science into decision-making by regulators. There should be greater transparency in environmental assessment processes that provide the variety of stakeholders involved in coastal development, including the private sector, managers, locals and environmentalists, increased access to information and broader review of the scientific assessments. Moreover, there is an increasing need to consider individual projects in the context of the cumulative effects of all impacts, including those in the past, present and future. In addition to this, more effective science advice could be given if the experts understood the context of their advice in terms of the governing laws and processes for environmental assessment.

On the issue of cumulative effects, the GBRMPA Strategic Assessment of the Great Barrier Reef Authority called for new, science-based accounting for the cumulative effects of multiple impacts to be incorporated into both the authority's policies, and national policy guiding Commonwealth's decisions arising from the national environmental law, the *Environment Protection and Biodiversity Conservation Act 1999*. The 2018 Cumulative Impact Management Policy was approved by the GBRMPA and the Commonwealth Department of the Environment soon after the dredge spoil disposal ban was enacted.

Protection and wise use are no longer enough

The scientific reporting of the massive extent of coral mortality on the GBR in 2015 led by Terry Hughes from James Cook University's ARC Centre of Excellence for Coral Reef Studies, including several other institutions under the collective title of the National Coral Bleaching Taskforce, produced many excellent scientific papers [e.g. 16, 19, 20]. The timely reporting throughout the multi-year event also informed the Reef managers, the GBRMPA and the Queensland National Parks and Wildlife Service, the public, the governments and the international community, who all wanted to understand the future of the GBR and coral reefs everywhere.

After the global marine heat stress in 2015–16, the GBRMPA convened the 2017 Reef Summit in Townsville, attended by 70 regional, national and international delegates representing marine park managers, Traditional Owners, government agencies, research institutions, industry groups, Reef users and other stakeholders. The primary output of the Reef Summit was the GBRMPA's *Great Barrier Reef Blueprint for Resilience*. An inclusive and transparent discussion had emerged from this summit, which identified climate change, driven by the burning of fossil fuels, as a major threat to the Reef and canvassed all the available pathways to build resilience for the Reef in the face of climate change. The blueprint highlighted the need to restore and maintain damaged ecosystems, adding an important component to the original GBRMPA goal set out in 1989: 'To provide for the protection, wise use, understanding and enjoyment of the Great Barrier Reef in perpetuity through the care and development of the Great Barrier Reef Marine Park.'

In the face of the existential challenge posed by climate change, the old adage of protection and wise use is no longer enough to sustain the GBR. In addition to building Reef resilience by local actions for the ecosystem, all available actions to reduce greenhouse gas emissions should be taken and championed by all those who care about the fate of the GBR and coral reefs worldwide.

The Intergovernmental Panel on Climate Change (IPCC)
Ove Hoegh-Guldberg

As described in many places of this book, coral reefs have been at the centre of the scientific evidence that rapid human-driven climate change is dramatically changing our world. Reefs were also one of the first ecosystems to demonstrate the severe ramifications if emissions of greenhouse gases such as CO_2 are not reduced dramatically. This narrative provides a transparent causal chain between climate change, impacts on ecosystems, and decreasing support from these ecosystems for human communities and development [95]. Nevertheless, non-science debates have sown the seeds of doubt about climate change over the past 40 years and, as a result, have dangerously reduced the rate of efforts to mitigate greenhouse gases [96].

Getting to the nub of the issue: the Intergovernmental Panel on Climate Change

A critical response to these challenges was to develop a robust scientific consensus of expert knowledge of climate change in the past, present and future contexts. The IPCC was established by the United Nations Environment Programme (UNEP) and the World Meteorological Organization (WMO), which the UN General Assembly endorsed in 1988. The IPCC currently has 195 member states. The aim was to assess the degree of consensus on critical issues relating to climate change, feeding into the UNFCCC.

The IPCC process revolves around a series of assessment reports. So far, there have been five assessment cycles, each cycle involving three assessment reports, which cover (1) physical science basis; (2) impacts, adaptation, and vulnerability; and (3) mitigation of climate change. Each assessment cycle also includes several special reports that involve 'deep' dives into emerging issues on topics such as (1) the implications of global warming of 1.5°C above pre-industrial levels; (2) the role of the ocean and cryosphere in a changing climate; and (3) climate change, desertification, land degradation, sustainable land management, food security and greenhouse gas fluxes in terrestrial ecosystems.

The expert chapter teams are supported by a technical support unit, with countries putting forward experts selected for their scientific excellence, advanced field knowledge and geographical inclusiveness. These experts volunteer their time for at least one 5–6-year cycle, which involve multiple internal and external reviews open to experts, industry, government and public sectors. For example, the 'Special report on global warming of 1.5°C' involved 91 authors from 40 countries and received 42 001 expert and government comments from thousands of qualified individuals, with each comment being rigorously responded to and publicly available (Table 8.2).

The scientific knowledge generated by the IPCC process feeds into the Conference of the Parties (COP) associated with the UNFCCC. Crucially, knowledge about climate change must be policy-relevant, but not policy-prescriptive. The IPCC process gathers the expert consensus required by governments and their agencies to make informed decisions and develop effective policies. It does not tell governments what to do. This knowledge consensus is transparent and objective, covering the full range of topics from the physical basis, impacts and vulnerability, to mitigation of climate change. It provides critical insights on climate change for governments to understand and respond to.

The Intergovernmental Panel on Climate Change perspective on climate risks, impacts, and coral reef conservation

Not surprisingly, the IPCC process has played a central role in developing consensus in the complex area of climate change. The 2015 United Nations Climate Change Conference, or Conference of the Parties (COP) 21, meeting in Paris was a turning point for effective action on climate change. With two exceptions, the international community signed on to article 1(a), which defined the temperature target that was required to keep climate change to a manageable level. Article 1(a) stipulates, 'holding the increase in the global average temperature to well below 2.0°C above pre-industrial levels and pursuing efforts to limit the temperature increase to 1.5°C above pre-industrial levels', recognising that this would significantly reduce the risks and impacts of climate change.

Table 8.2.   Author and reviewer statistics from the 'Special report on global warming of 1.5°C' (example of the review process for a typical IPCC report; https://www.ipcc.ch/).

Authors	91 authors from 44 citizenships and 40 countries 14 coordinating lead authors 60 lead authors 17 review editors 133 contributing authors (CAs)
Cited literature	Over 6000 cited references
Expert comments	A total of 42 001 expert and government review comments received and responded to: • first order draft (12 895) • second-order draft (25 476) • final government draft (3630).

At this point, one might heave a sigh of relief and imagine that the pointy end of climate change was some way off. However, as outlined elsewhere [9, 10], climate change has a sensitivity that is far higher than many might at first expect. As evidence of this, the IPCC Assessment Report 1 'found the increased incidence of bleaching to be consistent with ocean warming' [97] while the special report on 1.5°C published in 2018 [98] (40 years later) concluded that 'multiple lines of evidence indicate that the majority (70–90%) of warm water (tropical) coral reefs will disappear even if global warming is constrained to 1.5°C (very high confidence …)' [99]. As our understanding of the bleaching impacts of temperature has developed, so has our understanding of several other important climate impacts on coral reefs. Other factors such as ocean acidification, storm intensity, SLR, changing currents, and decreasing ocean oxygen levels interact with non-climate stressors, such as coastal pollution and overharvesting, to drive fundamental changes to levels of impact (in some cases) not seen for tens of millions of years [101].

Confronting the reality of a dangerously changing world

The IPCC plays a crucial role in providing scientific and expert knowledge to help us understand the past and, therefore, the future. For example, multiple lines of evidence reveal that 70–90 per cent of today's corals and coral reefs will disappear even if global warming is constrained to 1.5°C (very high confidence) [102]. This situation might be daunting to many people. However, rather than focus on the 70–90 per cent, we should focus on the 10–30 per cent that is left behind [100, 101]. There is evidence that some coral reefs are not as exposed to climate change as others and should therefore be conservation targets against local threats (e.g. pollution, overfishing), allowing them to function as regeneration centres for future coral reefs after climate stabilisation. This strategy is the basis for the 50 Reefs project [100, 101] and World Wildlife Fund's Coral Reef Rescue Initiative [102]. By combining rapid action on climate change with deep localised action on non-climate threats such as coastal pollution and overfishing, it is hoped that the future for the world's most diverse marine ecosystems will be preserved.

Advancing Australian reef governance into the Anthropocene: the Great Barrier Reef at the next frontier
Tiffany H. Morrison

Australia's regime of governance for the GBR is widely regarded as one of the most innovative environmental governance systems in the world [103, 104]. For more than 40 years, multiple stakeholders have worked together to create knowledge about the GBR, prioritise issues, formulate policy, delegate responsibility, and make decisions about how to sustain the Reef.

The creation of the *Great Barrier Reef Marine Park Act* and the GBRMPA in 1975 marked the first major governance innovations. These innovations were reinforced through listing of the GBR as a World Heritage site in 1981. Through the 1980s and early 1990s, additional land, water and coastal policies were introduced in Queensland. In 2004, a best practice GBR zoning plan was designed and implemented, based on the latest reserve design software, spatial analysis tools, and extensive public participation. A national and state Reef Water Quality Program and associated Reefplan followed, combining partnerships, incentives, and voluntary action to improve the management of diffuse terrestrial pollution from agriculture.

By the late 2000s, management of the GBR had received international acclaim, with the rezoning process receiving 19 international, national, and local awards (Fig. 8.24). The GBRMPA was regarded as the Trojan horse of sustainable marine governance, exporting its model to the world. These successes, and their challenges, inspired and informed a new agenda in environmental governance research and practice globally (see 'Zoning the Great Barrier Reef' in Chapter 5).

Fig. 8.24.   (A) Joint meeting of the Marine Park Authority and the Great Barrier Reef Consultative Committee on 13 August 2003. (B) Executive Director of the GBRMPA John Tanzer, the Honourable Senator Ian Campbell and Chairman of the GBRMPA and the Honourable Virginia Chadwick AO accept the World Wildlife Fund's Gift to the Earth award in October 2005 from WWF International Director General designate, James Leap. (C) Dr Fanny Douvere (UNESCO) announces the decision of the 21-country World Heritage Committee to postpone an 'in danger' decision until 2022, pending action on climate change and water quality. (Image credits: © Commonwealth of Australia (A) and (B), © UNESCO 2021 (C))

New challenges overturn governance assumptions

The paradigm underpinning the GBR's early governance success was one of ecosystem-based management. Ecosystem-based management assumes that strong *local* management can maintain ecological and social resilience by reducing proximate stressors, such as overharvesting and pollution, and undertaking local restoration. Under this paradigm, GBR governance has successfully focused on the goal of maintaining biodiversity and restoring the Reef to recent historical baselines. The primary targets of intervention have been the fishing, tourism and farming industries operating within the GBR and its adjacent coastal land catchments.

However, the escalating impacts of mining development and climate change have now rendered this paradigm untenable for coral reefs and many other ecosystems [105]. Proposals to expand several ports along the Queensland coastline, including dredging and dumping sediments, have underscored water quality issues, while the emergence of recurrent mass coral bleaching events has reinforced the limitations of conventional ecosystem governance for the GBR (see 'The UNESCO List of World Heritage in Danger: a lever for reform' in Chapter 7). Ecosystem-based management alone is clearly not able to withstand the growing biophysical and political pressures on the reef [106].

Growing threats also shine a spotlight on the increasingly complex systems for managing those threats. Ecosystem-based management of the Great Barrier Reef Marine Park is the responsibility of the Australian Government, primarily through the statutory GBRMPA. A highly collaborative working relationship, dating back to the 'Emerald Agreement' of 1979, exists with the State of Queensland, involving complementary marine, land, water, and coastal arrangements established over four decades. UNESCO provides important international oversight as a consequence of the 1981 World Heritage listing.

As the climate change and coral-bleaching crisis continue to unfold, this increasingly complex system for governing the Reef is becoming less effective [107]. Multiple additional stresses, including economic crises, resource industry pressure, and local political backlashes against conservation, have all combined to impact different aspects of the system [108]. In some cases, successive governments have continued to announce new laws, programs, funds and plans while simultaneously dismantling pre-existing laws, departments and funding. Low-visibility examples include the introduction of an offsets policy that encourages developers who want to build on or near the reef to make an offset payment into a Reef trust. While this policy funds activity to improve water quality, it has also made getting consent for development easier, because there is no mechanism to minimise the potential for undue industry influence. The federal Department of Water and Environment grants approval for developments, and also oversees the offset fund into which the developers pay. In other more visible cases, Queensland coasts and catchments legislation has been repealed (2013), and Australian climate law and policy dismantled (2014).

Responding to new challenges

Coral reef ecologists have shown how climate change, pollution, and overfishing have collectively degraded reefs. Biological and climate scientists have also documented that most drivers of ecological change on coral reefs are increasing at regional and global scales. Environmental social scientists have demonstrated that people, institutions and politics are critical to effective governance. A developing research and policy agenda is beginning to extend these perspectives to incorporate new governance challenges and innovations; consequently, a new reef governance paradigm is emerging, placing decarbonisation and adaptation at its centre ([108], Fig. 8.25).

Decarbonisation

Recurrent and severe coral bleaching means that decarbonisation rather than conservation is now understood as the defining challenge for coral reefs. In the past, the ecosystem-based management paradigm emphasised reefs and reef-dependent peoples as the single arena for action. Today, the challenge of decarbonisation is opening up a much wider variety of multiscale strategies for governing coral reefs, including political, economic, technological and cultural approaches. Radical interventions in

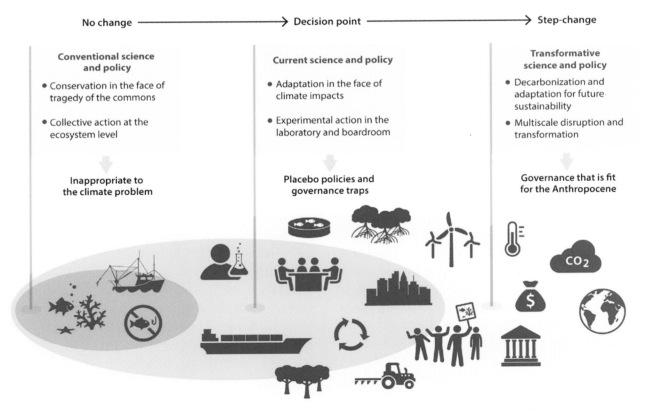

Fig. 8.25. Changing governance paradigms for coral reefs in the Anthropocene. (Image credit: Tiffany Morrison, reproduced from [106])

other ecosystems, such as forests, have shown how interlocking social and political approaches can limit carbon emissions, lift Indigenous peoples and communities out of poverty, and prevent wider ecosystem degradation. Other coral reef countries are already ahead of the curve. In Fiji, low-cost renewable energy technologies (e.g. solar photovoltaics, LED lighting) are already being distributed through local marine cooperatives to increase the uptake of low-carbon fuel sources [109]. Ambitious renewable energy development in Small Island Developing States (SIDS) of the Pacific has also strengthened the position of Pacific SIDS in global climate change negotiations [110].

In Australia, by contrast, a regional decarbonisation plan has been proposed to transition the coal-dependent GBR catchment to alternative energy sources, economies and livelihoods [111], but such a plan has yet to receive bipartisan support. While many business elites, politicians and senior bureaucrats continue to oppose regional decarbonisation, a new breed of activists is exploring the transformative role of sectors operating at higher scales, such as finance, health, energy and transport. Such work is extending the traditional and disciplinary boundaries of coral reef governance and relieving the pressure on environmental ministries and local reef managers, who are often the least powerful in terms of bringing about decarbonisation on behalf of the ecosystems within their own purview.

Adaptation

While decarbonisation is now the defining challenge for coral reefs, reefs and reef-dependent communities must also learn how to adapt to more intense global change. Today, the GBR is at a crossroads with respect to technology-led adaptation and community-based adaptation. Technology-led adaptation is mainly focused on biophysical interventions in the water (e.g. coral shading, AE and coral-seeding robots) [112]. Community-based adaptation, by contrast, entails meaningful social change in the fisheries, tourism, mining and agricultural communities, which depend on, or interact

with, the GBR. Such social change includes alternative industries and livelihoods, enabled by innovations in education, agriculture and energy. Australia has successfully implemented similar large-scale change previously [113]. In the 1990s, for example, the Australian Government successfully transitioned more than 12 forest ecosystems towards more sustainable use and livelihoods through a series of regional forest agreements and structural adjustments.

To date, substantial funding has been directed towards technology-led adaptation, while community-based adaptation is lagging. Technology-led approaches may be novel and potentially adaptive; however, they also carry unknown evolutionary and ecological risks, as well as pose serious ethical challenges [114]. Many technology-led approaches also falsely imply that it is possible to address ecosystem decline by curbing the symptoms of rising temperatures without dealing with the underlying drivers of rising greenhouse gas emissions. However, community-based adaptation approaches must also be tested for their political legitimacy and effectiveness to both measure the political feasibility of such interventions and modify them accordingly, and to guide the development of other, as yet unanticipated, interventions that might also be social and political, rather than scientific and technological. More work needs to be done across both the ecological and social sciences to assess and improve the legitimacy, efficacy and cost of these different adaptation solutions at meaningful scales.

Moving forward for the Great Barrier Reef and the world

Australian reefs are not just an environmental asset – they are also a governance problem with much at stake. Despite significant resources, time, effort and salience, their status remains disputed and their governance is stuck in a constant state of policy controversy. To liberate future coral reefs from today's policy controversy, scientists and policy-makers are now embracing a new governance paradigm, one that is fit for the Anthropocene.

Understanding how to manipulate ecological, social, and political dynamics at a variety of spatial and temporal scales is now integral to addressing the escalating problems that confront coral reefs. Indeed, securing a future for coral reefs under climate change is as much a political challenge as it is an ecological or a social one. Although the scientific hurdles of interdisciplinarity, complexity and urgency that are embedded within the new paradigm are challenging, the benefits are potentially global. If we succeed, efforts to effectively govern Australian reefs could continue to set an example for sustaining ecosystems and people into the next century and beyond.

References

1. McCulloch M, Fallon S, Wyndham T, Hendy E, Lough J, Barnes D (2003) Coral record of increased sediment flux to the inner Great Barrier Reef since European settlement. *Nature* **421**, 727–730. doi:10.1038/nature01361

2. Glynn PW (1984) Widespread coral mortality and the 1982–83 El Niño warming event. *Environmental Conservation* **11**, 133–146. doi:10.1017/S0376892900013825

3. Glynn PW, D'Croz L (1990) Experimental evidence for high temperature stress as the cause of El Niño-coincident coral mortality. *Coral Reefs* **8**, 181–191. doi:10.1007/BF00265009

4. Hoegh-Guldberg O, Smith GJ (1989) The effect of sudden changes in temperature, light and salinity on the population density and export of zooxanthellae from the reef corals *Stylophora pistillata* Esper and *Seriatopora hystrix* Dana. *Journal of Experimental Marine Biology and Ecology* **129**, 279–303. doi:10.1016/0022-0981(89)90109-3

5. Hoegh-Guldberg O, Salvat B (1995) Periodic mass-bleaching and elevated sea temperatures: bleaching of outer reef slope communities in Moorea, French Polynesia. *Marine Ecology Progress Series* **121**, 181–190. doi:10.3354/meps121181

6. Hoegh-Guldberg O, Berkelmans R, Oliver J (1997) Coral bleaching: implications for the Great Barrier Reef Marine Park. In *Proceedings of The Great Barrier Reef Science, Use and Management Conference.* 25–29 November, 1996, Townsville. (Ed. D Yellowlees) pp. 210–224.

7. Jones RJ, Hoegh-Guldberg O, Larkum AWD, Schreiber U (1998) Temperature-induced bleaching of corals begins with impairment of the CO_2 fixation mechanism in zooxanthellae. *Plant, Cell & Environment* **21**, 1219–1230. doi:10.1046/j.1365-3040.1998.00345.x

8. Marshall PA, Baird AH (2000) Bleaching of corals on the Great Barrier Reef: differential susceptibilities among taxa. *Coral Reefs* **19**, 155–163. doi:10.1007/s003380000086

9. Hoegh-Guldberg O (1999) Climate change, coral bleaching and the future of the world's coral reefs. *Marine and Freshwater Research* **50**, 839–866. doi:10.1071/MF99078

10. Hughes TP, Kerry JT, Álvarez-Noriega M, Álvarez-Romero JG, Anderson KD, Baird AH, Babcock RC, Beger M, Bellwood DR, Berkelmans R, *et al.* (2017) Global warming and recurrent mass bleaching of corals. *Nature* **543**, 373–377. doi:10.1038/nature21707

11. Hoegh-Guldberg O (2011) Drowning out the truth about the Great Barrier Reef. *The Conversation*, 30 August, <https://theconversation.com/drowning-out-the-truth-about-the-great-barrier-reef-2644>.

12. Kleypas JA, Buddemeier RW, Archer D, Gattuso J-P, Langdon C, Opdyke CBN (1999) Geochemical consequences of increased atmospheric carbon dioxide on coral reefs. *Science* **284**, 118–120. doi:10.1126/science.284.5411.118

13. Dove SG, Kline DI, Pantos O, Angly FE, Tyson GW, Hoegh-Guldberg O (2013) Future reef decalcification under a business-as-usual CO2 emission scenario. *Proceedings of the National Academy of Sciences of the United States of America* **110**, 15342–15347. doi:10.1073/pnas.1302701110

14. International Panel of Climate Change (2019) 'Special report on the ocean and cryosphere in a changing climate', <https://www.ipcc.ch/srocc/>.

15. Great Barrier Reef Marine Park Authority (2014) 'Great Barrier Reef region strategic assessment: strategic assessment report', <http://elibrary.gbrmpa.gov.au/jspui/handle/11017/2861>.

16. Hughes TP, Barnes ML, Bellwood DR, Cinne JE, Cumming GS, Jackson JBC, Kleypas J, van de

Leemput IA, Lough JM, Morrison TH, et al. (2017) Coral Reefs in the Anthropocene. *Nature* **546**, 82–90. doi:10.1038/nature22901

17. Dietzel A, Connolly SR, Hughes TP, Bode M (2021) The spatial footprint and patchiness of large-scale disturbances on coral reefs. *Global Change Biology* **27**, 4825–4838. doi:10.1111/gcb.15805

18. Hughes TP, Kerry JT, Connolly SR, Álvarez-Romero JG, Eakin CM, Heron SF, Moneghetti J (2021) Emergent properties in the responses of tropical corals to recurrent climate extremes. *Current Biology*, **31**(23), 5393–5399. doi:10.1016/j.cub.2021.10.046

19. Hughes TP, Kerry JT, Baird AH, Connolly SR, Dietzel A, Eakin CM, Heron SF, Hoey AS, Hoogenboom MO, Liu G, *et al.* (2018) Global warming transforms coral reef ecosystems. *Nature* **556**, 492–496. doi:10.1038/s41586-018-0041-2

20. Hughes TP, Kerry JT, Baird AB, Connelly SR, Chase TJ, Dietzel A, Hill T, Hoey AS, Hoogenboom MO, Jacobson M, *et al.* (2019) Global warming impairs stock-recruitment dynamics of corals. *Nature* **568**, 387–390. doi:10.1038/s41586-019-1081-y

21. Roelfsema CM, Kovacs EM, Ortz J-C, Callagham DP, Hock K, Mongin M, Johansen K, Mumby P, Wettle M, Ronan M, *et al.* (2020) Habitat maps to enhance monitoring and management of the Great Barrier Reef. *Coral Reefs* **39**, 1039–1054. doi:10.1007/s00338-020-01929-3

22. Booth DJ, Beretta GA, Brown L, Figueira WF (2018) Predicting success of range-expanding coral reef fish in temperate habitats using temperature-abundance relationships. *Frontiers in Marine Science* doi:10.3389/fmars.2018.00031.

23. Smith SM, Fox RJ, Booth DJ, Donelson JM (2018) Stick with your own kind, or hang with the locals?' Implications of shoaling strategy for tropical reef fish on a range-expansion frontline. *Global Change Biology* **24**, 1663–1672. doi:10.1111/gcb.14016

24. Baird AH, Sommer B, Madin JS (2012) Pole-ward range expansion of *Acropora* spp. along the east coast of Australia. *Coral Reefs* **31**, 1063. doi:10.1007/s00338-012-0928-6

25. Booth DJ, Sear J (2018) Coral expansion in Sydney and associated coral-reef fishes. *Coral Reefs* **37**, 995. doi:10.1007/s00338-018-1727-5

26. Booth DJ (2020) Opposing climate-change impacts on poleward-shifting coral-reef fishes. *Coral Reefs* **39**, 577–581. doi:10.1007/s00338-020-01919-5

27. Coni EO, Nagelkerken I, Ferreira CM, Connell SD, Booth DJ (2021) Ocean acidification may slow the pace of tropicalization of temperate fish communities. *Nature Climate Change* **11**, 249–256. doi:10.1038/s41558-020-00980-w

28. Pauly D (1995) Anecdotes and the shifting baseline syndrome of fisheries. *Trends in Ecology & Evolution* **10**, 430. doi:10.1016/S0169-5347(00)89171-5

29. Thurstan RH, Buckley SM, Pandolfi JM (2016) Oral histories: informing natural resource management using perceptions of the past. In *Perspectives on Oceans Past: A Handbook of Marine Environmental History*. (Eds K Schwerdtner Máñez and B Poulsen) pp. 155–173. Springer Verlag, Dordrecht, Germany.

30. Pandolfi JM, Bradbury RH, Sala E, Hughes TP, Bjorndal KA, Cooke RG, McArdle D, McClenachan L, Newman MJH, Paredes G, *et al.* (2003) Global trajectories of the long-term decline of coral reef ecosystems. *Science* **301**, 955–958. doi:10.1126/science.1085706

31. Pandolfi JM, Jackson JBC, Baron N, Bradbury RH, Guzman HM, Hughes TP, Kappel CV, Micheli F, Ogden JC, Possingham HP, Sala E (2005) Are US coral reefs on the slippery slope to slime? *Science* **307**, 1725–1726. doi:10.1126/science.1104258

32. Daley B (2014) *The Great Barrier Reef: An Environmental History*. Routledge, Taylor & Francis Group, London and New York.

33. Lewis SE, Bartley R, Wilkinson SN, Bainbridge ZT, Henderson AE, James CS, Irvine SA, Brodie JE (2021) Land use change in the river basins of the Great Barrier Reef, 1860 to 2019: a foundation for understanding environmental history across the catchment to reef continuum. *Marine Pollution Bulletin* **166**, 112193. doi:10.1016/j.marpolbul.2021.112193

34. Roff G, Clark TR, Reymond CE, Zhao JX, Feng Y, McCook LJ, Done TJ, Pandolfi JM (2013) Palaeoecological evidence of a historical collapse of corals at Pelorus Island, inshore Great Barrier Reef, following European settlement. *Proceedings of the Royal Society B – Biological Sciences* **280**. doi:10.1098/rspb.2012.2100

35. Buckley SM, Thurstan RH, Tobin A, Pandolfi JM (2017) Historical spatial reconstruction of a spawning-aggregation fishery. *Conservation Biology* **31**, 1322–1332. doi:10.1111/cobi.12940

36. Lybolt M, Neil D, Zhao J, Feng Y, Yu KF, Pandolfi JM (2011) Instability in a marginal coral reef: the shift from natural variability to a human-dominated seascape. *Frontiers in Ecology and the Environment* **9**, 154–160. doi:10.1890/090176

37. Thurstan RH, Campbell AB, Pandolfi JM (2016) Nineteenth century narratives reveal historic catch rates for Australian Snapper (*Pagrus auratus*). *Fish and Fisheries* **17**, 210–225. doi:10.1111/faf.12103

38. Budd AF, Pandolfi JM (2010) Evolutionary novelty is concentrated at the edge of coral species distributions. *Science* **328**, 1558–1561. doi:10.1126/science.1188947

39. Kleypas J, Eakin CM (2007) Scientists' perceptions of threats to coral reefs: results of a survey of coral reef researchers. *Bulletin of Marine Science* **80**, 419–436.

40. Veron JEN (2011) Ocean acidification and coral reefs: an emerging big picture. *Diversity (Basel)* **3**(2), 262–274. doi:10.3390/d3020262

41. Kleypas JA, Buddemeier RW, Gattuso JP (2001) The future of coral reefs in an age of global change. *International Journal of Earth Sciences* **90**, 426–437. doi:10.1007/s005310000125

42. Mollica NR, Guo W, Cohen AL, Huang KF, Foster GL, Donald HK, Solow AR (2018) Ocean acidification affects coral growth by reducing skeletal density. *Proceedings of the National Academy of Sciences of the United States of America* **115**, 1754–1759. doi:10.1073/pnas.1712806115

43. Furla P, Galgani I, Durand I, Allemand D (2000) Sources and mechanisms of inorganic carbon transport for coral calcification and photosynthesis. *The Journal of Experimental Biology* **203**, 3445–3457. doi:10.1242/jeb.203.22.3445

44. Fine M, Tchernov D (2007) Scleractinian coral species survive and recover from decalcification. *Science* **315**, (5280), 1811. doi:10.1126/science.1137094

45. Zickfeld K, Eby M, Weaver AJ, Alexander K, Crespin E, Edwards NR, Eliseev AV, Feulner G, Fichefet T, Forest CE, Friedlingstein P (2013) Long-term climate change commitment and reversibility: An EMIC intercomparison. *Journal of Climate* **26**, 5782–5809. doi:10.1175/JCLI-D-12-00584.1

46. Eyre BD, Cyronak T, Drupp P, De Carlo EH, Sachs JP, Andersson AJ (2018) Coral reefs will transition to net dissolving before end of century. *Science* **359**, 908–911. doi:10.1126/science.aao1118

47. Richter C, Wunsch M, Rasheed M, Kötter I, Badran MI (2001) Endoscopic exploration of Red Sea coral reefs reveals dense populations of cavity-dwelling sponges. *Nature* **413**, 726–730. doi:10.1038/35099547

48. Dove SG, Brown KT, Van Den Heuvel A, Chai A, Hoegh-Guldberg O (2020) Ocean warming and acidification uncouple calcification from calcifier biomass which accelerates coral reef decline. *Communications Earth & Environment* **1**(1), 55. doi:10.1038/s43247-020-00054-x

49. Langdon C, Takahashi T, Sweeney C, Chipman D, Goddard J, Marubini F, Aceves H, Barnett H, Atkinson MJ (2000) Effect of calcium carbonate saturation state on the calcification rate of an experimental coral reef. *Global Biogeochemical Cycles* **14**, 639–654. doi:10.1029/1999GB001195

50. Boström-Einarsson L, Babcock RC, Bayraktarov E, Ceccarelli D, Cook N, Ferse SC, Hancock B, Harrison P, Hein M, Shaver E, Smith A (2020) Coral restoration – a systematic review of current methods, successes, failures and future directions. *PLoS One* **15**, e0226631. doi:10.1371/journal.pone.0226631

51. Van Oppen MJ, Gates RD, Blackall LL, Cantin N, Chakravarti LJ, Chan WY, Cormick C, Crean A, Damjanovic K, Epstein H, Harrison PL (2017) Shifting paradigms in restoration of the world's coral reefs. *Global Change Biology* **23**, 3437–3448. doi:10.1111/gcb.13647

52. Epstein HE, Smith HA, Torda G, van Oppen MJ (2019) Microbiome engineering: enhancing climate resilience in corals. *Frontiers in Ecology and the Environment* **17**, 100–108. doi:10.1002/fee.2001

53. Quigley KM, Bay LK, van Oppen MJH (2019) The active spread of adaptive variation for reef resilience. *Ecology and Evolution* **9**, 11122–11135. doi:10.1002/ece3.5616

54. Anthony K, Bay LK, Costanza R, Firn J, Gunn J, Harrison P, Heyward A, Lundgren P, Mead D,

Moore T, Mumby PJ (2017) New interventions are needed to save coral reefs. *Nature Ecology & Evolution* **1**, 1420–1422. doi:10.1038/s41559-017-0313-5

55. Howlett L, Camp EF, Edmondson J, Henderson N, Suggett DJ (2021) Coral growth, survivorship and return-on-effort within nurseries at high-value sites on the Great Barrier Reef. *PLoS One* **16**, e0244961. doi:10.1371/journal.pone.0244961

56. Williams SL, Sur C, Janetski N, Hollarsmith JA, Rapi S, Barron L, Heatwole SJ, Yusuf AM, Yusuf S, Jompa J, Mars F (2019) Large-scale coral reef rehabilitation after blast fishing in Indonesia. *Restoration Ecology* **27**, 447–456. doi:10.1111/rec.12866

57. Hein MY, McLeod IM, Shaver EC, Vardi T, Pioch S, Boström-Einarsson L, Ahmed M, Grimsditch G (2020) *Coral Reef Restoration as a Strategy to Improve Ecosystem Services – A Guide to Coral Restoration Methods.* United Nations Environment Program, Nairobi, Kenya.

58. Suggett DJ, Camp EF, Edmondson J, Boström-Einarsson L, Ramler V, Lohr K, Patterson JT (2019) Optimizing return-on-effort for coral nursery and outplanting practices to aid restoration of the Great Barrier Reef. *Restoration Ecology* **27**(3), 683–693. doi:10.1111/rec.12916

59. Suggett DJ, Edmondson J, Howlett L, Camp EF (2020) Coralclip®: a low-cost solution for rapid and targeted out-planting of coral at scale. *Restoration Ecology* **28**(2), 289–296. doi:10.1111/rec.13070

60. Baums IB, Baker AC, Davies SW, Grottoli AG, Kenkel CD, Kitchen SA, Kuffner IB, LaJeunesse TC, Matz MV, Miller MW, *et al.* (2019) Considerations for maximizing the adaptive potential of restored coral populations in the western Atlantic. *Ecological Applications* **29**(8), 1–23. doi:10.1002/eap.1978.

61. Randall CJ, Negri AP, Quigley KM, Foster T, Ricardo GF, Webster NS, Bay LK, Harrison PL, Babcock RC, Heyward A (2020) Sexual production of corals for reef restoration in the Anthropocene. *Marine Ecology Progress Series* **635**, 203–232. doi:10.3354/meps13206

62. Baird AH, Guest JR, Willis BL (2009) Systematic and biogeographical patterns in the reproductive biology of scleractinian corals. *Annual Review of Ecology, Evolution, and Systematics* **40**(1), 551–571. doi:10.1146/annurev.ecolsys.110308.120220

63. Babcock RC, Heyward AJ (1986) Larval development of certain gamete-spawning scleractinian corals. *Coral Reefs* **5**(3), 111–116. doi:10.1007/BF002981781

64. Baird A, Guest JR, Edwards AJ, Bauman AG, Bouwmeester J, Mera H, Abrego D, Alvarez-Noriega M, Babcock RC, Barbosa MB, Bonito V (2021) An Indo-Pacific coral spawning database. *Scientific Data* **8**, 35. doi:10.1038/s41597-020-00793-8

65. Kojis BL, Quinn NJ (1982) Reproductive ecology of two faviid corals (Coelenterata: Scleractinia). *Marine Ecology Progress Series* **8**, 251–255. doi:10.3354/meps008251

66. Harrison P, Babcock RC, Bull GD, Oliver JK, Wallace CC, Willis BL (1984) Mass spawning in tropical reef corals. *Science* **223**, 1186–1189. doi:10.1126/science.223.4641.1186

67. Oliver JK, Willis BL (1987) Coral-spawn slicks in the Great Barrier Reef: preliminary observations. *Marine Biology* **94**, 521–529. doi:10.1007/BF00431398

68. Heyward AJ, Smith LD, Rees M, Field SN (2002) Enhancement of coral recruitment by in situ mass culture of coral larvae. *Marine Ecology Progress Series* **230**, 113–118. doi:10.3354/meps230113

69. Cruz DW, Harrison PL (2017) Enhanced larval supply and recruitment can replenish reef corals on degraded reefs. *Scientific Reports* **7**, 13985. doi:10.1038/s41598-017-14546-y

70. Omori M, Shibata S, Yokokawa M, Aota T, Watanuki A, Iwao K (2007) Survivorship and vertical distribution of coral embryos and planula larvae in floating rearing ponds. *Galaxea* **8**, 77–81. doi:10.3755/jcrs.8.77

71. Harrison PL (2018) 'Coral larval restoration on the GBR (final report)'. Great Barrier Reef Foundation, Brisbane.

72. Doropoulos C, Vons F, Elzinga J, ter Hofstede R, Salee K, van Koningsveld M, Babcock RC (2019) Testing industrial-scale coral restoration techniques: harvesting and culturing wild coral-spawn slicks. *Frontiers in Marine Science* **6**. doi:10.3389/fmars.2019.00658

73. van Hooidonk R, Maynard J, Tamelander J, Gove J, Ahmadia G, Raymundo L, Williams G, Heron SF, Planes S (2016) Local-scale projections of coral reef futures and implications of the Paris Agreement. *Scientific Reports* **6**(1), 39666. doi:10.1038/srep39666

74. Sully S, Burkepile DE, Donovan MK, Hodgson G, van Woesik R (2019) A global analysis of coral bleaching over the past two decades. *Nature Communications* **10**(1), 1–5. doi:10.1038/s41467-019-09238-2

75. van Oppen MJH, Oliver J, Putman H, Gates R. (2015) Building coral reef resilience through assisted evolution. *Proceedings of the National Academy of Sciences of the United States of America* **112**(8), 2307–2313. doi:10.1073/pnas.1422301112

76. Oliver TA, Palumbi SR (2011) Do fluctuating temperature environments elevate coral thermal tolerance? *Coral Reefs* **30**(2), 429–440. doi:10.1007/s00338-011-0721-y

77. Putnam HM (2021) Avenues of reef-building coral acclimatization in response to rapid environmental change. *The Journal of Experimental Biology* **224**(Suppl 1), jeb239319. doi:10.1242/jeb.239319

78. Bellantuono AJ, *et al.* (2012) Coral thermal tolerance: tuning gene expression to resist thermal stress. *PLoS One* **7**(11), e50685. doi:10.1371/journal.pone.0050685

79. Quigley KM, Willis BL, Kenkel CD (2019) Transgenerational inheritance of shuffled symbiont communities in the coral *Montipora digitate. Scientific Reports* **9**, 1–11. doi:10.1038/s41598-019-50045-y

80. Quigley KM, Randall CJ, van Oppen, MJH, Bay LK (2020) Assessing the role of historical temperature regime and algal symbionts on the heat tolerance of coral juveniles. *Biology Open* **9**(1), 1–11. doi:10.1242/bio.047316

81. Chan WY, Peplow LM, Menendez P, Hoffmann AA, van Oppen MJH (2018) Interspecific hybridization may provide novel opportunities for coral reef restoration. *Frontiers in Marine Science* **5**. doi:10.3389/fmars.2018.00160

82. Damjanovic K, Blackall LL, Webster NS, van Oppen MJ(2017) The contribution of microbial biotechnology to mitigating coral reef degradation. *Microbial Biotechnology* **10**(5), 1236–1243. doi:10.1111/1751-7915.12769

83. Rosado PM, Leite DCA, Duarte GAS, Chaloub RM, Jospin G, da Rocha UN, Saraiva JP, Dini-Andreote F, Eisen JA, Bourne DG, *et al.* (2019) Marine probiotics: increasing coral resistance to bleaching through microbiome manipulation. *The ISME Journal* **13**(4), 921–936. doi:10.1038/s41396-018-0323-6.

84. Morgans CA, Hung JY, Bourne DG, Quigley KM (2020) Symbiodiniaceae probiotics for use in bleaching recovery. *Restoration Ecology* **28**(2), 282–288. doi:10.1111/rec.13069

85. Buerger P, Alvarez-Roa C, Coppin CW, Pearce SL, Chakravarti LJ, Oakeshott JG, Edwards OR, Van Oppen MJ (2020) Heat-evolved microalgal symbionts increase coral bleaching tolerance. *Science Advances* **6**(20), eaba2498. doi:10.1126/sciadv.aba2498

86. Camp EF, Schoepf V, Mumby PJ, Hardtke LA, Rodolfo Metalpa R, Smith DJ, Suggett DJ (2018) The future of coral reefs subject to rapid climate change: lessons from natural extreme environments. *Frontiers in Marine Science* **5**, 4. doi:10.3389/fmars.2018.00004

87. Keppel G, Wardell-Johnson GW (2012) Refugia: keys to climate change management. *Global Change Biology* **18**, 2389–2391. doi:10.1111/j.1365-2486.2012.02729.x

88. Morikawa MK, Palumbi SR (2019) Using naturally occurring climate resilient corals to construct bleaching-resistant nurseries. *Proceedings of the National Academy of Sciences of the United States of America* **116**, 10586–10591. doi:10.1073/pnas.1721415116

89. Camp EF, Edmonson J, Doheny A, Rumney J, Grima AJ, Huete A, Suggett DJ (2019) Mangrove lagoons of the Great Barrier Reef support coral populations persisting under extreme environmental conditions. *Marine Ecology Progress Series* **625**, 1–14. doi:10.3354/meps13073

90. Morgan KM, Perry CT, Johnson JA, Smithers SG (2017) Nearshore turbid-zone corals exhibit high bleaching tolerance on the Great Barrier Reef following the 2016 ocean warming event. *Frontiers in Marine Science* **4**(Jul). doi:10.3389/fmars.2017.00224

91. Frade PR, Bongaerts P, Englebert N, Rogers N, Gonzalez-Rivero M, Hoegh-Guldberg O (2018) Deep reefs of the Great Barrier Reef offer limited thermal refuge during mass coral bleaching. *Nature Communications* **9**(1), 3447. doi:10.1038/s41467-018-05741-0

92. Cornwall CE, Comeau S, DeCarlo TM, Moore B, D'Alexis Q, McCulloch MT (2018) Resistance of corals and coralline algae to ocean acidification: physiological control of calcification under natural pH variability. *Proceedings of the Royal Society B: Biological Sciences* **285**(1884), 20181168. doi:10.1098/rspb.2018.1168.

93. Schoepf V, Jung EMU, McCulloch, MT, White NE, Stat M, Thomas L (2020) Thermally variable, macrotidal reef habitats promote rapid recovery from mass coral bleaching. *Frontiers in Marine Science* **7**, 245. doi:10.3389/fmars.2020.00245

94. Schoepf V, Carrion SA, Pfeifer SM, Naugle M, Dugal L, Bruyn J, McCulloch MT (2019) Stress-resistant corals may not acclimatize to ocean warming but maintain heat tolerance under cooler temperatures. *Nature Communications* **10**, 4031. doi:10.1038/s41467-019-12065-0

95. Walther GR, Post E, Convey P, Menzel A, Parmesan C, Beebee TJ, Fromentin JM, Hoegh-Guldberg O, Bairlein F (2002) Ecological responses to recent climate change. *Nature* **416**, 389–395. doi:10.1038/416389a

96. Oreskes N, Conway EM (2011) *Merchants of Doubt: How a Handful of Scientists Obscured the Truth on Issues from Tobacco Smoke to Global Warming.* Bloomsbury Publishing, New York NY.

97. Intergovernmental Panel on Climate Change (1990) *Climate Change. The IPCC Scientific Assessment.* IPCC, Geneva, Switzerland.

98. Hoegh-Guldberg O, Jacob D, Taylor M, Bindi M, Brown S, Camilloni I, Diedhiou A, Djalante R, Ebi KL, Engelbrecht F, *et al.* (2018) Impacts of 1.5°C of global warming on natural and human systems. In *Global Warming of 1.5°C*. pp. 175–311. IPCC (Intergovernmental Panel on Climate Change).

99. Hoegh-Guldberg O, Cai R, Poloczanska ESS, Brewer PGG, Sundby S, Hilmi K, Fabry VJJ, Jung S (2014) The ocean. In *Climate Change Impacts, Adaptation, and Vulnerability. Part B: Regional Aspects. Contribution of Working Group II to the Fifth Assessment Report of the Intergovernmental Panel on Climate Change.* pp. 1655–1731. Cambridge University Press, Cambridge, UK.

100. Beyer HL, Kennedy EV, Beger M, Chen CA, Cinner JE, Darling ES, Eakin CM, Gates RD, Heron SF, Knowlton N, Obura DO (2018) Risk-sensitive planning for conserving coral reefs under rapid climate change. *Conservation Letters* **11**, e12587. doi:10.1111/conl.12587

101. Hoegh-Guldberg O, Kennedy EV, Beyer HL, McClennen C, Possingham HP (2018) Securing a long-term future for coral reefs. *Trends in Ecology & Evolution* **33**, 936–944. doi:10.1016/j.tree.2018.09.006

102. CRRI (2021) Coral Reef Rescue Initiative, <https://coralreefrescueinitiative.org/>.

103. McCay BJ, Jones PJS (2011) Marine protected areas and the governance of marine ecosystems and fisheries. *Conservation Biology* **25**, 1130–1133. doi:10.1111/j.1523-1739.2011.01771.x

104. Olsson P, Folke C, Hughes TP (2008) Navigating the transition to ecosystem-based management of the Great Barrier Reef, Australia. *Proceedings of the National Academy of Sciences of the United States of America* **105**, 9489–9494. doi:10.1073/pnas.0706905105

105. Morrison TH, Hughes TP, Adger WN, Brown K, Barnett J, Lemos MC (2019) Save reefs to rescue all ecosystems. *Nature* **573**, 333–336. doi:10.1038/d41586-019-02737-8

106. Morrison TH, Adger N, Barnett J, Brown K, Possingham H, Hughes T (2020) Advancing coral reef governance into the Anthropocene. *One Earth* **2**, 64–74. doi:10.1016/j.oneear.2019.12.014

107. Morrison TH (2017) Evolving polycentric governance of the Great Barrier Reef. *Proceedings of the National Academy of Sciences of the United States of America* **114**, E3013–3021. doi:10.1073/pnas.1620830114

108. Craik AM (2017) 'Independent review of governance of the Great Barrier Reef Marine Park Authority'. Canberra, Department of the Environment and Energy.

109. Republic of Fiji (2018) *COP23 Talanoa Dialogue Submission: 'Where are We?'* UN Framework Convention on Climate Change, <https://cop23.unfccc.int/sites/default/files/resource/107_Talanoa%20dialogue_How%20Do%20We%20Get%20There.pdf>.

110. Dornan M (2015) Renewable energy development in small island developing states of the Pacific. *Resources* **4**, 490–506. doi:10.3390/resources4030490

111. Whittlesea E (2021) 'Central Queensland Energy Futures Summit report'. A report prepared by The Next Economy, Brisbane.

112. Condie SA, Anthony KR, Babcock RC, Baird ME, Beeden R, Fletcher CS, Gorton R, Harrison D, Hobday AJ, Plagányi ÉE, Westcott DA (2021) Large-scale interventions may delay decline of the Great

Barrier Reef. *Royal Society Open Science* **8**, 201296. doi:10.1098/rsos.201296

113. Morrison TH, Lane MB, Hibbard M (2015) Planning, governance and rural futures in Australia and the USA: revisiting the case for rural regional planning. *Journal of Environmental Planning and Management* **58**, 1601–1616. doi:10.1080/09640568.2014.940514

114. Filbee-Dexter K, Smajdor A (2019) Ethics of assisted evolution in marine conservation. *Frontiers in Marine Science* **6**, 20. doi:10.3389/fmars.2019.00020

EPILOGUE: THE EYE OF THE BEHOLDER

Sarah Hamylton, Pat Hutchings and Ove Hoegh-Guldberg

Australia's coral reefs traverse extensive continental shelves, meander along the Ningaloo and Kimberley coastlines, fringe much of the Queensland coastline, encircle offshore volcanic peaks of Lord Howe Island, and form labyrinths in the Torres Strait and Coral Sea. Just as these reefs have distinctive local environments, they also have unique histories of human occupation and interaction. As their physical structures have been gradually rearranged over time, so they have also been moulded as products of human perception. Collectively, their histories weave a rich tapestry of stories that positions Australia's reefs, in the minds of many, as some of the most beautiful yet isolated and exciting places on Earth.

Sea country has deep spiritual and ceremonial significance for Aboriginal peoples and Torres Strait Islanders, who have lived alongside Australia's reefs for many thousands of years. The relationship between Indigenous Australians and sea country imparts critical information on such aspects as reef origins, ownership, harvesting and navigation. This relationship also recognises a fundamental kinship between the many Traditional Owner groups around Australia and their local reefs that translates into daily duties of caring for sea country.

The perceptions of Australia's coral reefs have shifted. The qualities for which reefs were once feared – their remote isolation and the capacity of corals to grow in a diverse range of forms – have become their most prized characteristics. Once, a lone coral reef emerging unexpectedly from the blue depths was a jagged mortal shipping hazard; now, it is a resilient outpost of vulnerable marine biodiversity.

Australia is also economically reliant on the reefs that thrive inside its marine estate, as can be seen in the tight coupling between the fortunes of families in coastal fishing villages that once thrived in the Houtman Abrolhos Islands and the boom-and-bust cycles of the rock lobster populations on the offshore reefs of Western Australia. Reefs attract tourists and are places where fish, bêche-de-mer, oysters and pearls are harvested for industries that stretch back several centuries and supply international trading networks. Australian reefs have a value that demands we think carefully about our lifestyles so that we can live sustainably alongside them.

Beginning with the observations of early global voyagers, the Great Barrier Reef has emerged as a geographical focal point for expanding our scientific understanding of Australian coral reefs, helped by several expeditions and a network of research stations. Geologists have drilled down to resolve the age of the underlying reef platform. Biologists have unravelled the complexity of corals and their web of intricate relationships with marine life to answer fundamental questions about what corals are, as well as how they feed, reproduce and grow.

Australia's coral reefs are managed by a diverse and expanding community of Traditional Owners, marine park authorities, island rangers and, increasingly, by a global community whose actions are determining outcomes for coral reefs. Zoning the Great Barrier Reef Marine Park was an exercise that included thousands of people, placing Australia at the forefront of global coral reef stewardship with participation from scientific, public, and political communities. Managing reefs in Australia demands of us a range of actions, from directly removing a crown-of-thorns starfish feeding on a coral to taking the more convoluted step of curbing carbon emissions to address the underwater marine heatwaves that cause corals to bleach. As inshore coastal water quality has deteriorated with coastal development, so we have become increasingly aware of the impacts that humans have through land-use change and the need to manage these changes. Helping manage 'local' threats such as water quality and over-fishing may buy important time as we wrestle with the serious 'global' threats that coral reefs face from escalating sea temperatures, acidifying seas, sea-level rise, and intensifying storms. How we choose to intervene will depend on the availability of strategies to confront the challenges from these threats at credible scales for reefs and regions. This book illustrates some promise, but provides no clear strategy that applies a new technology to the millions of hectares that are commensurate with the current impacts on coral reefs.

The implications of climate change are woven through every chapter of this book. Indeed, as we send the proofs off to the printer, we are in the midst of confirmation of coral bleaching on the Great Barrier Reef. Torres Strait Islanders have recognised the consequences of sea-level rise and extreme weather events with mounting anxiety as communities living on low-lying islands have been relocated. We have come to understand that marginal coral reefs living in extreme environmental settings possess unique ecological characteristics that, while still exposed to climate change, may enable them to act as refugia for stressed corals under climate change Reef managers also increasingly rely on real-time technology (e.g. satellites) to help them monitor changing sea temperatures and hence the gargantuan impacts of mass coral bleaching and mortality. Since the 1998 mass coral bleaching event, the reef science community has spoken out about the grave implications of climate change for Australian coral reefs and advocated for both federal and state policies to reduce carbon emissions.

Marine and coastal scientists have sought to shape management decisions, policies and legislation regarding coral reefs through status reports, expert panels, senate inquiries and royal commissions. The origins of Australia's coral reef conservation movement can be traced to underwater reef surveys carried out by graduate science students to document the marine life of Ellison reef, supporting a case in a Queensland Court that overturned a proposal to mine the reef for limestone. In the early 1970s, this entrained a series of events that, via a royal commission on oil drilling, led to a moratorium on oil and gas exploration and the formation of the Great Barrier Reef Marine Park (1975), which was eventually listed as a World Heritage site (1981). In the 1980s and 1990s, a broader network of marine parks were established around Australia's coastline, reflecting a growing awareness of the finite and vulnerable nature of coral reefs. At times, the urge to conserve has met with tension in Australia, particularly the concerns of the fishing community at ring-fencing large swathes of ocean and banning unsustainable extraction techniques. Establishing the boundaries of marine parks has been a hotly contested political issue in the Coral Sea, and in other areas.

Those concerned about protecting coral reefs have increasingly spoken out about addressing climate change as the number one threat to reef health in Australia. At times, this concern has manifested in local issues, such as the 'coal *v.* coral' battle at Abbot Point on the Great Barrier Reef and the tensions between the offshore oil and gas industry and those seeking to halt industrial development at Ningaloo. The question of whether to place the Great Barrier Reef on the List of World Heritage in Danger speaks to a broader need for the Australian Federal Government to move away from a fossil-fuel-based national energy policy.

Coral reefs hold a special place in the human imagination. Even the most devoted divers among us will only visit a tiny fraction of Australia's 50 000 km² of coral reef. As the American novelist Wallace Stegner once wrote in defence of wild land when he penned his famous *Wilderness Letter* [1]:

> *We simply need wild country available to us, even if we never do more than drive to its edge and look in. For it can be a means of reassuring ourselves of our sanity as creatures, a part of the geography of hope.*

Tim Winton echoes this sentiment in Chapter 7 when he writes of Ningaloo that, for our own sanity and honour, we humans need to know there are still some exceptional sites left intact; ecosystems whose richness and enduring health afford us hope for the future, even if we never get the chance to visit them ourselves.

The diversity of writers in this book reflects the multitude of ways in which Australian reefs hold importance: spiritually, traditionally, culturally, artistically, resourcefully, biologically, geologically, structurally, historically, economically, emotionally, socially and environmentally. These writers hold a common view that we must continue to protect Australia's coral reefs because of what we know that they are and, critically, what we are yet to discover they can be. The importance of knowing that such places exist and recognising the myriad ways in which they touch our lives cannot be overstated.

Reference

1. Stegner W, Snow VD (1995) *Wilderness Letter*. Red Butte Press, University of Utah, Marriott Library, Salt Lake City UT.

INDEX